ISLANDED

Britain, Sri Lanka, and the Bounds of an Indian Ocean Colony

SUJIT SIVASUNDARAM

THE UNIVERSITY OF CHICAGO PRESS

CHICAGO AND LONDON

SUJIT SIVASUNDARAM is Lecturer in World and Imperial History since 1500 at the University of Cambridge and fellow of Gonville and Caius College, Cambridge. He is the author of *Nature and the Godly Empire: Science and Evangelical Mission in the Pacific, 1795–1850*. He won a Philip Leverhulme Prize for History in 2012, awarded to scholars in the United Kingdom for outstanding contributions to research.

The University of Chicago Press, Chicago 60637
The University of Chicago Press, Ltd., London
© 2013 by The University of Chicago
All rights reserved. Published 2013.
Printed in the United States of America

22 21 20 19 18 17 16 15 14 13   1 2 3 4 5

ISBN-13: 978-0-226-03822-3 (cloth)
ISBN-13: 978-0-226-03836-0 (e-book)

Library of Congress Cataloging-in-Publication Data
Sivasundaram, Sujit.
  Islanded : Britain, Sri Lanka, and the bounds of an Indian Ocean colony / Sujit Sivasundaram.
    pages ; cm
  Includes bibliographical references and index.
    ISBN 978-0-226-03822-3 (cloth: alk. paper) — ISBN 978-0-226-03836-0 (e-book)
  1. Sri Lanka—Colonization.   2. Sri Lanka—Politics and government—18th century.   3. Sri Lanka—Politics and government—19th century.   4. Sri Lanka—Relations—Great Britain.   5. Great Britain—Relations—Sri Lanka.   I. Title.
    DS489.7.S545 2013
    954.93'02—dc23

                                                                    2012045010

♾ This paper meets the requirements of ANSI/NISO z39.48-1992 (Permanence of Paper).

CONTENTS

ACKNOWLEDGMENTS

Sri Lanka is a richly complicated place. For me, the challenge of writing about it is compounded by the fact that I still call it my home and so feel a great attachment to it. When I first turned to doctoral work I weighed up the idea of writing on Sri Lanka, but was given a very good piece of advice, namely, that it takes too much to write about a place where one has seen an ethnic conflict unfold. So now I have finally completed the book that I postponed in researching the history of the Pacific Ocean as a graduate student. The intellectual space and distance of being distracted by another region has been helpful. In writing this book I have come to know my home all over again.

Everyone in Sri Lanka is a historian of sorts: everyone wishes to put you right when you tell a historical story. I have greatly enjoyed debating the terms of this book over rice and curry on various visits back to the island. Yet one thing that I really did not expect was the generosity of almost all the people who sat down with me over the past years in Sri Lanka: collectors of palm-leaf manuscripts, antiquarians, archivists, scientists, colleagues in Sri Lankan universities, retired academics, public intellectuals, and friends of all kinds. Whereas academic knowledge is guarded and owned as intellectual property in the West, I found the willingness of Sri Lankans to share information and even access to materials in their possession exemplary.

If this book was born out of a return to Sri Lanka, it has also been born out of close engagement with the research and teaching of South Asia at the University of Cambridge. It is an attempt to break Sri Lankan studies out of its isolation from broader theoretical and methodological issues in the wide literature on South Asia. The book took an agonistic turn when I realized how marginalized the study of Sri Lanka is in the West, and how South Asia often equates with India in the historical literature. The scholar whose work has most affected the shape of this project has to be Christopher A. Bayly, who has read many versions and provided careful and thought-provoking suggestions. I'd also like to thank my other colleagues at Cambridge in South Asian studies, as well as world history, who have provided me with intellectual encouragement, especially Shruti Kapila, Tim Harper, Megan Vaughan, Joya Chatterji, and Emma Hunter. I have continued to draw inspiration from the work of James A. Secord and Simon Schaffer in the Department of History and Philosophy of Science. I'd also like to place on record my gratitude for the welcome extended to me at the London School of Economics in the two years I spent there lecturing in South Asian history. More recently, I have enjoyed teaching about Sri Lanka, among other topics, to a range of bright undergraduates and graduates at Cambridge, from whom I have learned much. In particular I thank Eleanor Harding for her work as a research assistant for a few weeks as I got the last pieces of the book together.

The book benefited from the assistance and support of the American Institute for Sri Lankan Studies on countless occasions. I am indebted in particular to John Rogers, who has read various essays in the past—as well as the full manuscript—with an eye for detail that I have not met in any other reader of academic work. It has been a delight to join a younger generation of Sri Lankan historians, and I thank in particular Alan Strathern, Mark Frost, Zoltán Biedermann, Alicia Schrikker, and Sandagomi Coperahewa for their critical companionship. I also thank Nira Wickramasinghe for reading the entire manuscript and making valuable suggestions and Gananath Obeyesekere for the inspiration of his own work and for his lively engagement with various pieces I sent to him. Udaya Meddegama did me a great service in supplementing my language skills by translating various palm-leaf manuscripts for me. Over the years, and particularly at the inception of this project, a range of other scholars and authors working on Sri Lanka have taken time to discuss my ideas and helped in practical ways; among them are Senake Bandaranayake, K. M. De Silva, Lorna Dewaraja, Nirmal Dewasiri, James Duncan, Charles Hallisey, Kumari Jayawardena, Patrick Peebles, Ismeth Raheem, Michael Roberts, Jonathan Spencer, SinhaRaja Tammita Delgoda, and C. G. Uragoda. The staff of the National Archives of Sri Lanka, the Colombo Museum, the library of the Royal Asiatic Society of Sri Lanka, the library of the University of Peradeniya, the National Archives of the United Kingdom at Kew, and the British Library deserve my thanks. I very much appreciated the assistance of all those at the University of Chicago Press, and especially, Alan Thomas,

Randolph Petilos, and Richard Allen, who enthusiastically saw the book to print.

The research for this book was first undertaken when I became a research fellow at Gonville and Caius College. The master and fellows of the college greatly honored me in electing me to this post. I still recall my delight, in the first weeks of the research fellowship, when I took out the drawer for Sri Lanka in the card catalog of the Royal Commonwealth Society collections in the Cambridge University Library to begin work on this project. It was invaluable to have that time away from the full course of teaching with which I am now very acquainted. Gonville and Caius has also provided me with a vibrant intellectual community, and I thank my colleagues in history in particular. I should also record my gratitude to the Centre for Research in Arts, Social Sciences, and the Humanities who appointed me to an early career research fellowship for one term, which was useful in finalizing a couple of the chapters which form this book. Papers connected with this project were presented at various venues in South Asia, Europe, and the United States, and I benefited from all who contributed by their questions, commentary, and hospitality.

This book is dedicated to my grandmother, Mano Muthu Krishna, and my step-grandfather, George Candappa, who lived in Colombo. It is a great joy that my grandmother Mano, who has had such an inspirational life as a journalist and educationalist, is with us to see the book in print, and that I can at last answer her question, "When is it coming out?" If there is anyone who should be blamed for the fact that I became a historian (and not the scientist I was intended to be), she is that person. Rather unintentionally the lives of my grandparents, including my maternal grandfather from Ambalangoda in southern Sri Lanka and my paternal grandparents who lived outside Jaffna in northern Sri Lanka, are reflected in the argument mounted in this book. Their lives have been the lives of travelers caught between the memory of India and Sri Lanka's historic links and the ghost of those links in the troubled present. They themselves and their ancestors traversed the waters that separate the island from the mainland. I also thank my parents, Ramola and Siva, for coping with my regular invasions on visits to the archives and conferences in Sri Lanka, and my sister Renu and her family for putting up with all the history. My parents-in-law, Hazel and Paul, through their help with child care have made an important contribution to my research. The book would never have been completed without Caroline, Toby Tarun, and Anjali Alice. By a fitting coincidence I write this very last bit of it just days before the expected arrival of another traveler.

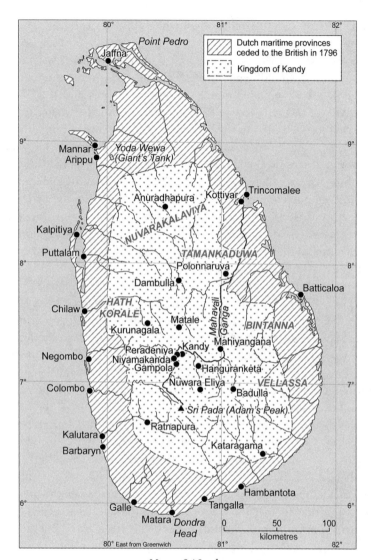

Map 2. Sri Lanka

ISLANDED

Fig. o.1. "Town of Kandy from Castle Hill." From William Lyttleton, *A Set of Views in the Island of Ceylon* (London, 1819). Reproduced by kind permission of the Syndics of Cambridge University Library.

Paths through Mountains and Seas

Hidden within a ring of hills, the city of Kandy is best appreciated from above. At its center is the lake, built by the last king to rule in the island of Sri Lanka, at the start of the nineteenth century, and first named "The Ocean of Milk" in allegorical reference to the cosmic ocean that lies at the foot of Mount Meru in the center of the universe. On one side of the lake lie the Royal Palace and the temple of Buddha's Tooth, signifying that ruling power has been linked to the patronage of Buddhism and ownership of Buddha's Tooth Relic. Extending outward from the lake are the modern city's busy thoroughfares, which link it to all corners of the island.[1] From the highland plain of Kandy, it takes about three hours to journey down to Colombo, along a road that winds its way through a landscape which descends in altitude in dramatic fashion. Tea and rubber plantations slowly give way to fields of rice, and coconut trees appear in abundance to herald the closeness of the sea. Yet this road, from Kandy down to Colombo, hides the historical processes that have made the Kandyan kingdom and its legacy part of a nation. Until the nineteenth century, "the great city" or *Mahanuvara* in the "country on the hills" or *Kandaudarata*, which was Anglicized as Kandy, was the nerve center of a spiritual and political entity.[2] This was a South Asian kingdom that resisted European encroachment in the most sustained manner.[3]

1. For a reading of Kandy's landscape history, see James Duncan, *The City as Text: The Politics of Landscape Interpretation in the Kandyan Kingdom.* For an account of Kandy in 1815, see L.J.B. Turner, "The Town of Kandy about the year 1815 A.D."

2. *Kandaudarata* is itself a short form for *kandaudapasrata,* "the five regions on the hills."

3. For the Kandyan kingdom, see Lorna Dewaraja, *A Study of the Political,*

3

Stories of the expansion of Europe in South Asia sometimes cast early modern empires as water-borne parasites, having command of the sea but simply taking up small stretches of territory and building factories along the coasts. In this picture, the British win out against their European rivals at sea and then start forming alliances with South Asian states around a system of military fiscalism; they work closely with land-based Indian financiers and merchants and protect the rights of landowners. It is from this ground of mutual interest that British conquest takes shape. In this spirit, David Washbrook writes of South India: "If we are seeking 'difference' between European and Indian cultures one obvious one lies at the shoreline."[4] When thinking of Kandy and Lankan history, it is also easy to slip into this narrative, so that Kandy constitutes the last remaining badge of indigenous kingship; it is a landlocked kingdom set irrevocably against the maritime prowess of Europe on the coast, and it is in decline. Europeans, firstly the Portuguese (r. 1594–1658) and then the Dutch (r. 1640–1796), throttle the kingdom of Kandy, so that by the early nineteenth century it implodes internally into factionalism and court intrigue; British attempts to forge an alliance, on the model of those in India between successor states to the Mughals and the East India Company, lead nowhere, and Kandy is finally taken over by the British in 1815.[5] In contrast, along the coastal belt of the island of Sri Lanka, successive waves of European colonists have brought about social change, so that there is an essential difference by the British period between "lowlanders" and the untouched warrior Kandyans in the highlands.

One primary aim of this work is to take issue with this narrative, by theorizing the period that cuts across the fall of Kandy and the consolidation of British rule over the whole island, or from roughly the eighteenth

---

Administrative, and Social Structure of the Kandyan Kingdom of Ceylon, 1707–1760 and the second edition of this work, Lorna Dewaraja, *The Kandyan Kingdom of Sri Lanka*.

4. David Washbrook, "South India, 1770–1840: The Colonial Transition," 512.

5. For the thesis of the decline of the kingdom of Kandy, see for instance Ralph Pieris, *Sinhalese Social Organization: The Kandyan Period*. For the standard account stressing intrigue and isolation, see "The Fall of the Kingdom of Kandy," chapter 17 in K. M. De Silva, *A History of Sri Lanka*. For a more recent history of Sri Lanka, focusing in on the modern period, see Nira Wickramasinghe, *Sri Lanka in the Modern Age: A History of Contested Identities*. For recent work on the Portuguese and Dutch in Sri Lanka, see Alan Strathern, *Kingship and Conversion in Sixteenth-Century Sri Lanka: Portuguese Imperialism in a Buddhist Land*, and Alicia Schrikker, *Dutch and British Colonial Intervention in Sri Lanka, 1780–1815: Expansion and Reform*.

century to the middle of the nineteenth century, and by so doing to open up new ways of thinking about the consolidation of colonialism in South Asia. This was a key interval in Sri Lankan history, for it saw the end of the kingly line, the imposition of European control over the whole island, and the attempt to reform and create a modern colony. Though the hold of Europe over Kandy was certainly underway before the British arrival, it is incorrect to write the kingdom off as being in terminal decline in the eighteenth century. To begin this enterprise it is important to avoid the rigid dichotomies of the *kingdom* and the *colonial state*, the *indigenous* and the *colonial*, and the *highland* and the *coastal*.

The tactic of distancing and separating Europe from the Kandyans, and exaggerating the kingdom's isolation, emerged partly out of colonial rhetoric. In the words of a mid-nineteenth-century historian of Ceylon, the Kandyan kingdom was protected by a "species of natural circular fortification" which allowed the Kandyans to defy European modes of warfare for three centuries.[6] Writing in 1841, Lieutenant De Butts noted that the "physiognomy of mountaineers is influenced by the bold scenery amid which they reside, and which is supposed to impart somewhat of *hardiesse* to their manners and aspect." This physiognomic difference was said to map on to a divergence in character, evident in the "servility" and "effeminate" nature of the lowlanders, which contrasted with the elevated manliness of the highlanders.[7] In a popular commentary, Robert Percival wrote of how Europeans who were brought into contact with the climate of Kandy fell ill with debilitating "hill or jungle fever."[8] Added to this was the trope of the oriental despot, which the British quickly attached to the last king of Kandy, Sri Vickrama Rajasimha. One tale that the British publicized was how the king allegedly slaughtered the family of a fleeing minister, Ahalepola, by ordering the heads of Ahalepola's children to be put into a mortar and pounded with a pestle by their mother.[9]

The difficulty in holding up this contrast, of Kandy in the hills and the Europeans on the coast separated by culture, topography, climate, good rule, and the fabric of modernity, is illustrated again by considering the

---

6. Henry Charles Sirr, *Ceylon and the Cingalese* (1850), 1:219.

7. Augustus De Butts, *Rambles in Ceylon* (1841), 140–41.

8. Robert Percival, *An Account of the Island of Ceylon containing its History, Geography, Natural History, with the Manners and Customs of its Various Inhabitants* (1803), 239.

9. For a discussion of the veracity of all the sources connected with the slaying of the family of Ahalepola, see Paul E. Pieris, *Tri Sinhala: The Last Phase, 1796–1815,* appendix H.

road. An intriguing palm-leaf manuscript, *Kolamba sita mahanuvarata mahaparaak tanima* ("The Laying of a Road from Colombo to the Great City"), enrolls the road the British built in the 1820s in a genre of pilgrim poems, so that the colonists appear as Buddhist kings, upon whom merit can be loaded for the act of benefaction they have undertaken in building a road to Kandy.[10] Rather than seeing the road as a symbol of the expansion of colonial political economy or the rolling out of the modern state, this poet notes how pilgrims now worship at the Tooth Relic "peacefully" and how "the English are doing things surpassing the achievements of ancient kings." In India, by the nineteenth century, the legacy of the Mughal Empire was controversial, and a series of successor states and kingdoms gave way in piecemeal fashion to the colonial state. At the southern end of India, some large princely states remained in place under British protection. In Lanka, however, there was just one kingdom that remained by the seventeenth century, and it had disappeared by 1815. In this context, it was possible for there to be an evocative similarity between the kingdom of Kandy and the modern state, because they operated in the same field and because there was a regal vacuum. This symmetry allowed the British to be seen as inheritors of the mantle of Buddhist kingship, with its connected discourses of patronage and lineage.[11] It was an inheritance that the British were willing to cultivate for the political legitimacy it provided.

Unlike any of the European powers who preceded them, the British could be seen to stand in the lineage of Buddhist kings, because they took into captivity the last king of Kandy and into possession the Tooth Relic of the Buddha, the sacred signifier of their right to rule. Turning again to understudied palm-leaf texts, there are ceremonial verses written in honor of the British that utilize the same forms of description as were common in palm-leaf texts written in honor of kings. For instance, in *Jorgi Astaka*, written in honor of King George, the British monarch is "Bearer of all wealth, born in the solar clan, glorious like the sun with a chest which is the dwelling place of Goddess Lakshmi, who is like a beauty spot of other kings, like unto a house for Goddess Saraswati, and like a lion to the elephant-like enemy kings."[12] These lines may be interpreted alongside the well-established South Asian tradition of *prasasti* poetry, liter-

---

10. *Kolamba sita mahanuvarata mahaparaak tanima.* Palm-leaf manuscript in the Wellcome Library, WS 165 (v), University of London.

11. I first presented this argument in Sujit Sivasundaram, "Buddhist Kingship, British Archaeology and Historical Narratives in Sri Lanka, c. 1750–1850."

12. *Jorgi Astaka.* Palm-leaf manuscript in the British Library, Or 6601 (11).

ally "praise of kings." Other verses honor British officers in Ceylon, such as the orientalist and judge Alexander Johnston, or John D'Oyly, the first Resident of Kandy, who occupied the Royal Palace, or the Governor who oversaw the fall of Kandy, Robert Brownrigg. These individuals are lauded for being of the "solar caste," likened to a "lion," and linked to a "divine clan," all signs of kingly blood.[13] In another genre of texts, which give histories of the Tooth Relic of the Buddha, there is an interesting description of John D'Oyly. He is said to be "like the ancient kings" and to have been given the "three countries," which refers to the Tri Simhala, a signifier of the entire island, and to have arranged for sacrifices to the Tooth Relic. "He is a good man, fulfilling perfections to become a Buddha."[14] In a third category of texts, namely the *vamsas*, which narrate historical events, we find British governors slotted into a line which starts with the kings of the island. For instance, in a text titled *Sinhala Rajavamsa*, which has seen additions in the nineteenth century, the taking of Kandy is characterized as an act of the English king.[15] The text moves from the doings of Sinhala kings to describe the feats accomplished by British governors, such as the building of roads and bridges, and later the introduction of railways, clock towers, mail coaches, and electricity. The death of Charles Elliott, an Irish newspaper editor and an important figure in the rebellion of 1848 in Ceylon, with which this book ends, is also noted. There is also interest in the commemoration of the governors' good rule, in portraits and statues to honor their names. This idea of the British as new kings of the island, and as doers of righteous acts like the Buddhist kings of the past, only went into abeyance as the state disassociated itself from the official patronage of Buddhism in the mid-nineteenth century.

Yet the difficulty of separating Kandy from Europe must also be appreciated from the perspective of Kandy. This kingdom did not see itself as bound by the highlands. Though it kept knowledge of how to access the kingdom privileged, it saw itself as lord of the entire island.[16] The

13. See other manuscripts under classmark Or 6601 (11), including *Jonston Astaka*, *John Armour Astaka*, *Deanwood Astaka*, and *Sir Robert Brownrigg Astaka*. For D'Oyly, see *Dalada Astaka*, Or 6602 (8).

14. *Dathagotrapradipaya*. Palm-leaf manuscript in the British Library, Or. 6606 (27).

15. *Sinhala Rajavamsa*. Palm-leaf manuscript at the Colombo Museum Library, Sri Lanka, A.R. 14.

16. The need to engage with the ritual and symbolic history of Kandy and to see it as a tributary overlord is well made in Michael Roberts, *Sinhala Consciousness in the Kandyan Period, 1590s to 1815*. See also John Holt, *The Religious World of Kirti Sri: Buddhism, Art, and Politics in Late Medieval Sri Lanka*.

Kandyan kings thought that the island was a territory specially sancti-
fied by the Buddha, who had appeared magically three times on the island
after his enlightenment. Hence the way of referring to the entire island as
Tri Simhala—which takes up the idea that the territory was divided into
three historic kingdoms under one umbrella. Ballads concerned with geography and boundaries specifically took up the term "Sri Lamkadvipaye"
[island of Sri Lanka]. The Buddhist monks of the lowlands, such as Karatota Dhammarama (1737–1827), who is a key figure for this study, attached
significance to the patronage of the Kandyan king and those monks with
close relations with him. Even as he did this Karatota sought for the patronage of the Dutch and the British in turn. If the Kandyans could see
themselves as universal lords of what became the entire territory of the
colonial state, and the British could be cast as Buddhist kings, another key
route through which the kingdom was enmeshed with the modern state
lies in its placement vis-à-vis indigeneity. It is not the case that Kandy was
simply a Sinhala kingdom, to take up the label of the majority ethnicity of
the island, or that British rule of the island was exclusively British.

This kingdom was characterized by an attention to both the indigenous
and the cosmopolitan, and the ideal of universal kingship which operated
at court sought to draw in diverse communities, individuals, and European
powers as vassals to the lord who ruled.[17] These ranged from Muslim traders
and physicians to Malay soldiers and Englishmen. One of the key riddles of
this period is how the last kings, despite sponsoring fantastic spectacles of
Buddhist piety and learning, could at the same time have been born as Hindus of South Indian descent, who took the European label "Malabar," the
earlier term for people who later became identified as Tamil.[18] When victorious against the British in the first Anglo-Kandyan war of 1803, the ruling
king could be lauded as a splendid specimen of Sinhalaness. However, once

17. This claim complicates further the sense of Roberts' arguments, by urging that
Sinhalaness, though present, was not the only ideological orientation. There were ways
in which "others" were integrated into the structures of the state, so that it could be
cosmopolitan, even while Sinhalaness was kept alive.

18. The status of these kings who constituted the "Nayakkar line" has served as
a focus of debate; see K.N.O. Dharmadasa, "The Sinhala-Buddhist Identity and the
Nayakkar Dynasty in the Politics of the Kandyan Kingdom, 1793–1815," and H. L.
Seneviratne, "'The Alien King.'" For more recent contributions to this debate, see
R.A.L.H. Gunawardana, "Colonialism, Ethnicity and the Construction of the Past: The
Changing 'Ethnic Identity' of the Last Four Kings of the Kandyan Kingdom"; Michael
Roberts, *Sinhala Consciousness*, 46–52; and Anne M. Blackburn, *Buddhist Learning
and Textual Practice in Eighteenth-century Lankan Monastic Culture*, 32–35.

the kingdom had fallen by 1815, the same king could become, in the words of another palm-leaf text, the *Vadiga Hatana*, a "Demala [Tamil] eunuch, a heretic wicked and vile," or alternatively, "a large earth-worm that came to dig the ground day and night."[19] The growing number of this king's relatives in the city were described in this latter poem as "crazed devils" who ignored "Sinhala men and caused distress to them." This king's rule, according to the poet, was a direct threat to Buddhism; he demolished stupas, and, close to the shrine room of the Tooth Relic, built an octagon which rose above it and in which he enjoyed the pleasure of women.

The need to complicate the binary of precolonial and colonial, dependent as it is on a simple sense of what counts as indigenous and foreign, is also evident in relation to British rule. When the kingdom of Kandy fell to the British, the Convention that marked the event issued this remarkable promise: "The religion of Boodhoo, professed by the chiefs and inhabitants of these provinces is declared inviolable, and its rights, ministers, and places of worship are to be maintained and protected."[20] Even before this promise was made, British orientalism and exploration followed the intellectual pathways of Kandy by turning for texts and images to a class of monks, who emerged out of a reformation of Buddhism undertaken under the patronage of the Kandyan kings in the eighteenth century. This became a program of reformation which sought to resuscitate knowledge of Pali texts, a language that the British orientalists saw as holding the key to the Sinhala past. At the same time, British interest in exploration and antiquarianism followed the Kandyan traditions of pilgrimage and religious travel, so much so that at times some ordinary Britons could be thought of as pilgrims rather than explorers, despite their scientific instruments. In the case of botany and agricultural knowledge, which was critical to the emergence of Ceylon as a plantation colony by the 1840s, the major botanical garden at Peradeniya, outside Kandy, was in fact built on the site of a temple garden.

The entanglement of the Dutch in this story provides another direction of complication, beyond the binary of Kandy versus Britain. For the structures of the state, in medicine, surveying, and other technical departments, which were key to projecting public benefit in the new colony, be-

---

19. M. E. Fernando, ed., *Vadiga Hatana* (also titled *Ahalepola Varnava*).

20. "Official Declaration of the Settlement of the Kandyan Provinces," in John Davy, *An Account of the Interior of Ceylon, and of its Inhabitants*, appendix 1, 500–501. For controversy over this clause, see Paul E. Pieris, *Sinhale and the Patriots, 1815–1818*.

gan to rely heavily on Dutch intermediaries and Portuguese descendants, in addition to the odd German, Italian, and Frenchman. In fact British botany drew from Dutch traditions and also from the work of a range of mixed-ancestry individuals. This was a region of Asia where multiple layers of Europeans had traveled and settled, and their descendants mattered for the nature of British rule. Even the Americans entered the island as missionaries in Jaffna from 1813. As the decades of the nineteenth century rolled past, the British, having turned to the Buddhist monks at first for the information they needed to govern and legislate, turned now in an age of liberal reform from the 1830s to other Europeans who could prop up the expansion of their state bureaucracy, thus creating a hierarchy of assistants and races. How then can this regime be seen as British? Working across the Anglo-Dutch fault line is in keeping with a new literature in Dutch imperial historiography which stresses similarities and inheritances in the making of imperial structures, and in terms of people and policy.[21]

In appreciating the impossibility of separating the authentic Kandyans from the British, or even of isolating the British as a monolith, one gains a more nuanced view of the appearance of the modern colonial state. Yet it cannot be the intention of this book to tell this story from all the geographic vantage points in the island. Since Kandy is often taken as indicating the "indigenous" heritage in contemporary Sri Lanka, this argument focuses on its entanglement with the nineteenth-century British colonial state. Following on from this, neither the coast nor the highland is inherently more "indigenous." At various points in the book other centers of political power and heritage come into brief focus, such as Colombo, Galle, or Jaffna. It would have been plausible to reject Kandy more wholly and to take the coastal belt as indicative of a baseline for assessment of the colonial takeover. After all, Kandy in the eighteenth century did not represent the majority of the island's population or wealth. However, the focus on Kandy is defensible for the fact that it represented the lineage of kingship, which was vital to symbolic politics, and it was strategically and

---

21. For this argument with respect to Sri Lanka, see Schrikker, *Dutch and British Colonial Intervention in Sri Lanka*. For other work on the interface of Dutch and other imperial structures, see Leonard Blussé, *Visible Cities: Canton, Nagasaki and Batavia and the Coming of the Americans*; Kerry Ward, *Networks of Empire: Forced Migration in the Dutch East India Company*; and Eric Tagliacozzo, *Secret Trades: Smuggling and States Along a Southeast Asian Frontier, 1865–1915*.

economically important for the British to subdue it. Yet, this concern with Kandy must not imply that Kandy can be equated with Lanka. Rather, in addition to Kandy and the British, there were other political, religious, and intellectual currents at play in Lanka, and such communities articulated alternative conceptions of the island's placement in the sea.

❀

Debates about the colonial transition inevitably fall into questions of continuity and change, and impact and rupture.[22] Looking from South Asia, these debates are about the ascendance of the British Raj in place of the successor regimes to the Mughals, but in Lanka the sense of colonial transition is more complicated in that there were several transitions in this period, from the Dutch to the British, from Kandy to the British, and by the colonization of the whole island for the first time. This necessitates a wider usage for the phrase "colonial transition" as applied to eighteenth- and nineteenth-century Lanka, and this is the more inclusive sense in which this term is used here. In addition, by bringing Kandy alongside the British, and by drawing on palm-leaf manuscripts, temple murals, and the colonial archive, this argument takes for granted the fact that sources and voices from multiple perspectives need to be studied. In a place such as Sri Lanka, which has witnessed repeated waves of migration, colonization, and longstanding trade contact, the idea of impact, with its sense of division between colonizer and colonized, cannot be upheld.[23]

Instead of deciding between continuity or change, the argument offered here stresses *recycling* and *movement*. I have thought of this claim in the manner of a wheel on a bullock cart taking the road from Kandy to Colombo. After opening the road, Governor Barnes wrote in 1827 that "the number of carts in Colombo has doubled since the carriage road to

22. For debates about colonial transition from multiple perspectives, see Seema Alavi, ed., *The Eighteenth Century in India*; C. A. Bayly, *The Origins of Nationality in India: Patriotism and Ethical Government in the Making of Modern India*; Partha Chatterjee, *The Nation and its Fragments: Colonial and Postcolonial Histories*; Nicholas Dirks, *Castes of Mind: Colonialism and the Making of Modern India*; Peter Marshall, ed., *The Eighteenth Century in Indian History: Evolution or Revolution?* and Gyan Prakash, "Writing Post-Orientalist Histories of the Third World: Perspectives from Indian Historiography."

23. For a critique of the idea of colonial impact, and also of colonial transition, see Jon Wilson, "Early Colonial India beyond Empire."

Kandy was opened and is daily increasing."[24] Like the wheel of such a cart, the process of colonial transition saw a forward momentum onward to the colonial state, but it picked up on prior knowledge and pathways and spun round on itself. Traveling on a cart involves a journey which is rather bumpy, and the bullocks regularly need to be reined in by the driver and reminded of the correct route and destination, or else it is possible to end up in a very different place, such as a rice field by the roadside. Unlike the alternatives of continuity and change, the terms of recycling and movement gesture toward the way in which colonial transition is not singular in time or well defined in terms of path or space. There isn't one colonial transition or even one axis against which colonial transition can be measured. Change is constant and every change is changed in turn, and continuity is there but in that continuity the very idea of what came in the past, namely the precolonial or the indigenous, is repackaged and redefined.

Accordingly, in this view of colonial transition the definitional mutability of the "indigenous" within the transformations of global empire and capital is manifest. This follows recent work in South Asian intellectual history by Shruti Kapila, Andrew Sartori, and others, which scrutinizes the history and reassembly of concepts, such as the self or culture, rather than positing them as givens or tracing the life of a given concept in practices of imitation or diffusion.[25] In this book, indigeneity is cast as working alongside cosmopolitanism. These two terms have separate meanings, but they are not opposites as much as concepts which feed off each other. Indigeneity is taken to include a variety of changeable claims to and definitions of belonging in the island space, linked to descent, lineage, status, residence, culture, ethnicity, and religiosity. Meanwhile, cosmopolitanism is understood to include many different processes of universalizing reason and attachment, which stretch beyond the boundaries of the polity. Yet there were local cosmopolitanisms and assertions of the indigenous which sought to naturalize the seemingly foreign, and this makes a clear-cut distinction between these two terms impossible. The visibility of indigeneity and cosmopolitanism is predicated on the contexts of their operation, for example, the state and the religious and intellectual order. The argument of this work is that the changing state in Lanka sought to define the indigenous over and over again, and yet it had to come to terms with the cos-

24. Dispatch from Edward Barnes, dated 27 April 1827, CO 54/97, The National Archives, Kew, U.K.; hereafter TNA.

25. See Shruti Kapila, ed., "An Intellectual History for India," and Andrew Sartori, *Bengal in Global Concept History: Culturalism in the Age of Capital.*

mopolitan in doing so. This manner of tracing the co-constitution of the indigenous and cosmopolitan follows recent theoretical works, according to which the desire for cosmopolitanism undoubtedly comes from wanting to engage with otherness, but it begins from a position of rootedness and self. Both terms have boundaries which are permeable.[26]

The particularity of the Lankan story also needs to be appreciated and contextualized in relation to the Indian Ocean more broadly.[27] Kandy was not a landlocked kingdom: it had connections through migration and patterns of religious, cultural, and economic life to South Asia as well as Southeast Asia. Before 1796, the king's traders and monks at times evaded Dutch control, and at other times utilized Dutch vessels for the express purpose of keeping links alive. Buddhist monks traveled to and from Burma and Siam, taking and bringing texts, and the reformation of Buddhism in eighteenth-century Kandy depended on the reintroduction of higher ordination by Siamese monks in 1753. Muslim and "Malabar" traders, in addition to Chetties, also connected the Kandyan court and rural villages to the sea, and kept alive Kandy's historic ownership of five ports along the coast for its trade. The kingdom's ownership of these ports had been interrupted when the Dutch proclaimed their right of rule over a continuous strip of land around the island by 1766. Nonetheless, Kandy saw itself not only as lord over the island, but with the power to speak and connect across the sea. This does not take away from the fact that its economy was heavily dependent on service tenures and that land mattered deeply to it; rather, it makes the case that it was a polity which saw itself as a center, and that sense of centricity could encompass the ocean as well as the land. Even as the last king sought to conjure the ocean through his lake, and laid his capital around the ocean lake, the ocean beyond was also no absolute barrier to Kandy's reach.

Thinking from the Indian Ocean provides another way of working out the relations, during the period of colonial transition, between land and sea. The division of highlands versus coastal belt did not simply mark Europe against Kandy, but neither must the shape of the island as land within

26. See, for instance, Steven Vertovec and Robin Cohen, eds., *Conceiving Cosmopolitanism: Theory, Context, and Practice*, and, for a long perspective, Sheldon Pollock, "Cosmopolitan and Vernacular in History," and also Walter D. Mignolo, "The Many Faces of Cosmo-polis: Border Thinking and Critical Cosmopolitanism."

27. For recent works on the Indian Ocean which stress the role of India, see Thomas Metcalf, *Imperial Connections: India in the Indian Ocean Arena, 1860–1920*, and Sugata Bose, *A Hundred Horizons: The Indian Ocean in an Age of Global Empire*.

sea be taken at face value. Another two words might be used to summarize the argument of this book about colonial transition, namely *islanding* and *partitioning*. Although the intervention of the British and the takeover of the whole island did not displace the legacy of Kandy, it nevertheless set in process a journey where every change and counter-change was directed in a nonlinear sense to the colonial state. Unlike India, Ceylon was a Crown colony, and initially a garrison state under military governors. This is important, because the rivalry between the East India Company in India and Crown in Ceylon meant that, in governmental terms, the island was cast off from the mainland by the 1830s. It became a separable island colony, and there was a concerted attempt to dredge a channel between the island and the mainland to prevent Company vessels from needing to go around the island in traveling between Bombay and Calcutta. The apparatus through which Ceylon was unified under a centralized regime of what Indian observers expressly declared to be "colonialism" set in motion a discursive and intellectual way of thinking and writing of this space as a romanticized and sexualized island, a lost Eden, and a place which was very different to the barren and Hindu mainland. The island's Buddhism was seen to hold a key to the mainland's past, and this religious system was thought to have lessened the force of some of the norms of society in India, such as caste or gender oppression. In an important revisionist work, Jon Wilson writes: "The categories of 'colonial discourse' that Britons used to justify colonial domination emerged *within* rather than before the process of state formation in India."[28] It was through the colonial state that Lanka was repositioned in the ocean as an island. While Ceylon's British rule shared many parallels with British India, this act of disconnection meant that it served as a different laboratory for forms of state-making, following a separate chronology and leaving a different legacy, for instance in relation to ethnic identities.[29]

Partition is a loaded word in South Asian historiography, associated with the cataclysmic events of 1947 that saw the birth of independent Pakistan and India, and its history has been impacted by the nationalist

28. Wilson, "Early Colonial India," 968.

29. For other works that stress the need to consider the status of islands critically, see Rod Edmond and Vanessa Smith, eds., *Islands in History and Representation*; Richard Grove, *Green Imperialism: Colonial Expansion, Tropical Island Edens and the Origins of Environmentalism, 1600–1860*; and Greg Dening, *Islands and Beaches: Discourses on a Silent Land, Marquesas, 1774–1880*.

Fig. 0.2. "India and Ceylon" (London, 1831), a map published by Baldwin and Cradock for the Society for the Diffusion of Useful Knowledge, showing South India and Ceylon. Author's photograph.

ideology of these two post-colonial states.[30] The argument here emphasizes partitioning rather than partition to highlight that what happened to the island's relationship to the mainland was a process rather than an event—and an incomplete one. Partitioning might be conceived of in at least three different senses for the period at the start of the nineteenth century, which saw the rise of British rule in Ceylon, and these relate to the policing of flows of peoples, connections of lands, and the exchange of commodities. It is important to emphasize that this partitioning was not a radical dislocation of past patterns, and in this sense it did not come purely from British initiatives. The legacy of kingly understandings of the unity

30. See, for instance, Ayesha Jalal, *The Sole Spokesman: Jinnah, the Muslim League and the Demand for Pakistan*, and Joya Chatterji, *Bengal Divided: Hindu Communalism and Partition, 1932–1947*.

of the island—and the idea of Tri Simhala—is significant here. There is evidence that what it meant to be "Sinhala" was hardening in the eighteenth century prior to the arrival of British colonization and its restatement of "Sinhala." In addition, British policy with respect to the separable status of the island followed the Dutch, and the Dutch in turn had a keen sense of the utility of bounded sovereignties. At the very start of British rule, and briefly, there was an experiment with Company rule, and so not a knee-jerk reaction against all things Indian.[31] Further, this partitioning certainly did not leave fully formed nations or even stable colonies. But in using the word "partitioning," it is possible to point to the intent of British colonialism to create a separate state, with separate channels of accountability and a distinct idea of space. This intent seeded the structural and ideological forms which in the long run gave rise to Sri Lanka.[32]

This argument pluralizes the terrain and stretches the periodization of accounts of partition in South Asia. Unlike its more famous counterpart, the bottleneck of the partitioning of the island and the mainland occurred at the start of the British rule rather than at its end. Extrapolating from this forgotten partitioning, it is time to link together the fragmentations of territory in different parts of "Greater India," with their different chronologies, encompassing Nepal, Burma, Afghanistan, Bhutan, and Tibet, among others.[33] It is also important to work across the axis of South and Southeast Asian history, by comparing events in Lanka to the Straits settlements, later Malaysia and Singapore. This will yield a rather more complicated and politically meaningful history of partition and provide a deeper understanding of the connection of the territorial form to notions of subjecthood and indigeneity. Like 1947 in India, the partitioning of Sri Lanka from India was not a success. The control over the migration of peoples, the making of separate channels of navigation, and the establish-

31. This follows the work of Michael Roberts, *Sinhala Consciousness*, and Alicia Sckrikker, *Dutch and British Colonial Intervention in Sri Lanka, 1780–1815*.

32. This takes further previous studies of the early British period of Sri Lankan history: Colvin R. de Silva, *Ceylon under British Occupation, 1795–1833*, and U. C. Wickremeratne, *The Conservative Nature of the British Rule of Sri Lanka: With Particular Emphasis on the Period, 1796–1802*. See also the recent articles by John Rogers, "Caste as a Social Category and Identity in Colonial Lanka," and "Early British Rule and Social Classification in Colonial Lanka."

33. The call to think of partition in a new transnational space is evident in the work of Faisal Devji; see his recent Kingsley Martin Memorial Lecture in Cambridge, 2009, entitled "Muslim Zion." For the forgotten partitions of Burma, see Thant Myint-U, *Where China Meets India: Burma and the New Crossroads of Asia*.

ment of a thriving independent economy for the island did not take their intended shape. The contemporary legacy of British colonization in South Asia lies therefore in the limited and yet disruptive attempts to reorganize identity, land, and trade. However, historiographically speaking at least, the motivation of British colonists to make a separate territory has until now won the day. For Sri Lanka is roundly ignored by Indian historians, even as sources relevant to Lankan history have been housed in different places to those relevant to Indian history.[34] At the same time, it is not seen to be part of the history of Southeast Asia. In London for instance, the sources for Indian history are kept primarily at the British Library, separated by a tiresome journey from the sources for Lanka in the National Archives.

Viewing British state-making in Lanka as *islanding* and *partitioning* must be put together with the idea of colonial transition as *recycling* and *movement*. Both the making of the island and the partition of the island from the mainland were processes, things that worked with aims rather than simply arriving at their destinations. In this sense a separable state— islanded and partitioned—was in formation rather than a fully formed object. Among the consequences of this process of state-making, one of the most critical comes into view in chapters 1 and 8, namely ethnicity.

⚬

The origins of Sri Lanka's ethnic division between the majority Sinhalese and minority Tamils has quite naturally, in the context of the recent war, served as a key question in the historiography on Sri Lanka, and it is appropriate that this study uses this theme as a start and an end.[35] Recently, the literature has shifted to a post-Saidian position by suggesting that the advent of the British in the early nineteenth century is important but not

34. As a way of illustrating this, I note that in Sugata Bose and Ayesha Jalal's *Modern South Asia: History, Culture and Political Economy*, which is a very useful textbook, with a title that consciously seeks to subvert the national story of India, there is only the briefest mention of the island.

35. See, for instance, Jonathan Spencer, ed., *Sri Lanka: History and the Roots of Conflict*, and Pradeep Jeganathan and Qadri Ismail, eds., *Unmaking the Nation: The Politics of Identity and History in Modern Sri Lanka*. The seminal article on the formation of identity in the island also emphasizes the advent of colonial categories as definitive. It has been reprinted twice in 1984 and 1990 since its first appearance: R.A.L.H. Gunawaradana, "The People of the Lion: Sinhala Identity and Ideology in History and Historiography."

all-defining, and this is particularly marked in the work of John Rogers.[36] According to this line of argument, the consolidation of ethnicity is usually dated to the period after the 1830s, which saw a liberal age of reform that dispensed with caste differences, leaving ethnicity as the prevalent form of colonial social categorization; this contrasted with the continued use of caste in India. This new direction of argument emphasizes that early colonial categorizations before 1830 were slippery and ill-formed and that their power was restricted, and that they should not be taken as a baseline. Yet Rogers hints at the significance of islanding and partitioning: "Why did race or nationality emerge as the central social category in colonial Lanka . . . ? The British decision to govern Lanka separately from the mainland created conditions favourable to the development of distinct patterns of identity formation on the island."[37]

Thinking critically about this process of separation, it is possible to argue that British state-making in the early nineteenth century raised the question first of "Who belongs?" which turned by the 1830s into "Who represents?" Chapter 1 is concerned with the first of these questions and shows how Britons sought to see "Malabars," who later were termed "Tamil," as belonging in mainland India, in contrast to the true indigenes of the island, who were Sinhalese and Crown subjects. In this they misunderstood the way in which indigeneity was couched within the idea of universal kingship in Kandy. This means that the period prior to 1830 is critical. Chapter 8 shows how belonging could become institutionalized in an age of reform, with the establishment of a Legislative Council with "native" members and with the expansion of schools and the press. This new battery of initiatives generated a further set of anxieties on the island. In particular, there were rival quests to take control of the emerging public spheres, and this led to the provision of a structural home for notions of ethnicity, language, and European descent and status. This was a contradictory consequence of the reformers' strategy of unity and their impulse to liberalize the people's engagement with government. But even as questions of "native-ness" were central to the convolutions of the state and the rise of divergent publics, the resonance of kingship carried on into the 1848 rebellion, which marks the end point of the book.

36. John D. Rogers, "Colonial Perceptions of Ethnicity and Culture in Early Nineteenth-century Sri Lanka," and Rogers, "Early British Rule and Social Classification in Lanka." See also, K.N.O. Dharmadasa, *Language, Religion and Ethnic Assertiveness: The Growth of Sinhalese Nationalism in Sri Lanka.*

37. Rogers, "Early British Rule," 646.

Just after the 1848 rebellion, a very short letter to the editor appeared in *The Observer* from "A Tamil." The correspondent hit on an issue which illustrates the argument about ethnicity rather well. Why, the correspondent asked, did the Europeans on the island call the Tamils of Ceylon "Malabars" rather than "Tamils," when they didn't use this label for the Tamils of the mainland? Did this denote an attachment to the theory that they had arrived in the island from the "Malabar coast"?[38] Questions of ancestry and community were also central in other arenas, such as the intellectual societies of the island. The Royal Asiatic Society of Ceylon, in addition to providing a context for the study of the island's history, also served as a venue for discussion of the lineage and traditions of different communities. The first president, Justice John Stark, addressed the society in 1845, the year of its formation, specifically laying out the need to trace the history of the island in terms of waves of migration, in relation to the Sinhalese and "Malabars," as well as the Portuguese, Dutch, and English.[39] *The Examiner* took particular interest in the society's discussions about the relative "improvability" of the various "native" communities. The first in a series of articles devoted to the society's early discussions provided a contrasting diagnosis of the plight of the Sinhalese vis-à-vis other communities: "The greatest strides in native improvements of any kind have been accomplished by foreigners, such as Moormen, Malabars &c. We have never yet heard of a Sinhalese becoming really a capitalist. All the native ship-owners are either of the Burgher class or Moormen. The Government contracts are in nearly every case taken by others than natives of Ceylon."[40] Taking the argument further, the journal urged that if the Sinhalese were to be improved, they needed to be taxed; and this was advice given prior to the rebellion, where taxation was a key issue. All of this bears out the signs of ethnicity in the emerging public spheres, and of the relationship between discussions of ethnicity and histories of mobility and origin. Yet the point of relevance to the argument is that these debates

38. "Malabars and Tamils," *The Observer*, 13 December 1848. Another correspondent, taking the title of Kusta-raja, the name of the statue of the coconut king, to be discussed in chapter 5, clarified the question by pointing to the Portuguese ancestry of the term. "Malabars and Tamils," *The Observer*, 21 December 1848.

39. "Address by the Hon Justice Stark, Delivered at the Opening of the General Meeting of the Asiatic Society of Ceylon, Thursday May 1, 1845," 1.

40. "The Advancement of the Singhalese," *The Examiner*, 28 May 1847. As an aside, it is worth noting that in fact there was a Sinhalese capitalist class developing in this period.

were framed within the contortions of the colonial state. The *Examiner's* views on race take up discussions of maritime commerce and free trade.

Between the first and the last chapter, this book is organized according to themes. Though it moves from eighteenth-century Kandy to mid-nineteenth-century rebellion, its chronology must only be taken loosely, in the spirit of the argument about colonial transition not having a linear track. The themes that are studied from chapters 2 to 7 are trade, orientalism, exploration and antiquarianism, botany and landscape, geography, cartography and engineering, and medicine. This list signals the centrality of the question of the relationship between knowledge and power to debates about colonial transition. In a seminal work, Christopher A. Bayly writes of South Asia: "The conquerors needed to reach into and manipulate the indigenous system of communication in new colonies."[41] In this view the British were able to collect information which was critical to rule India through the spread of schools and missions, and the expansion of the bureaucracy and the press. But Bayly's argument is as much about the limits of the colonial state, in its reliance on Indian runners, spies, and holy men in order to gain access to this knowledge. These limits were also apparent in the state's ability to run aground in moments of rebellion, such as in 1857–59, for instance, when there were "knowledge panics" on the part of the British. One implication of Bayly's work is that colonial knowledge should not be seen in abstract terms as what is held in the head by the British, nor should it be posited as a unit or divided into two categories of practice, across the axis of colonizer and colonized. Rather knowledge was what it took to govern, and it had a rich spectrum of manifestations. Indeed, in South Asia there were variegated and multiple communities who held specialist knowledge, who utilized oral and printed modes of thought, and who picked upon divergent historic traditions of theorizing the body, language, and political debate. At the interface of these rival strands of knowledge, or rival "orders" to use Bayly's language, there was a great deal of conflict, arising at times from the exchange or theft of knowledge, and these episodes were often tied to exercises of political legitimacy. While the accumulation of knowledge has recently been posited as one of the chief means through which an empire's power was made manifest, moments where the British "did not know," or where they relied on others, bear out how an emphasis on power in the literature needs to be balanced by attention to weakness and failure.

41. C. A. Bayly, *Empire and Information: Intelligence Gathering and Social Communication in India, 1780–1870*, 6.

Bayly's argument works especially well for Ceylon. Because of Lanka's prior history of colonialism, the range of mixed-race communities and European descendants was greater than the average in South Asia; at the same time, the vibrancy of Buddhist, Hindu, Christian, and Muslim traditions of knowledge provided greater intellectual breadth. To cite Percival again: "There is no part of the world where so many different languages are spoken, or which contains such a mixture of nations, manners and religions." This mixture included "Moors of every class, Malabars, Travancorians, Malays, Hindoos, Gentoos, Chinese, Persians, Arabians, Turks, Maldivians, Javians, and natives of all the Asiatic isles; Parsees . . . [and] also a number of Africans, Caffres, Buganese, a mixed race of Africans and Asiatics" (fig. 0.3).[42] Further, Ceylon's Crown state was more penetrative than the Company's indirect rule in India. By the third and fourth decades of the nineteenth century, there arose a sphere of public debate between the governing classes, other Europeans, and local elites, akin to that in North India, but this sphere was marked by the military culture of British presence on the island. Military governors sought to intervene in wars conducted in the press; and the means of dealing with epidemics and disease bore the imprint of the military context, as did programs of road- or bridge-building. Though Buddhist monks may be compared with the *pandits* and *munshis* who worked alongside the British orientalists and administrators of India, the political position of monks in Lanka was rather different. For the British had taken the *sasana* in Kandy under its direct patronage, and in doing this had respected it as a center of political power, in the absence of a king. In these ways, the island provides a lens through which to view India, but where contesting traditions of knowledge in the nineteenth century take an even greater political charge and fragment into an even more eclectic range of possibilities of attachment and heritage.

The island's heterogeneity notwithstanding, the British turned to narrow categories to define its culture and traditions, even as they recycled extant traditions. One key area where this is apparent is in relation to Buddhism.[43] When British orientalists sought to "discover" sacred sites,

42. Percival, *An Account of the Island of Ceylon*, 143.

43. It is worth noting that there is a very good literature on Buddhist studies of Lanka, and the argument of this book intervenes in that literature too. See, for instance, Anne M. Blackburn, *Buddhist Learning and Textual Practice in Eighteenth-century Lankan Monastic Culture*, and Blackburn, *Locations of Buddhism: Colonialism and Modernity in Sri Lanka*; also K. Malalgoda, *Buddhism in Sinhalese Society, 1750–1900: A Study in Religious Revival and Change*, and Richard Francis Gombrich and Gananath Obeyesekere, *Buddhism Transformed: Religious Change in Sri Lanka*.

Fig. 0.3. "Priest of Buddha, Malay(?), Moorman Constable, and Kandyan Woman," illustrative of the variety of peoples of the island, from a volume of drawings entitled *Sinhala Drawings*. Courtesy Lindley Library, Royal Horticultural Society, London.

they fitted into a tradition of kingship, for the Kandyan kings had done the same. Following Charles Hallisey, there was an "intercultural mimesis" in orientalists' views of Buddhism in Ceylon. By this he signals how the "culture of a subjectified people, and its patterns of change, influenced the investigator to represent that culture in a certain manner."[44] However, even as kingly Buddhist patronage influenced orientalist study, Buddhism was cast as a discrete religion to be studied by attention to a particular set of texts, images, and sites. It became de-linked from Hinduism in British thought, and this was in keeping with Lanka's partitioning from the mainland. Ceylon became a signifier of Buddhism on the global stage, as priests and artifacts of all kinds were taken to Britain for study and display. The link between London and the island redefined what counted as indigenous.

Even as London needs to be brought into the permutations of Lanka's intellectual history, the spaces in which knowledge operates may also be deconstructed by unraveling the island itself, rather than seeing it as a whole. The journey from Kandy to Colombo is a useful starting point for the book, but it is necessary to leave this road, taking up other routes within the island. Chapters 4 and 5 seek to break the island up, by showing the work that went into making it a whole unit. Chapter 4 moves into the peripheries of the old Kandyan kingdom by looking at the changing status of Anuradhapura, the ruined city, and Sri Pada or Adam's Peak, the mythologized mountain. By placing these peripheral localities at center stage, it is possible to see how both Kandyan kings and British colonizers sought to bring the furthest localities under their authority, but also how often such spaces sustained rival religious, political, and cultural identities that undercut those which were emanating from the center and which sought to make meaning for them. Chapter 5 considers the representational politics of the division between the coastal belt and the highlands, which mapped on to the quintessential colonial divide between the tropical and the temperate, by looking at two of the icons which were central to these spaces, the coconut tree and the hill station. This division, which took further shape even as the British penetrated the furthest highlands, was a means of mobilizing colonial power, and it recycled extant meanings, traditions, and encounters with the land, which were kingly.

Yet the slippages and entanglements between knowledge traditions of all kinds does not mean that this was a nonviolent story. Chapter 7, which

44. Charles Hallisey, "Roads Taken and Not Taken in the Study of Theravada Buddhism," 33.

considers medicine, brings this out clearly. Medical practice was pivotal to the making of the island state. The British saw diseases such as smallpox and cholera as having a maritime history, and the carriers of disease were said to be migrants. Often "Malabars" and Muslims, and merchant families and wandering holy men, were suspects in the hunt for the sources of disease. This tracing of disease pathways set the context for discourses of class and race. Measures of surveillance were not only directed outward to arrivals from India and elsewhere, but also inward, most visibly in the layout of the capital of Colombo. Colombo's various districts came under close scrutiny in the midst of epidemics, and these measures served to entrench their character either as European or "native" residential areas. Indeed because medicine pertained to the interstices between the state and body, it was a prime means whereby the category of public benefit was first defined in Ceylon, and also where resistance to the new state structures was articulated, for instance by those who resisted vaccination.

The making of the garrison state of British Ceylon therefore necessitated scrutiny of the boundaries of sea and land in the Indian Ocean. Until quite recently Indian Ocean historiography has oriented itself to the period prior to 1750, and to the overriding question of the impact of early modern imperialism on extant systems of trade and exchange. This great sea has been divided into sub-zones, for instance, the Arabian Sea and the Bay of Bengal, or the southwest Indian Ocean. Yet Lanka does not belong exclusively in a particular zone, for it lies in the middle of the Indian Ocean even as it sustained connections to the southwest at the moment of the advent of the British. To understand this, it is helpful to take up Sugata Bose's call for a more creative use of "space and time" in this "inter-regional arena," in order to bring to life more horizons in the Indian Ocean, and to pluralize the spaces of water in which ideas and structures moved at particular times but not at others.[45] In this regard, it is important to keep in mind that Ceylon was made into a Crown colony in the turbulent global age of revolutions, characterized as it was by contradictory commitments to rights as well as state power, to free trade as well as indentured labor.[46] In this narrative, Ceylon may be interpreted alongside lands along the far southwest Indian Ocean, as Mauritius and the Cape were also taken into British Crown control in the context of the Napoleonic wars. Later the Commission of Eastern Inquiry, which proposed a

45. Bose, *A Hundred Horizons*, chap. 1.
46. David Armitage and Sanjay Subrahmanyam, eds., *The Age of Revolutions in a Global Context, c. 1760–1840*.

wide-scale program of reforms in Ceylon in the 1830s, came to the island after investigating these two other Crown colonies.

Moreover, Lanka's place in the web of British Crown colonialism with Mauritius and the Cape was overlaid upon how the Dutch had previously connected southern Africa to Lanka in the century that had just passed. In Lanka, as in the southwest Indian Ocean, the British had to come to terms with how to handle creolization and what it took to make a successful plantation colony.[47] By the time of the 1848 rebellion, Indian indentured laborers were beginning to change the nature of labor across this world, and Indians thought themselves to be "overseas" in all of these three places. Imperial careering and traveling also linked Mauritius, the Cape, and Ceylon.

Indian Ocean historiography is characterized by a spirit of liberation, where the Indian Ocean is nobody's property, and it is not constrained by the scholarly protocols of area studies, of East Africa, South Asia, or Southeast Asia for instance.[48] Yet it is undeniable that this concept of the Indian Ocean is itself a construction that needs to be critiqued, and on its worst showing in the literature it comes to stand for the easy movement of people, ideas, and things, in constant motion across vast distance. Such a view can at times take individual and unrepresentative biographies as indicative of a larger narrative, or focus on port cities to the exclusion of the hinterland. Rather, the methodology of this work is to return Indian Ocean studies to the local, without reifying a new area in another's stead. The island as a locality might be one mode of interpretation that could be applied more widely to this sea. Its microcosmic scale is useful for historical study of broader processes. Yet, to understand "islanding," it is important not to take the physical geography of islands at face value, or to assume that the localness and boundedness of the island is natural. Instead, by scrutinizing the making of islands, through discourses and as states, and as intensive spaces of colonialism, it is possible to move away from

47. For recent work on the southwest Indian Ocean, see Pier M. Larson, *Ocean of Letters: Language and Creolization in an Indian Ocean Diaspora*; Kerry Ward, *Networks of Empire*; and Megan Vaughan, *Creating the Creole Island: Slavery in Eighteenth-Century Mauritius*.

48. For analysis of the Indian Ocean as a historiographical method, see Markus P. M. Vink, "Indian Ocean Studies and the New Thalassology," and Isabel Hofmeyr, "The Black Atlantic Meets the Indian Ocean: Forging New Paradigms of Transnationalism for the Global South." For an important groundbreaking example of Indian Ocean historiography, see Engseng Ho, *The Graves of Tarim: Genealogy and Mobility Across the Indian Ocean*.

the dominance of large landmasses in world history. By keeping a secure footing in a particular place-in-the-making, a unit in the imperial system, the danger of being stranded in the sea, without a sense of perspective or without a limit to the local, is avoided. People and things can move, and Lanka can be outward looking, but not without restraints.

The road between Colombo and Kandy deserves a parting glance. When its chief enthusiast, Governor Barnes, was accompanying the visiting Bishop of Calcutta, Rev. Heber, on a trip to Kandy in 1825, he took the opportunity of showcasing the recently finished road. Barnes drove the bishop in a bandy (a bullock-drawn carriage), while Mrs. Heber accompanied two of the main architects of the road and several other European gentlemen in a palanquin carriage. Half way from Colombo, Mrs. Heber wrote: "We were met by an extraordinary personage, the second Adigar of Candy, followed by a numerous retinue, and preceded by one man carrying a crooked silver rod, and by another with a long whip, which he cracked at times with great vehemence; this is considered a mark of dignity among the Candians."[49] The Adigar or chief minister was elaborately dressed, and Heber went into detail about this. Barnes and the bishop dismounted from their bandy and exchanged pleasantries with the Adigar. It is striking that both British colonizers and Kandyan elites could use the new road like this, as a site of procession. Yet the irony is that the bishop was also taken through a tunnel that Barnes had built on this road, specifically to respond to a legend amongst the Sinhalese that the island would not be taken over by outsiders until a hole was made through the mountains of Kandy. When they went through this tunnel, Barnes made "the Caffres set up a yell, which reverberating against its roof and sides, had a most savage wild effect."[50]

The presence of African-descent attendants gestures to how the practices of prior colonizers, particularly the Portuguese, who had brought laborers and recruited regiments from the southwest Indian Ocean, continued into the 1820s.[51] Yet the usefulness of this snapshot lies in how it points to the constant redeployment and redefinition of the authentic, and

49. Reginald and Amelia Heber, *Narrative of a Journey through the Upper Provinces of India from Calcutta to Bombay, 1824–1825, with Notes upon Ceylon Written by Mrs Heber* (1828), 2:244.

50. Ibid., 2:255.

51. For commentary on how the British recruited a regiment of seven hundred African-descent soldiers from Portuguese Goa, see James Cordiner, *A Description of Ceylon: Containing an Account of the Country, Inhabitants and Natural Productions* (1807), 1:65.

also the recourse on the part of the British to the ideal of the king who receives homage from his ministers. Despite these recyclings, it is undeniable that the British state intruded into Lanka. Its presence on the island and its power to redefine pathways through mountains and seas was an act of violent confidence as much as limitation and failure.

Peoples

Sri Lanka has been a nodal point of migration at the center of the Indian Ocean, attracting waves of traders, conquerors, and holy men from India and also from West and East Asia.[1] Even though this long-running history of cosmopolitanism is undeniable, in the modern period the island's people have preferred to emphasize matters of "nativeness" in arbitrating among themselves. Of critical concern in this chapter is the nature of British colonialism, and the manner in which it consolidated a separate and unified territory of governance in the island and set in motion a colonial policing of the movement of peoples, so that belonging on the island equated with a different identity from that of coming from the mainland or elsewhere in Asia. The British undertook a process of islanding and partitioning Lankans, particularly from Indians.

This colonial program recontextualized some elements of pre-British social structure. As the kingdom of Kandy engaged in repeated battles with Europeans, identities within the kingdom started to contort, so that outsiders were cast as separate from the "Sinhalese."[2] This ran in parallel with the continued arrival of foreigners within Kandy, resulting, for instance, from warfare with Europeans. The British misunderstood this balance between attention to indigeneity and cosmopolitanism. After they took over the kingdom, they sought to repatriate to India the "Malabars"

---

1. A previous version of this chapter appeared as Sujit Sivasundaram, "Ethnicity, Indigeneity and Migration in the Advent of British Rule to Ceylon." For the long-running status of Lanka as a hub of migration, see, for instance, Michael Roberts, "From Southern India to Lanka."

2. For senses of Sinhala in this period, see R.A.L.H. Gunawardana, "People of the Lion," and K.N.O. Dharmadasa, "'The People of the Lion.'"

of Kandy, an ethnic label that later became "Tamil." They narrowed pre-existing patterns of identity in Kandy and sought after the truly indigenous in their newly acquired territory, so that it could be marked as distinct from India.

To understand the islanding of Lankan ethnicities, it is important to think beyond the island, to transcolonial processes and structures.[3] Sri Lankan ethnicities emerged in the context of the movement of peoples between India, the island, and the wider region and in the colonial state's attempt to impose new norms and meanings on those movements. To follow Arjun Appadurai's provocative work, in the contemporary world the difference between majorities and minorities has been crystallized out of the entanglement of state-building with globalization. In the colonial era as well the friction between the making of the colony and the flow of peoples shifted and created distinctions between those who belonged and who did not, and this is evident in the period that saw the advent of British rule to Sri Lanka.[4]

It is easy to consider island-making and its turn to indigeneity and ethnicity as a matter of discourse alone. Yet structural interventions are critical—at least because the political organization of the island underwent dramatic changes in the last decades of the eighteenth century and first decades of the nineteenth. The separation of islanders and mainlanders came about partly because of the structural irritations between the different arms of the British Empire, namely Crown and Company, and their need to define separate populations of subjects. When Indians came to the island for the purposes of war and labor, a new regime of distinction and segregation dictated their treatment. Following Sanjay Subrahmanyam, the structure of the state in Asia was in flux at the end of the eighteenth century, and this affected discourses of identity.[5]

The discussion begins by examining the different axes of Kandy's overseas relations at the end of the eighteenth century and how those relations impacted on the composition and self-awareness of the kingdom of

3. Rogers' careful work, "Caste as a Social Category and Identity in Colonial Lanka," makes the point that Sri Lanka should be contextualized as a region within South Asia. The relationship between the mainland and the island also features in Sanjay Subrahmanyam's article, "Noble Harvest: Managing the Pearl Fishery of Mannar, 1500–1925."

4. Arjun Appadurai, *Fear of Small Numbers: An Essay on the Geography of Anger*, 45.

5. Sanjay Subrahmanyam, *Penumbral Visions: Making Polities in Early Modern South India.*

Kandy. The main body of the argument looks at the arrival of the British and the emergence of a regime of partitioning islanders from mainlanders, and in particular at an attempt to repatriate Kandyan Malabars. The end of the chapter looks at the opposite direction of travel, by considering how the British regulated the movement of Indians from the continent to the island.

TRANSREGIONAL KINGSHIP

The last king of Kandy, Sri Vickrama Rajasimha, died in captivity on 30 January 1832 in the fort at Vellore in the Madras Presidency (fig. 1.1). The family of Tipu Sultan, who had ruled Mysore and was vilified as an oriental despot by the British, was also kept captive in this fort. The European surgeon who attended Sri Vickrama Rajasimha found him to be "affected generally with the dropsy"; but the king also asked to be attended by a "native medical practitioner," who was probably an ayurvedic practitioner who accompanied him from Kandy, and in his last hours preferred

The late King of Kandy, from a drawing by a Native.

Fig. 1.1. "The Late King of Kandy, from a drawing by a Native." From John Davy, *An Account of the Interior of Ceylon with Its Inhabitants with Travels in That Island* (London, 1821). Author's photograph.

the latter.[6] He asked his keeper to burn his body on a plot of ground assigned for the purpose, which would be sufficient to erect a "kind of tomb
being built over the ashes . . . a small garden being formed and a small
Chattry being erected for the accommodation of a superintendent Bramin
[sic] and water to travellers."[7] In making this request the king pointed his
keeper to a drawing of the family tombs at Kandy to show the building
that he hoped would be raised over his ashes. Yet in asking for a "Chattry," the king seemed to have had in mind a dome-shaped Hindu funerary
monument typical for instance of the Rajputs of India.[8]

Despite the kingdom of Kandy's Buddhist heritage, in this conversation its memory is forged within Hindu norms. Understanding why Sri
Vickrama Rajasimha was taken from the highlands of Ceylon to South
India and why he adopted Hindu symbols might serve as a point of entry
in discussing how the British intervened in the economy of migration between the mainland and the island.

Sri Vickrama Rajasimha marked the end of the Nayakkar royal line,
which is said to have commenced with the ascension of Sri Viyaya Rajasimha (r. 1739–47), and which marked its origins from South India.[9] In the
context of the fall of the coastal polities of the island to invading Europeans, there was a dearth of suitable brides of solar caste for the Kandyan
monarchy in the interior of the island. In South India, meanwhile, a class
of settlers, including military adventurers and governors, called the Nayaks, had broken away from the nominal overlordship of the Vijayanagara
empire. There was thus a congruence of interest between the Kandyans'
need to procure brides who could be presented as belonging to the solar
caste and the Nayaks' need in South India for a stabilizing of their for-
tunes. The Nayaks thus married into the Kandyan royal line, and eventually took it over when a Kandyan monarch was childless from his Nayakkar queens.

6. For the details of the ayurvedic practitioner, see J.A.W. Perera, "Sri Wickrama
Rajasinha's Exile in India, 1816–1832," 427.

7. Letter dated 20 January 1831, from Lieut. Col. F. P. Stewart, Paymaster of Stipends at Vellore, to Richard Clive, Acting Secretary of Government, and letter dated
9 January 1832 from Stewart to Henry Chamier, Chief Secretary of Government, file
F/4/1461, India Office Records, Asia, Pacific and Africa Collections, British Library,
London (hereafter IOC).

8. See R. L. Mishra, *The Mortuary Monuments in Ancient and Medieval India*, 95.

9. For details of the rise and fall of the Nayakkar line, see L. Dewaraja, *A Study of
the Political, Administrative, and Social Structure of the Kandyan Kingdom*, and the
second edition of this work, *The Kandyan Kingdom of Sri Lanka*.

Given the entrenched debate on the character of the Nayakkar line, the following question is critical: were they always perceived as foreigners from South India, or were they internalized?[10] They certainly portrayed themselves as pious Buddhists in keeping with Kandy's religious ethos, and they were tutored in Sinhala and Pali by Buddhist priests while overseeing a period of cultural renaissance in the interior. On the other hand, the plot of 1760 to depose the Nayakkar monarch, Kirti Sri Rajasimha (r. 1747–81), may have been prompted by the sight of his adherence to a Hindu custom, of anointing himself with ash.[11] When the Nayaks multiplied, they were set apart in Kandy, having for their use a separate street, which after the British invasion was called "Malabar Street."[12] It is useful to see the Nayaks as being both excluded from and included by what it meant to be Sinhala and Buddhist, where the sense of these categories is taken to indicate the period's meanings. The traditional idea of bounded or static identity is unhelpful in coming to terms with the shifts in both the self-presentation of these monarchs and in how they were viewed by their courts. At the same time, the political and cultural import of being Sinhala was not equivalent; it was possible at times to be a Sinhala king even while not being Sinhala in cultural terms.

For the purposes of the argument here, what is more important is an issue that has attracted far less attention from scholars, namely how the Nayakkar line forged their place in the politics and culture of the wider region. The kingdom of Kandy continued to have linkages with Southeast Asia and South India prior to its conquest by the British, and this involved the passage of peoples and the formation of a sense of regional community. Thinking towards Southeast Asia, it is interesting to note how the late eighteenth century saw the centralization and integration of a number of polities.[13] Of particular importance for its marked similarity with the island is the reconstruction of cultural practices across Burma and Siam, which, like in the island, involved a new kingly patronage of Thera-

10. For references to this debate, see the Introduction. None of the extant literature follows the story of the repatriation of the last king of Kandy to Vellore Fort.

11. Michael Roberts, *Sinhala Consciousness in the Kandyan Period, 1590s to 1815,* 49. Other historians disagree and point to the economic motivations of this plot; see Dewaraja, *A Study,* 108.

12. Roberts, *Sinhala Consciousness,* 51.

13. Anthony Reid, ed., *The Last Stand of Asian Autonomies: Responses to Modernity in the Diverse States of South-east Asia and Korea, 1750–1900,* introduction and chap. 1, and Victor Lieberman, *Strange Parallels: South-east Asia in Global Context, c. 800–1830.*

vada Buddhism, the restoration of scholarly monks, and kingly interest in works of scholarship, translation, history, and art.[14]

In the centuries prior to Kandy's fall, various island kings had established links with Burma and Siam. Particularly noteworthy is the rather exaggerated claim in the twelfth-century section of the Buddhist chronicle the *Mahavamsa* of how the monarch Parakramabahu I, who ascended the throne in 1153 AD, waged a victorious war against the king of Ramanna, equipped with a navy which "sailed forth in the midst of the ocean . . . like a swimming island."[15] Ramanna may be taken to be Ra-manya, or what later became lower Burma.[16] This was exceptional: for the most part relations with Burma were friendly and beneficial for both sides, and were cemented by the shared bond of Theravada Buddhism and the passage of monks between the two territories. When the need arose for religious revival, Burma looked to the island. Similarly, the Kandyan kingdom's rulers, who observed the need to reestablish higher ordination for the Buddhist clergy, looked to Burma. At the tail ends of both the sixteenth and seventeenth centuries, monks were brought from Arakan to Kandy. Bang on time at the end of the next century, in 1799, a monk journeyed to Burma with five novices to gain higher ordination there, this time to the British territories on the coast. On returning to Ceylon in 1803, this group set up a new fraternity, the Amarapura Nikaya, which continues to this date.[17] The Amarapura Nikaya provides a successful example of the localization of an imported heritage of Theravada Buddhism.[18]

Religious reformation also underpinned Kandy's connections with Siam. There were two failed attempts to reformulate the island's Buddhist *sangha* by contact with Siam during the reign of Sri Viyaja Rajasimha (r. 1739–47); these were followed by two successful attempts in the reign of his successor Kirti Sri Rajasimha (r. 1747–82). Twenty-five monks from

14. For more on this, see Sujit Sivasundaram, "Appropriation to Supremacy: Ideas of the 'native' in the rise of British imperial heritage.'

15. W. Geiger, trans., *Culavamsa, Being the Most Recent Part of the Mahavamsa*, 2:76, line 56.

16. For scrutiny of this claim, see Sirima Wickremasinghe, "Ceylon's Relations with South-east Asia, with Special Reference to Burma."

17. For the passage of monks back and forth, see Malalgoda, *Buddhism in Sinhalese Society, 1750–1900: A Study of Religious Revival and Change*, 56–57, 97–98.

18. For the argument about successful importation, see Anne M. Blackburn, "Localizing Lineage: Importing Higher Education in Theravadin South and Southeast Asia."

Siam arrived in 1753, while a second group arrived in 1756.[19] According to one palm-leaf source, five hundred monks were conferred with higher ordination in 1753 by the Siamese delegation, and then taught the Pali language and literature.[20] The *Narendra Caritavalokana Pradipikava*, a historical chronicle on palm leaf, which was allegedly written at the invitation of Governor Edward Barnes, nineteen years after the fall of Kandy, notes the following about how Kirti Sri established and then celebrated the reintroduction of higher ordination from Siam:

> As there was not a single monk with higher ordination in Lanka, the king knowing that the Sasana was on the path of decline, having decided "I must support the sublime and wonderful Buddha Sasana," having prepared a great deal of royal gifts and objects as offerings, having treated well the Sinhala Ministers and the Dutch ministers, handed over the gifts and royal messages to them, and sent them away to Siam by ship. Then all of them, without experiencing any danger, arrived at the city of Ayodhya, and having met with King Dharmika of Siam, presented him with the gifts and the royal message. Then that king, having treated those ministers well, having read the letter and knowing that the Buddha Sasana had disappeared from Lanka, became overwhelmed with grief. He decided to reestablish the religion of Buddha in Lanka. Having summoned and gathered senior monks of Siam led by the Sangaraja [chief of monks], having explained to them the situation in Lanka, he sent all of these into the ship to be taken to Lanka: elder Sthavira Upali and other monks of ten categories, text books that were not available at the time in Lanka, gold statues and golden books and many other gifts to ambassadors of Lanka, special gifts to the king and a royal letter to the king of Lanka, a few ministers of Ayodhya who were capable of protecting the monks, and a valuable gem-set casket for keeping the Tooth Relic. He entrusted everything to the ministers who had gone to Siam from Lanka, and granted them leave to return. Then they arrived by ship at the port of Trincomalee [in Lanka]. When King Kirti Sri heard of their arrival he informed the citizens of Kandy, ordered his ministers to repair and decorate the way from Trincomalee

---

19. Malalgoda, *Buddhism in Sinhalese Society*, 61–67.

20. See Nandasena Mudiyanse, "Correspondence between Siam and Sri Lanka in the 18th century." This includes printed Sinhala versions of two palm-leaf manuscripts connecting the Siamese and Kandyan courts.

to Kandy, and built temples at many places, and sent forth his ministers to receive them.[21]

The strength of this connection with Siam is exemplified in the inner
history of the 1760 plot against Kirti Sri: some local Buddhist monks, who
had benefited from Kirti Sri's religious reformation, sought to assassinate
the king and replace him with a Siamese prince.[22] The plotters designed
an elaborate plan to kill Kirti Sri. They set a pit of sharp spikes under
Kirti Sri's chair at a religious ceremony. Having heard of the plot, the monarch arrived at the ceremony, exposed the pit, and the event carried on as
if nothing had happened. The Siamese prince and monks were sent back
home.[23]

Kandy's relations with the outside world followed the geographical contours of Dutch colonialism. The Dutch shipped ambassadors and
monks on their vessels. For instance, Kirti Sri's embassy to Siam in 1750
went via Aceh, Sumatra, and Malacca. Throughout the journey the vessel flew the Dutch flag and only lowered it and replaced it with the "Lion
Flag of Lanka" when it approached Siam.[24] The Dutch for their own part
were deeply suspicious of the external relations of Kandy, and in particular
of the connections with the wider world that the Nayakkar line brought
with them, as immigrants from South India. They provided assistance in
procuring brides and monks in order to control these relations as much as
possible.[25]

The movements of peoples and the sense of community were not only
directed towards Southeast Asia and were not purely religious in this period. A series of intriguing letters between the kingdom of Kandy and a
South Indian coastal polity, possibly Arcot, and also between Kandy and
the French based in Pondicherry, suggest the need to place the highland

---

21. *Narendra Caritavalokana Pradipikava* [The lamp to look into the history of
Kings] by Yatanvala Mahathera of Asgiri Vihara. Translation undertaken by Prof. Udaya
Meddegama of the University of Peradeniya, Sri Lanka, from Sinhala to English, on the
basis of the version published by Mahabodhi Press, Colombo, 1926, and edited by P. A.
Hewavitarana.

22. Malalgoda, *Buddhism in Sinhalese Society*, 66; for a full discussion of this plot,
see Dewaraja, *The Kandyan Kingdom of Sri Lanka*, 119–24.

23. Dewaraja, *The Kandyan Kingdom*, 121–24.

24. Paul E. Pieris, trans., "An Account of Kirti Sri's Embassy to Siam in 1672 Saka
[1750 A.D.]," 22.

25. Dewaraja, *The Kandyan Kingdom of Sri Lanka*, 99–103.

state in the context of the Indian mainland.[26] The Nayakkars did not forget their mainland connections upon coming to power in Kandy; indeed, they spoke to the mainland with a sense of authority and with the expectation of respect. In doing this they did not necessarily forge a new relationship to India. But their correspondence is striking, given the traditional view of how landlocked Kandy was by European colonists who had taken the coasts.

A letter, possibly from the Kandyan monarch Rajadhi Rajasimha, was drafted in gold characters and wrapped in two muslin handkerchiefs and put into a large bag of gold tinsel cloth and wrapped again in a handkerchief. The text includes the notice of gifts: "We are in receipt of the set of golden garment [sic] which you with your good will sent unto us. In return We are gracefully sending a set of golden garments, a letter bearing our seal and two elephants, one a she-elephant and the other a baby."[27] Accompanying this letter was another of the same date from "Divaka Wickramasinghe, the General of His Most Gracious Majesty (the Beneficent Great Court), the Lord of Sri Lanka." Wickramasinghe heaped praise on the character of his enlightened monarch, who was "resplendent with multitudinous glory as clear and excessively white as snow, kunda flowers, sandal paste, autumnal Moon, milk, white lotus, celestial elephant, stars, pearl necklace. . . ."[28] He pointed out that ambassadors from "many countries" had visited Kandy. Having drawn attention to Kandy's greatness, he then meted out his criticism of Arcot. He noted how the ambassadors had not followed the etiquette of mutual gift giving in a way that was consistent with the honor of the court: "Some forms of etiquette observed in the island of Lanka may appear disrespectful to you and some of yours may appear disrespectful to us. . . . Therefore do not send such Ambassadors. If such are sent we shall not receive them nor talk to them."

British soldiers retrieved these letters between Kandy and Arcot from

26. S. Rasanayagam, trans., "Tamil Documents in the Government Archives." This includes five translated letters. I refer to three of them in this paragraph. In addition to this relationship with Arcot, Kandy also had close links with the Tevar of Ramnad, whose territory was separated from the island by the narrowest stretch of sea, being on the opposite coast of the mainland. See Dewaraja, The Kandyan Kingdom of Sri Lanka, 97–99.

27. "Draft of the King's Reply to the Nawab of Arcot, dated 4 November 1786," in Rasanyagam, "Tamil Documents," 10.

28. "Draft of a Letter from the Chief Minister of the King to the Minister of the Nawab" in Rasanayagam, "Tamil Documents," 10–11.

the palace at Kandy after the city's invasion by the British in 1815. Another of the letters was written in Arabic-Tamil, by Magdom Lebbe or Magdom Ismail from Ramnad, on the east coast of South India, and was addressed to the chief minister at Kandy. It carries news from the mainland: there is mention of a war in Madras and how "owing to storms and floods, sloops and boats are in great distress."[29] Most of the contents are devoted to matters of maritime trade. The need to write in Arabic-Tamil is spelt out: it is to prevent information falling into the hands of the British. The importance of the straits surrounding the island of Mannar, on the northwest coast of Ceylon, are also explained: "There is a house near the Sundresa Aiyer Chattiram [traveler's bungalow] in the Isthmus of Pamban near Kovilgramamam village at the confluence of the southern and northern seas which is a convenient place for going and coming." The agent communicates his plan of housing his son at this point. He tells of the possibility of trading in cloth from the mainland: "If after a year or two of business in clothes [sic] we find it profitable we can always do that business. If the present samples are approved of, we can send clothes [sic] and shell bangles."[30]

In the decades before the British conquest, the kingdom of Kandy therefore had at least two axes of relations with the outside world: toward Burma and Siam, and also toward South India. These relations suggest a formation of identity and kingship which drew on regional patterns, and which was at the same time grasping for independence and respect. The identity of the king, and the religion and politics of the kingdom, were forged in a larger sphere, and the Nayaks maintained these outside connections.

COSMOPOLITAN SINHALANESS

These political, religious, and economic links shaped the character of the kingdom of Kandy and, as will be evident below, its discourse of Sinhalaness. They influenced the composition of Kandy and the way it conceived of its place in a wider sphere of relations.

War poems serve as an intriguing set of sources for the history of the

---

29. "An Arabic-Tamil Letter Written by Magdom Lebbe alias Magdom Ismail, an Envoy from Pondicherry, to the Chief Minister at Kandy from Ramanadapuram on or about 1 October 1799," in Rasanayagam, "Tamil Documents," 12–13.

30. For more on the role of Muslim traders within the Kandyan kingdom, see Lorna Dewaraja, *The Muslims of Sri Lanka: One Thousand Years of Ethnic Harmony, 900–1915.*

island.[31] One called *Ingrisi Hatana*, which translates literally as the "English Battle," commemorates the victory that the Kandyans won over the British in 1803 (figs. 1.2 and 1.3). It may have been the case that war poems were recited to inspire troops or, when recited before the king, to celebrate victory, and the preserved copy of *Ingrisi Hatana* may be a written transcription of an oral ballad. Such a use is consistent with the metric forms of these poems. The *Ingrisi Hatana* shows how warfare with the British, like previous wars, served to strengthen Kandyan confidence and identity. Its rendition of the defeated English is gory, and their defeat is consistently linked to the strength of Sinhalaness:

Behold! how the Sinhala troops showed their might on the battle field, cutting, slashing, flinging on the ground, chasing behind the enemy; beating, tying up, uttering abusive words, asking "How are you doing?," snatching what they were wearing and looting, without showing any compassion.

Some men in the Sinhala army bearing large and strong clubs as weapons, chasing behind the enemy and beating on their heads, killed them. Some others snatched from their hands their weapons such as lances, and umbrellas and flags, while some others took away their elephants, horses, and buffaloes.

Some men in the Sinhala army, by throwing on the ground the enemy soldiers, tore off the red armor they were wearing; some others smashed on the ground pots of rice and hoppers [a kind of pancake] etc. they were cooking, some others took away boxes of money, glasses and barrels of arrack, rum etc.

Some clever fighters, jumping into the middle of the fight, beheaded the enemy soldiers, some others having subdued the enemy threw them on the ground having bound their hands from behind. Some others tell them, "If you are so smart, get out from here alive." Then some enemy soldiers ran away unable to go through all this.

Thus the Sinhala army, not scared of war, showed their might. Then some of the enemy soldiers dropped their weapons. They were

31. The discussion of war poems in this paragraph draws on the analytical claims of Michael Roberts, *Sinhala Consciousness*. The citation from the *Ingrisi Hatana* is from a full translation of this manuscript undertaken from Sinhala into English in collaboration with Prof. Udaya Meddegama of the University of Peradeniya, Sri Lanka. Copies of the *Ingrisi Hatana* are available to view in the Museum Library, Colombo; see, for instance, K.11.

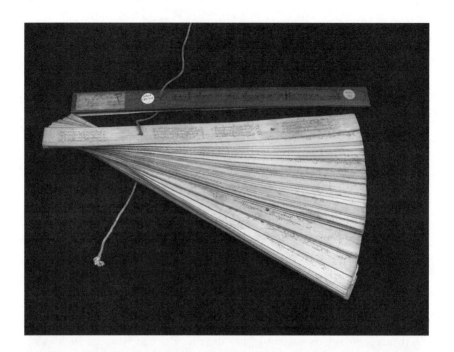

Fig. 1.2 and 1.3. A copy of the *Ingrisi Hatana* at the Colombo Museum,
Sri Lanka, and detail from a leaf of the same copy. By permission of
the Director of the Colombo Museum. Author's photograph.

taken to be shown to the king. Others in the English army ran away defeated in the war.

The bulk of the *Ingrisi Hatana* is devoted to the attitude and status of Sri Vickrama Rajasimha, who was later exiled to South India by the British. The king's actions fit within the lineage of Sinahalaness. Repeatedly, he is said to have united the three Sinhalas—Ruhunu, Pihiti, and Maya—as three separate historic principalities of the one Sinhalese kingdom. "As brave as a lion that rips open the heads of elephant-like enemies, King Sri Vickrama Rajasimha, glorious and majestic, shines like a bright light in the 'three Simhala.'" At the same time, the poet pointed to the spread of the king's fame beyond the realms of the island: "Having brought Sri Lanka under one parasol [as a united kingdom], spreading glory in many other countries, His Majesty King Sri Vickrama, may you exist enjoying pleasure like God-king Sakra!" The "whole of India" was said to be "shining with the splendor" of this king. The victory would be known in "the world" for "five thousand years." This act of uniting the land in victory was said to have made the king "a lion-king that displays its splendor in the middle of elephants."

This mode of representing kingly triumph was well established in the literature of the island. The "lion king" motif in particular signifies Sinhalaness, and points to a persistent myth of origin of the Sinhala people, according to which it has at times been claimed that they descended from a lion.[32] It is also striking that the poet casts the king as having united the whole of Lanka by his victory, pointing to how the Kandyans thought that they were universal lords of the whole island.

Yet there is another side to the kingdom of Kandy that needs to be kept in view. In reality, despite the discourse of Sinhalaness, the Kandyan kingdom was cosmopolitan at least at the elite level; it contained significant elements of diversity within itself. In addition to the monarchs, the Kandyan court and kingdom contained a number of important functionaries who came from or who traced their descent from elsewhere, and who kept their identities intact while being integrated into the structures of the state. In 1810, John D'Oyly, the first Resident of Kandy, who famously depended on spies disguised as monks and traders to discover the workings of Kandy, noted that the paid soldiers of the king of Kandy consisted of two hundred and fifty to three hundred Malays, two hundred "Kaffirs"

32. For the origin myth, see R.A.L.H. Gunawardana, "The People of the Lion: Sinhala Identity and Ideology in History and Historiography."

or troops from Africa, twenty sepoys from India, two hundred and fifty Muslims, and one hundred "Malabars."[33] It is likely that a good number of these soldiers were deserters who had worked within the ranks of European armies on the island, or their descendants. Throughout its wars with Europeans, the Kandyans also utilized a number of Europeans in their own ranks. In 1803, the British were shocked at the defection of one of their artillerymen, Benson, to the side of the Kandyans. Benson then took charge of the production of gunpowder in Kandy.[34]

Kandy's cosmopolitanism was not only reflected in the manner of its defense but also carried through to matters of trade. Here Kandy relied very heavily on a number of Moors or Muslims, who were forced into the interior in the context of persecution by the Portuguese. These Moors became the prime advisers on matters of trade to the Kandyan kings, and they formed part of the carriage bullock department. They had charge, as well, of the royal monopoly over the areca nut trade, and their trading practice will be discussed in chapter 2. The facility with which Moors were integrated into the Kandyan kingdom is also demonstrated by their taking up the task of cleaning the silver and gold vessels used at the temple of the Buddha's Tooth Relic.[35] European expansion therefore provided not only a context for the firming up of the identity of the Kandyan state as Sinhala but also prompted a series of migrations into Kandy that made it a cosmopolitan kingdom.

While these changes in composition and placement were taking place in the interior, the arrival of immigrants from South India went on outside the formal territory of Kandy as well, where the absorption of immigrants led to the formation of new castes.[36] For instance, immigrants from the weaver caste in South India became members of the new caste of cinnamon peelers. In the early British period connections between the island and the Coromandel coast were particularly strong. From there migrants were said to come and reside in the island "for years, carrying on a brisk trade and forming connections with families which are of the same caste as themselves."[37] One historian of immigration has argued that the pres-

33. Dewaraja, *The Kandyan Kingdom of Sri Lanka*, 201.

34. Channa Wickremasekera, *Kandy at War: Indigenous Military Resistance to European Expansion in Sri Lanka, 1594–1818*, 60.

35. All of this information is from Dewaraja, *The Muslims of Sri Lanka*, chapter 4.

36. Roberts, "From Southern India to Lanka." Also for a general analysis of castes in the island in relation to the mainland, see Rogers, "Caste as a Social Category."

37. Cited in Michael Roberts, "From Southern India to Lanka: The Traffic in Commodities, Bodies and Myths from the Thirteenth Century Onwards," 38.

ence of people of South Indian origin is a continuous feature of the island's history, and that even the tide of indentured laborers who worked on plantations in the later decades of the nineteenth century should be contextualized in relation to earlier patterns.[38]

Clearly, the relationship between the island's peoples and the wider world was complex: arrivals from the outside were assimilated even as they changed existent social distinctions. The structural patterns of Kandy's relations with the outside world and the wars with Europeans had an impact on the kingdom's organization and composition, and in turn on its discourse of Sinhalaness. But these structural patterns were contradictory—they led to a sense both of indigeneity and of foreignness.

IN SEARCH OF THE INDIGENOUS

The British certainly did not understand the complexity of these relations, and their policies eventually swung to a position where the island was said to be distinct from the mainland. For the British were in search of the truly indigenous in the island of Ceylon—which would then consolidate their new colony as a cohesive whole. This was particularly marked with respect to how they sought to repatriate those in Kandy who had recently come from India, and who were related to or associated with the royal family.

The first few years of British rule on the coastal belt of the island, before the taking of Kandy, provide a context for understanding how Britons recast the placement of islanders in the wider region. The British first took the Dutch territories of the coast in 1796 in fear that they would fall to the French and governed them under the East India Company's Madras Presidency. This arrangement led the early British officers in Ceylon to attempt to make its governance conform to the pattern of the Madras Presidency.[39] Robert Andrews, who was put in charge of revenues, grew apprehensive of the powers of the chief-headmen, through whom Europeans had governed prior to this period. In a proclamation of 1796, these men had their authority stripped; their duties were entrusted instead to officers from South India. In addition to these "Malabar" officials, a whole range of adventurers arrived in the island from South India who took up tax farming. There was widespread discontent at these changes, and a revisionist reading suggests that discontent pivoted on peasant feelings of anger at the for-

---

38. Patrick Peebles, *The Plantation Tamils of Ceylon.*

39. The material for this paragraph arises primarily from Colvin R. De Silva, *Ceylon under British Occupation, 1795–1833*, vol. 1, chap. 7.

eign agents of government, in addition to the standard explanations of the
imposition of a tax on coconut trees and the abolition of service tenures.
Revolt followed, and the Madras government lost its direct authority over
the island's territories.[40] By 1798, Frederick North became the first Governor of Ceylon. He was asked to report not to Madras but to Calcutta and
the Company's Court of Directors. North had control of civil and military
duties, while the Company had charge of commerce in this period of dual
control of the Company and Crown. Given the failures of Madras' rule of
Ceylon, North cultivated a disdain of the southern Presidency.[41] Though
he had at first hoped that Ceylon would be placed directly under Bengal as
a separate Presidency, in 1802, when it became a Crown Colony, he slowly
realized that this gave him a measure of extra authority.[42]

North's frosty relations with Madras also appear in his dealings with
Kandy. Throughout his career, and most particularly in his disastrous attempt to invade Kandy in 1803, he was motivated by the belief that the
Kandyan court was split by a "Malabar faction": namely that the king and
his relatives were resented by the Sinhala aristocracy on the basis of being
foreigners from South India.[43] In this view, he followed the Dutch percep-
tion of the Kandyan court. In the mid-eighteenth century one Dutchman
wrote of how he wished to see a "Kandyan prince" on the throne so that
"the pernicious coast Nayakkars, Malabars and Moorish scum" could
be thrown out.[44] North attempted at first to forge a subsidiary alliance

40. See also Alicia Schrikker, *Dutch and British Intervention in Sri Lanka, 1780–
1815,* 155ff. Very recently there have been further works pertaining to the string of
rebellions in early-nineteenth-century Lanka, including the events of these years; see
Nira Wickramasinghe, "Many Little Revolts or One Rebellion? The Maritime Provinces
of Ceylon/Sri Lanka between 1796 and 1800," and Kumari Jayawardene, *Perpetual Ferment: Popular Revolts in Sri Lanka in the Eighteenth and Nineteenth Centuries.*

41. See also U. C. Wickremeratne, "The English East India Company and Society
in the Maritime Provinces of Ceylon, 1796–1802," and Schrikker, *Dutch and British
Intervention in Sri Lanka,* 146.

42. Schrikker, *Dutch and British Intervention in Sri Lanka,* 155.

43. U. C. Wickremeratne, in "Lord North and the Kandyan Kingdom, 1798–1805,"
goes to the extent of suggesting that the factionalism in the Kandyan court was conceived and exaggerated by the British, while others such as James S. Duncan in *The
City as Text* and Alicia Schrikker, *Dutch and British Intervention in Sri Lanka,* 208,
have argued that there was factionalism in the Kandyan court and that it was economic. Regardless of the view we take of the intrigues of the Kandyan court, it is clear
from the evidence cited below that North privileged a view of "Malabars" as foreign.

44. From Dewaraja, *The Muslims of Sri Lanka,* 77. For more on Dutch views of the
Kandyan court, see K. W. Goonewardene, "The Accession of Sri Vijaya Rajasimha."

with Kandy, modeled on those common in India. Accordingly, he hoped
to station a British garrison in Kandy and in return to appropriate a part
of the kingdom's revenue to the British government.[45] He negotiated with
the chief minister, Pilima Talauve, and sought to place him on the throne
instead of the king. But North was unable to find an "ethnic" wedge between Pilima Talauve, who he perceived to be the chief Sinhala aristocrat,
and his king. In 1803, war ensued when negotiations broke down, and it resulted in the disaster recounted in the *Ingrisi Hatana*; the Nayakkar king
was restored.

Yet British ideas of the need to rid Kandy of "Malabar" influence began to have a slow effect. In 1812, the king executed Pilima Talauve for
treason, and Ahalepola was appointed as chief minister. Ahalepola in turn
cooperated with the British and eventually fled to the maritime provinces,
and his family was slain by the king. From there he commissioned texts
that provided an ethnicized critique of the Nayakkar line which played
up their Indian ancestry and Hindu leanings.[46] One example that bears
out this ethnic vocabulary is the *Kiralasandesaya*, which conforms to the
traditional genre of poems which trace the route of a messenger-bird. In
this case the name of the poem is literally "Lapwing's message," and it is
written in praise of Ahalepola's pretensions to the Kandyan throne; the
bird traces a path between temples to ask for blessings for Ahalepola. [47]
The *Kiralasandesaya* was possibly written in early 1815, and it contains
examples of anti-Tamil invective. The word translated "Tamil" is most
often *demala* in the original:

Of the deposed king:
Like a strong bull gone mad in rut
Trying to uproot the earth by attacking it with its horns
The Tamil rascal thought of suppressing
The noble people of Lanka, the descendants of high castes.
. . . .
The evil, ungrateful and lowly Tamil
Thus he began to ruin this Lanka.
Giving all kinds of trouble to the powerful and rich noblemen
By looting their villages and property.

45. For North's attempt at forging a subsidiary alliance, see Wickremeratne, "Lord
North and the Kandyan Kingdom."

46. Roberts, *Sinhala Consciousness*, 51.

47. C. E. Godakumbura, *Sinhalese Literature*, 204.

Of Ahalepola, who is given full credit for having deposed the king:

> Villages and lands that were occupied by the Tamil bandits
> This heroic king redeemed for the people
> Singing in sweet words of praise
> Such people come and go having worshipped the lotus of his feet.

It is relevant, in the context of the discussion to follow, to note how the poet describes the repatriation of the king and his relatives to India. He noted that Ahalepola "captured" these Tamils, who were "like bulls," and "deported them from the Sinhala country to go begging elsewhere." Though there is no evidence for this, it is suggestive to consider British ideologies of difference as providing a context for Ahalepola's invective. For it was not Ahalepola who conquered Kandy, or who sought to deport the Nayakkar line, but the British with whom he was cooperating.

After the fall of Kandy, Governor Robert Brownrigg noted the predicament of the "Malabars adhering to the King": they were caught between a loyalty to the monarch and the hope of returning back to India.[48] British benevolence dictated that they should all be repatriated:

> The Malabars from the Coast of Coromandel, as well as the Moors from the same quarter, are by their birth and parentage the natural subjects of His Britannick Majesty, and of the Hon. The East India Company. They are exhorted to keep in mind this bond of Allegiance—and to hold in view the hope of being able (as loyal subjects of the British Empire) to return with safety and protection to their families, relations, friends and cast, in their native countries, under the Hon. Company's Government. . . . Such safety and protection, with a passport to their country, and every reasonable assistance and support, is hereby offered to them—thus timely before they become involved in the guilt of actual hostility and armed opposition . . . neglecting which warning, they will incur the danger of being treated not only as enemies but as traitors.[49]

In stipulating why the king should go to South India, Brownrigg wrote that Sri Vickrama Rajasimha should be kept "amongst those of his own

---

48. See dispatch dated 25 February 1815, from Robert Brownrigg in Kandy reporting the capture of the king. CO/54/55, TNA.

49. Proclamation of His Excellency, Governor Robert Brownrigg, dated 11 February 1815. CO 54/55, TNA.

cast and consequently in or near that part of the country from whence his family originates." The Company, however, was worried that his presence in South India would "disturb the tranquility of our Districts." As a compromise it was agreed to house him in Vellore fort.[50] William Granville, who took the king and his entourage across to India in the ship *H.M.S. Cornwallis*, in 1816, left a rather colorful journal of the monarch's conduct on board. In this journal, the king is presented as being unable to come to terms with the loss of status. On 8 February 1816, for instance: "He fixed his eyes on the ocean before him, an element altogether new to him, and seemed to think on the mutability of power, and his own irreparable misfortunes."[51]

The last king of Kandy was set apart as a Malabar even in appearance. According to one commentator, he was a "stout good looking Malabar, with a peculiarly keen and rolling eye."[52] In being transported to India, he was described in contrasting ways. On the one hand, when without ceremonial clothing, he was said to appear of "no higher rank than any other native of caste in India." On the other hand, his appearance was also said to fit with his character of despotism, since he lost his temper at will and raised his voice "to the highest pitch" in addressing his inferiors.[53] While "Malabar" was used as a term for many people on the coast, the program of repatriation Brownrigg set in motion was confined to those in Kandy who were linked to the Nayakkar line. However, in the fort and pettah of Colombo, on the coast, the British did not allow Malabars and Moors to own houses or grounds, following a Dutch regulation.[54]

The Crown government of the island carefully classified the Nayakkar prisoners who moved across to India in this period. When the brig *Eliza Tutocoveen* took across thirty-three Nayakkar prisoners in May 1815, their details were recorded in tabular form: the Company was told which "cast" each individual belonged to, "what country" they belonged to, and

50. See letter dated 8 April 1815, from Robert Brownrigg to Hugh Elliot, Governor in Council. Also secret letter from Fort St. George dated 7 October 1815 and letter dated 26 April 1815 from Hugh Elliot to Robert Brownrigg. All in file F/4/515, IOC.

51. William Granville, "Deportation of Sri Vikrama Rajasinha," quotation on 497.

52. J. W. Bennett, *Ceylon and its Capabilities: An Account of its Natural Resources, Indigenous Productions and Commercial Facilities* (1843), 410.

53. Granville, "Deportation of Sri Vikrama Rajasinha," quotation on 544.

54. See dispatch dated 7 July 1832 from Governor Horton to London, CO/54/117, TNA, repealing the acts that were put in place banning "Moors and Malabars" owning land and houses in the Colombo fort and pettah.

also how long they had been resident in Kandy.[55] Over eighty people were imprisoned in Vellore fort with the king, while many others were dispersed in South India.[56] The entire contingent of repatriates was divided into several categories. First there was the king and his relations, and the eighty or so prisoners composed of this category. Second, there was a class of persons who had had resided in Kandy for a long period, and were "in some degree aliens in their native country." This class was said not to require treatment as prisoners, and the Ceylon government were unwilling to pay for their upkeep, except for a short period before they returned to their ordinary occupations. The third class of repatriates were those who according to the Ceylon government were "merely sojourners on this side of the Gulph." In addition to these three classes Ceylon also sent to India a small number of "Malays, Caffres and a few natives of Bengal."[57] While the Dutch had complained of the influence of people they deemed to be foreigners in Kandy, they never had the power to orchestrate this kind of eviction.

Nonetheless, one must take care in interpreting these events as a display of British power. For the British quickly found that their program of repatriation was difficult to carry out. The exiles did not see themselves to be residents of South India. The numerous petitions which they addressed to the British complained of being stranded in a foreign country.[58] In a case which is rather striking as an instance of resistance, seven prisoners disembarked at Cuddalore, and two of them insisted that "they [were] natives of Kandy and not of Malabar."[59] In another case, ten Kandyan families that arrived in Tanjore alleged that "their destination to Chalempalegam in Tondiman's country must have been founded on some mistake . . . that they know no such place in Tondiman's country [and] that they are with the exception of one of their number utter strangers to Tondiman's country."[60]

55. Among the "countries" listed were "Tanjore, Ramnda, Negapatam, Bendigalle, Puducotte, Madura, Trenevelly, Seveganta, Velantcheryy, Colleloor."

56. This figure comes from a letter dated 1 April 1816 from Charles Marriott to the Chief Secretary of Government at Fort St. George, F/4/515, IOC.

57. The different categories are laid out in Robert Brownrigg to Hugh Eliott, Governor in Council, Fort St. George, dated 8 April 1815, F/4/515 IOC.

58. See, for instance, F/4/880, IOC, for a batch of such petitions.

59. Letter dated 21 March 1816, from J. Macdonald to the Chief Secretary of Government, Fort St. George, F/4/515, IOC.

60. From the Resident of Tanjore to the Chief Secretary of Government Fort St. George, dated 1 October 1816, F/4/527, IOC.

The surprise of the prisoners arose in part because they did not see a difference between Crown and Company. One petitioner wrote to the Crown Governor of Ceylon: "I humbly beg leave to state that after the Honourable Company became masters of Candy, it has pleased the Honorable Company to send my late father and other families to these parts of country."[61] By 1816, the Crown government had to admit that some of the prisoners of war had returned to the island. In attempting to force its policy through, the government declared that it was illegal for anyone who was a "Malabar" resident in Kandy one year prior to the kingdom's takeover to remain in the island without "written permission" from a representative of the Governor.[62]

As might be expected, the king himself did not adjust to his new situation. Charles Marriott, who was in charge of him, wrote:

> To eradicate the kingly notions of a person (and that person by no means a wise one) who has by his account been seated on a throne about nineteen years, must be the work of time and infinite patience, and till these notions are eradicated it is useless to expect that ideas of private comfort will be planted or grown up.[63]

Sri Vickrama Rajasimha attempted to assemble his court while in exile, by calling his ministers at stated hours. He asked for a crown to be made out of beaten gold in his possession.[64] For his daughter's ear-ring feast he asked for 800 seers of raw rice, 2,500 young coconuts, 3,000 plantains, 200 candles, 1,500 limes, 30 jack fruits, 10 sheep, 200 eggs, 50 fowls, and 30 large fish. An important clue lies in his request that the fish be caught in inland water tanks, like those which were part of the Kandyan kingdom.[65] He expected similar extravagance for the marriage of his daughter. In response it was noted: "Independently of the objections on the score of expence it would obviously be very ill judged to indulge on his part or that of his

61. Petition to His Excellency the Governor in Council at Ceylon, the humble representation of Dince Swamy, the son of late Condesawmy Naicker, head brother in law of Istree Rajady Rajah Singa Maha Rajah, the king of Candy, F/4/1461, IOC.

62. Order of Council of Government of Ceylon, signed James Gay, Secretary of Council, dated 24 June 1816, F/4/527, IOC.

63. Letter dated 11 October 1816, from Charles Marriott to Chief Secretary of Government, Fort St. George, F/4/527, IOC.

64. See the correspondence in F/4/527, IOC. Also for some other details of the king's conduct in exile, see Perera, "Sri Wickrama Rajasinha's Exile in India."

65. Petition transmitted to Ceylon, dated 18 June 1821, F/4/880, IOC.

family a taste for the splendours of royalty."[66] The captive king was never resettled away from Vellore Fort. His attempt to merge Kandyan traditions with Hindu funerary rites, noted above, is therefore very fitting. It points to how the British idea that a "Malabar" could separate his identity from a "Sinhala" had failed to materialize. By 1834, the Ceylon government had to shift its policy and allow all the "Malabar" exiles, except the near relations of the family of the king, the option of returning to the island.[67]

The program of repatriation, and the partitioning of the mainland from the island, meant that the colonial state was particularly suspicious of the movement of people from India to Ceylon, and within Ceylon from the Kandyan territories to the lowlands, in the years after Kandy's fall. For instance, in 1816, one sitting magistrate suggested that, if "Malabars" wished to move in either direction between Kandy and Colombo, they should obtain a pass, which would only be granted after a personal interview. Sitting magistrates and police officers were instructed to carefully search after Malabars who moved between territories.[68] This also took on a further urgency in the context of the rebellion of 1817–18, when D'Oyly instructed that a watch be kept for "suspicious Malabar persons in the Maritime provinces." In particular, in 1817, he was keen to apprehend "three Malabars of whom the Principal is middle aged and middle sized, having the lower beard only shaved and remarkably good looking— another thin and middle sized having his forehead anointed with ash, the third a lad of 14 or 16 years old."[69] D'Oyly's general view in this period was that Malabars who had a "connection by marriage with Kandyans" and possessed lands in the Kandyan territories and who had been "separated by long absence from relations and some even unknown to their own nation in the Coast" might be allowed to stay, as long as they possessed a certificate. It is especially interesting to note the reason given for this issue of a certificate. It was to prevent "molestation."[70]

66. Chief Secretary Hill to the Officiating Paymaster of Stipends, Vellore, dated 23 September 1825, F/4/1013, IOC.

67. Extract from Political Letter from Fort St. George, dated 11 July 1834, F/4/1461, IOC.

68. Sitting Magistrates Office, to J. Sutherland, Secretary of the Kandyan Provinces, 31 July 1816, Lot 10/130, Sri Lanka National Archives, Colombo, Sri Lanka, hereafter SLNA.

69. Letter dated 24 October 1817 from John D'Oyly to James Sutherland, Lot 21/51B, Kandy branch of the Sri Lanka National Archives, Kandy, hereafter SLNAK.

70. Letter dated 31 January 1817 from John D'Oyly to James Sutherland, Lot 21/51B, SLNAK.

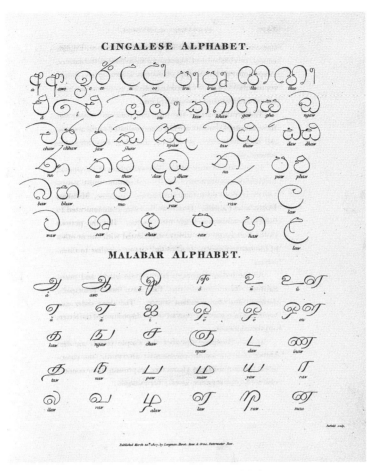

Fig. 1.4. "Cingalese Alphabet and Malabar Alphabet." From James
Cordiner, *A Description of Ceylon* (London, 1807). Reproduced by kind
permission of the Syndics of Cambridge University Library.

One class of people who came under colonial scrutiny as a result of
this directive were the mendicants, who lived an itinerant life across
South Asia. Eleven mendicants were detained on arriving in Mannar on
the northwest coast from India on September 1816, and were said from
their "appearance and manners" to be "exactly the sort of men whom it
is the desire of the Government to prevent penetrating into the interior"[71].
A magistrate also reported how he had apprehended in 1816 a "Malabar

71. Correspondence from Cutcherry, Mannar, dated 4 September 1816, Lot 10/130,
SLNA.

man who calls himself Cahilasen Poille and pretends to be a native of
Colombo and is going in search of medicinal herbs."[72] The sitting magistrate of Colombo documented the case of a man called Ramparasad
and sent on the transcript of an interview with him, of which an extract
follows:

Q. What is your Native Place?
A. Benares.
Q. What is your usual occupation?
A. I am Brahmin Beggar by profession.
Q. When were you last at Benares?
A. Four years ago, since that I was at Poonah and Kokam and Cochin
   and then I took a circuit on the Coast of Coromandel.
Q. What is your object in coming to Ceylon?
A. I came to perform a religious promise at Cataragam and to go to
   Adam's Peak.
Q. Why did you proceed without a passport?
A. I was not aware it was necessary—I landed at Mannar in company
   with three others vizt. Gooolapadoo and Iwat Ghirey, who are both
   gone away—the third is now with me, who is called Bederadesus.
   . . . .
Q. Did you converse with many Kandyan Malabars in the Kandyan
   Country?
A. I conversed at Cateragam with none. I can't speak Tamul or Cingalese. I understood from the Hindostanee Priest at Cateragam that
   many of the persons I saw were Kandyans, but they were not dressed
   like coast Malabars.[73]

The questioner determined that Ramparasad's route of travel encompassed
both north India and south India. In Ceylon, after landing in Mannar, the
man had spent six weeks in Jaffna in north Ceylon, from where he proceeded to Trincomalee on the east coast of Ceylon for a month, and then
south to Kataragama, which is sacred to both Hindus and Buddhists, and
Badulla, before climbing Adam's Peak, a site of religious pilgrimage which

72. Sitting Magistrates Court to Secretary to the Kandyan Department, Colombo,
dated 5 August 1816, Lot 10/130, SLNA.
73. Interview dated 13 September 1816, Colombo Sitting Magistrates Office,
Lot 10/130, SLNA.

will be studied in a later chapter. The questioner also displayed an interest in the connections between this man and the Malabars of Kandy and whether they spoke the same language, and whether they shared the same religious customs and festivals.

One case that presented some problems in this project of fixing people into categories, and of watching the movement of holy men, was that of a man by the name of Sinnasamy, who was described by D'Oyly as a "Malabar priest of Buddhu", and who in the context of the scrutiny of Malabars voluntarily appeared in Kandy from Matara "for the purpose of requesting permission to reside there without molestation." D'Oyly sent him off to Colombo for further investigation. After receiving reports from the Mudliyar of Matara, it was decided to approve his request. D'Oyly advised Colombo: "I perceive no objections to acquiescing in his wish to reside at some temple in the maritime provs. Provided it be at a distance from the limits [with the Kandyan territories] & under the care of some priest who can be confided in." In particular D'Oyly was keen to give the instruction that Sinnasamy should not wander into the Kandyan territories, for his "Malabar extraction might possibly give rise to alarming Rumours."[74]

There is some evidence that the extent to which the British policed the movement of Malabars in the years after the fall of Kandy led to new tensions between communities, which takes up D'Oyly's fear of the molestation of Malabars. This is evident in two petitions that arrived before Governor Robert Brownrigg, from "Weeraragoe of Candy, now at Colombo."[75] The petitioner, who identified himself as a man of Tanjore in South India, had arrived in Kandy forty-five years prior to the petitions and acted as a merchant with no connection to the royal family of Kandy. After the royal family and its relations were detained, the petitioner claimed that his house was attacked by a man named "Muddor, inhabitant of Candy in the accompany of some Cingalese men in the night time." In fear, the petitioner fled Kandy after obtaining a passport from John D'Oyly. After this he was detained on the coast in Colombo and asked to make a weekly appearance before a magistrate, and treated as a prisoner. His family was, in the petitioner's words, "reduced to insufferable indigence and starva-

74. The material related to the "Malabar priest of Buddha" is from letter dated 26 February 1817 from John D'Oyly to James Sutherland and letter dated 24 October 1817 from John D'Oyly to James Sutherland, Lot 21/51B, SLNAK.

75. Petitions to Sir Robert Brownrigg from Weeraragoe of Candy now at Colombo, Lot 10/130, SLNA.

tions without having any assistance nor means of support whatever in this strange place." In a further petition, this man complained of no response from Brownrigg and said that he been reduced to utter poverty, having even to sell his clothes, and that members of his family had fallen ill.

The process of structural experimentation and realignment, according to which the island was at first grafted on to the southern tip of India and then cut off, is vital to an interpretation of how the British attempted to classify the inhabitants of the island. The British first came to think of the island's peoples in relation to their territories in South India, which contrasts with how the Dutch facilitated Kandy's links with Southeast Asia. The British sought to formalize the identity of the island with respect to the mainland in the early years of dual control between Crown and Company. But when this strategy was found to be unsuccessful, and Ceylon grew apart from Madras, a new sense of identity was born that separated islanders and mainlanders.

PARTITIONING FORCES

The British policy with respect to the Malabars of Kandy might be contextualized in relation to the colonial state's handling of other movements of people after the fall of Kandy in the period up to the 1830s and beyond. The legacy of the partitioning of India and Ceylon is also evident in the regulation of the opposite direction of travel, of soldiers and plantation workers from India to Ceylon.

Britain's taking of the island's coasts was first articulated as a way of defending India. For instance, the politician George Pitt wrote that the island was "the most valuable colonial possession in the globe as giving to our Indian empire a security which it had not enjoyed from its first establishment."[76] Given the military significance of the island, the early Governors of Ceylon believed that India owed Ceylon the favor of supplying troops when they were required. Men from the mainland were used in the Kandyan wars of 1803 and 1815, and to quell the wide-scale rebellion that engulfed the interior in 1817–18. Yet the Governors of Ceylon had cause to complain about the delay and bureaucracy that the island experienced in the Company's handling of their requests for reinforcement. For instance, Brownrigg noted his "heavy disappointment" when one of his requests for a supporting force for the 1815 war, which eventually saw Kandy's fall, was denied because the Company's army was fully occupied

76. Cited in de Silva, *Ceylon under British Occupation*, 20.

on the mainland.[77] During the rebellion in 1817–18, he wrote again of the "anxiety and distress" he had labored under because the Madras Presidency did not dispatch the native troops he had expected. He noted that these additional troops would have enabled him to "occupy those parts of the Country, which being abandoned by [him] for want of troops, afforded secure retreats to the rebel chiefs, as well as resources to feed the flame of rebellion, which was expiring."[78]

The troops that did arrive also generated a point of discord. From the start, Governor Brownrigg complained about the "expensive Staff Establishments" sent with every regiment of troops arriving from India.[79] He noted that the Indian government's desire to organize a separate Commissariat for their troops would result in the "greatest confusion."[80] He was anxious that the Indian troops be placed securely under his command and that no intermediate officer should do injustice to Brownrigg's "Rank in His Majesty's Service."[81] The different structures under which the men from the mainland were organized eventually led to open discontent. In July 1818, Colonel Arthur Molesworth wrote to Brownrigg telling him of the great dissatisfaction prevailing among the troops from India at the rate of exchange which determined their pay:

> They embarked on this Service under the full conviction from former usage that they were to receive their Rations gratis and that they should be paid in Gold Pagodas or in Arcot Rupees exactly in the same manner as on the Coast, consequently they were inclined to leave on an average two thirds of their Pay with their families. . . .[82]

Instead these "coast sepoys" had been paid in Ceylon fanams at a depreciated rate of exchange, and this had meant that some very common articles, which they were able to buy while on service in the continent, were beyond their reach. Molesworth noted then that the sepoy or local troops

77. Dispatch dated 16 January 1815 from Robert Brownrigg, CO/54/55, TNA.

78. Letter dated 18 August 1818 from Robert Brownrigg, CO/54/71, TNA.

79. Letter dated 17 August 1818 from Robert Brownrigg, CO/54/71, TNA.

80. Letter dated 3 August 1818 from Robert Brownrigg to Vice President in Council, Fort William, CO/54/71, TNA.

81. Secret and Political Letter dated 18 August 1818 from Robert Brownrigg, CO 54/71, TNA.

82. Letter dated 16 July 1818, to Hugh Elliot, Governor in Council, Fort William, from Arthur Molesworth, Commander of the Madras Troops serving in Ceylon, CO 54/73, TNA.

"really cannot exist on the present rate of exchange." Brownrigg was
alarmed at the prospect of mutiny and intervened in declaring that the
rate of exchange of Fort St. George should apply to the payment of these
troops.[83] However, he pressed India on a question of principle. The confusion about the terms that applied to Indian troops arose from a distinction
between "field" service and "foreign" service: they were paid and rationed
as if they were on "field service" when they had expected to be treated as
if they were on "foreign service" when they embarked from India.[84] In effect he asked the Company whether it viewed Ceylon as "foreign." Having
acquired a complete description of the "resources of the Country, the extent of Supplies available for various Troops, and the rates at which those
Supplies, including various petty articles in common use with the natives
are procurable," the Company decided that they were unwilling to class
service on Ceylon as home service; instead the precedent of Indian troops
who served in 1795 in the Molucca Islands, now known as Maluku Islands
and part of the Malay archipelago, would apply to Ceylon.[85]

Given this exchange, it is unsurprising to discover how the body of
an Indian was cast as distinct to that of an islander. The Company's officers noted the peculiar propensity among the Indians troops of being
"affected by ulcers in the lower extremities," and how this resulted from
the climate of the interior.[86] In reply, all that one of the leading medical men of the island could do was to assert that "the persuasion which
appears to prevail at Fort St. George of the general unhealthiness of the
Interior of this Island . . . is by far too unqualified, and is taken up on
loose and vague grounds."[87] By the end of the 1817–18 rebellion, Brownrigg like North before him had learnt his lesson. He announced that he
would embark on an extensive program of recruitment amongst men on
the island, and so ensured that his dependency on the mainland would be
reined in:

83. General order dated 19 July 1818, from J. B. Gascoigne, Deputy Assistant Adjutant General, CO 54/73, TNA.

84. Letter dated 10 July 1818, from J. J. Wood, Military Secretary to Edward Wood,
Secretary of Government, Fort St. George, CO/54/73, TNA.

85. Letter dated 28 July 1818, from Edward Wood, Secretary of Government, Fort
St. George to the Commissary General, CO/54/73, TNA.

86. Letter dated 18 June 1818, from Alex Watson M.D., President Medical Board to
Hugh Elliott, Governor in Council, CO/54/73, TNA.

87. Letter dated 23 October 1818, from the Deputy Inspector of Hospitals to Secretary for the Kandyan Provinces, CO/54/73, TNA.

His Excellency the Governor and Commander of the Forces, considering it advisable to raise a Corps for the defence of the British Domin-
ions in the Island of Ceylon to consist of His Majesty's Native Cingalese Subjects, invites such Persons of the Class of Lascoreens of the
Vellale, Fisher and Chando Casts, as are willing to serve the Crown as
Soldiers in any part of Ceylon, and are able bodied, to offer themselves
for enlistment for a term of Three Years. . . .[88]

Later, however, once Ceylon had been consolidated as a colony, troops
were again moved between the two territories.[89] In 1825, troops from Ceylon were somewhat begrudgingly sent to Burma, which was termed "the
state adjacent," at the request of the Governor General in Council in India,
Edward Paget; in 1837, Governor Robert Horton sought to firm up the procedure whereby Ceylon could supply troops to the Madras Presidency at
times of emergency.[90]

The use of Indians in the island was evident in other contexts as well:
mainlanders were used for the purpose of labor on roads, transportation
of supplies in the course of war, the repair of irrigation tanks, and what
has been the most studied instance of their use, as plantation workers.[91]
Yet the correspondence on the use of Indian troops is important because it
provides a snapshot of one the earliest uses of Indian labor on the island,
and so reveals the mechanisms that came to dictate later uses of Indians.

From the 1830s, the British saw plantation laborers arriving from the
mainland as distinctly Indian and so a separate lot. This was despite the
fact that there is evidence that Indian plantation workers were sometimes

88. General Order, 27 January 1819, CO/54/73, TNA.

89. For the use of Indian troops overseas in the later period, see T. Metcalf, *Imperial
Connections: India in the Indian Ocean Arena, 1860–1920*, chapter 3.

90. Letter dated 17 January 1825 from Governor Edward Barnes to London, and
attached letters, CO 54/88, TNA, and letter dated 2 September 1837 from Governor
Horton to London, CO/54/155, letter dated 30 June 1838 from Governor Mackenzie to
London, CO/54/163, and letter dated 30 July 1838 from Governor General Auckland, in
India to Mackenzie, CO/54/164, TNA.

91. For the construction of roads, see Sujit Sivasundaram, "Tales of the Land: Brit-
ish Geography and Kandyan Resistance in Sri Lanka, c. 1803–1850." For other uses of
Indian labor, see Roland Wenzlhuemer, "Indian Labour Immigration and British Labour
Policy in Nineteenth-century Ceylon," and Peebles, *The Plantation Tamils of Ceylon*,
chapter 1. See also Ian H. Vanden Driesen, *The Long Walk: Indian Plantation Labour in
Sri Lanka in the Nineteenth Century*, 18–20.

recruited on the island rather than through agents sent to South India.[92] In the early years some traveled by sea to Colombo and then marched inland, while others crossed over by boats from Ramnad to Mannar and then traveled to the interior on foot. By both routes the walk by foot was about one hundred and fifty miles. At the start these workers were part of a floating community, traveling back and forth between the mainland and the island, though this changed with time as they became heavily indebted and effectively indentured. The official figures provide some evidence in support of this: in 1839 there were 2,719 arrivals and 2,202 departures, and in 1845 there were 73,401 arrivals and 24,804 departures.[93] Some allowance must be made here for the fact that departures were less easy to account for than arrivals, but even when this is taken into account it is clear that many workers met their deaths on the island and that for whatever reason a sizeable number never returned to India. While on the island, the workers very quickly became a separate community: they worked in districts with plantations and lived in poor accommodations "behind the line." By the middle of the nineteenth century, the plight of these workers served as an effective reminder of the distinction between the "Malabars" and the other inhabitants of the island. The British treated them as domiciled foreigners. Curiously, the British utilized the term "Malabar coolies" to refer to the plantation workers for most of the nineteenth century, even when they did not come from the Malabar coast.

Discussing the status of plantation workers takes us beyond the first three decades of the nineteenth century. Yet it makes the point that by this time there had come to pass a partitioning of islanders and mainlanders, and a partitioning of those on the island into distinct groups. These partitions arose out of the structural interventions of British colonization. Yet the term "partition" should not lead to the assumption that these colonial interventions were final or fully successful. The case of plantation workers is a good one to support such a qualification. The plantation workers were later termed "Indian Tamils" in contrast to "Ceylon Tamils," who lived on the coast, and in line with the slow replacement of "Malabar" with "Tamil." In 1964 and 1974, some were given Indian citizenship, while others were given Sri Lankan citizenship; however, a remnant remained stateless until receiving Sri Lankan citizenship in 1988 and 2003.[94] In this way the distinction between mainlanders and islanders continued to be a

92. Peebles, *The Plantation Tamils*, 26.
93. Vanden Driesen, *The Long Walk*, 20.
94. Peebles, *The Plantation Tamils*, 226.

potent one in the politics of post-independence Sri Lanka. In effect, this partitioning left quite a lot of unfinished business.

CONCLUSION

The islanding and partitioning of Sri Lanka involved a decisive attempt on the part of the British to intervene in the movements of people across the sea that divided what became separate territories. This process carried through in the manner of British governance and reform into the middle of the nineteenth century, as will be discussed in chapter 8. Popular discursive imaginings of territory need to be tied in securely to structural interventions.

It is not my claim that hybrid identities gave way to harder classifications with the arrival of the British. Such a statement is too simple, for British categories were themselves changeable. Such a statement also essentializes the character of the pre-British identities of the island. In using the term "cosmopolitan" for Kandy, I have done so with the meaning of a cosmopolitanism bounded by a sense of universal kingship and Sinhalaness. Rather than radically redrafting identities, the impact of British ideologies emerged as a consequence of the power of British colonialism to change the political organization of the island and its society. Crown rule intervened more powerfully than any previous external power had done in the evolving pattern of identity within the island and so narrowed the options for the future. In particular the British were concerned to track indigeneity and to exalt it as a determinant of difference, while isolating the foreign. This was partly a result of the need to stabilize the colony in political terms and to order it as a unit. The irritations between different forms of British governance meant that the Malabars, who later became Tamils, were said not to belong in a Crown territory but rather in mainland India.

In other colonial territories as well, the nature of colonialism may be linked to changing ideas of ethnicity. Following John Comaroff, who writes primarily of Africa, ethnicities are best understood not as things but relations and their content is wrought in the particularities of the ongoing historical construction of such connections.[95] From this it follows that ethnicities, like other identities, are about the placement of the self in relation to other peoples, in everyday life and in political and social

95. John Comaroff, "Ethnicity, Nationalism and the Politics of Difference in an Age of Revolution," 166.

processes. This means that studying shifts in relations leads to an understanding of changing conceptions of ethnicity. It is important to add that
such relations between peoples should not be localized too quickly, for it
is not only our age that has witnessed globalization or migration. Thinking of ethnicity as relational across distance sidelines a debate about when
ethnicity arose, and in particular the question of whether it was precolonial or colonial. Instead, what comes into view is the transnational context of the local and how the shifting sense of the transnational impacts
on the local. This interest in the shifting placement of the island, with the
onset of British colonialism, continues to be a theme in the next chapter,
which develops the argument about islanding and partitioning with respect to trade.

CHAPTER TWO

Trade

Sri Lanka lies in the middle of two historic webs of commerce: one to the west, stretching to West Asia and East Africa and down the coast of Africa, and another to the east, to Southeast Asia and beyond. The significance of Ceylon as a nodal point of trade and empire carried through into the nineteenth century, as the British reconfigured its placement in global networks of capitalist trade. This was not totally unlike what the Dutch had done before them; yet the legislative, structural, and bureaucratic dimensions of this intervention were more extensive in scale. The British program of resetting the trading bounds of the island included passes, papers, and arrangements that privileged trade with Europe and other Crown colonies over India. It was matched by an imaginative, discursive, and physical concern with maritime relations and access to waterways, so as to police the oceanic boundaries of this island colony. Islanding was thus a project that sought to regulate not only the land and its people but access to the sea as well.

In working with the period that cuts across the eighteenth and early nineteenth centuries, it is possible to connect scholarship on the history of the Indian Ocean with work on the British colonization of Ceylon. Following the panoramic writings of K. N. Chaudhuri, it has become a norm to treat the Indian Ocean world as an analytically coherent space, which is identifiable not as a spatial construct but more through the dynamics of social and economic relations. For Chaudhuri, the Battle of Plassey in 1757 was an important marker of how the unity of this world changed, leading to "technological and civilizational breaks" which gave rise to modern forms of empire and state-making.[1] At the same time he stresses that the

---

1. K. N. Chaudhuri, *Asia before Europe: Economy and Civilisation of the Indian Ocean from the Rise of Islam to 1750*, 41.

impact of European colonialism came through the "system of collective decision-making, through committees and councils with delegated authority," and he urges that this logic was of a different kind to that which had underpinned the less regularized activity of the merchants of the Indian Ocean world.[2] A similar track was followed by the distinguished historian from Sri Lanka, Sinnappah Arasaratnam, who also finished his studies of trade in the Indian Ocean with the end of the eighteenth century, after which period he posited a more intensive phase of modernization. Yet he noted that "intense exposure to western rule in the sixteenth to eighteenth centuries brought about developments in the island's society and economy which set it on a path different to the Indian subcontinent."[3] Both Chaudhuri and Arasaratnam, by ending in the eighteenth century, left it to other scholars of the nineteenth century to chart exactly how modern colonialism transformed the Indian Ocean world. This gave rise to some gaps. From the Indian side attention was paid to Bengal and north India, in creating a theory of colonial transition for the East India Company, leaving the Indian Ocean world somewhat in the background. The South Indian story of colonial transition paid no attention to Sri Lanka. From the side of Sri Lanka, scholars such as Vijaya Samaraweera produced important and detailed work which sought to document the legislative and commercial policies of British colonialism, but which did not have the aim of contextualizing the nineteenth century in terms of what came before.[4]

While the last chapter considered the placement of Kandy in the Indian Ocean world, this one adds a closer attention to Dutch policy with respect to Lanka's place in the sea. The first section of the chapter illustrates how the relationship between the Dutch and the Kandyans cannot be equated to sea- versus land-based power; a more organic and interlinked system of dependence characterized the seventeenth and eighteenth centuries. The remainder of the chapter documents how the British sought to bind the island as a maritime space, discursively, geographically, legislatively,

2. K. N. Chaudhuri, *Trade and Civilisation in the Indian Ocean: An Economic History from the Rise of Islam to 1750*, 83.

3. Sinnappah Arasaratnam, *Ceylon and the Dutch, 1600–1800: External Influences and Internal Change in Early Modern Sri Lanka*, 2.

4. For instance, see Vijaya Samaraweera, "Ceylon's Trade Relations with Coromandel during Early British times, 1796–1837"; Samaraweera, "The Cinnamon Trade of Ceylon"; and also Samaraweera, "Economic and Social Developments under the British, 1796–1832."

and commercially. Despite its partitioning of the island from the mainland, the interjection of British colonialism lay in the relocation of the island in this ocean rather than in an attempt to take Lanka away from the maritime sphere of this sea. This intervention in maritime trade and its structural and representational trappings was orchestrated via Crown colonialism, which narrowed existent pathways, as in the case of ethnicity, discussed in the previous chapter.

THE SEA AND SOVEREIGN POWER

Though the Kandyan kingdom looks landlocked on paper, its rivalry with the Dutch was also conducted by sea. In 1744, Narenappa Nayakkar, the father-in-law of the ruling king and the father of two other kings, sought to take two vessels under his charge across the sea from the port of Kalpitiya to the territory of the Tevar of Ramnad on the opposite coast. According to a Dutch historian of the eighteenth century, the Dutch East India Company (Vereenigde Oost-Indische Compagnie, VOC) was seeking at this stage to prevent smuggling by "cruising" off the coast "everywhere," but the Kandyan court sought to move "heaven and earth" to carry on a trade over the seas. In order to push through his plan, Narenappa arrested the Dutch master of horses from Jaffnapatnam, as well as two keepers of elephants, in addition to detaining their animals. He used the name of the king of Kandy to legitimize his actions. He even threatened to prevent the VOC's transfer of letters to and from Jaffnapatnam if he was not allowed to trade across the sea. "As nothing could be done with this turbulent man, a company of solders was sent by sea to prevent him from molesting the inhabitants: should he use force, they were to resist him with force and search his vessels." In the end Narenappa gave up this maritime stand-off and returned to Kandy, hearing of the death of one of the queens.[5]

In developing the economic aspect of Kandy's engagement with the sea and making sense of Narenappa's actions, it is pertinent to pay attention to the question of access to ports in the diplomatic wrangles between the Dutch and the Kandyans. The Kandyans had control over five ports at the time that the Dutch succeeded the Portuguese on the coastal belt in

5. All quotations for this paragraph are taken from F. De Vos, ed. and trans., "A Short History of the Principal Events that Occurred in the Island of Ceilon, Since the Arrival of the First Netherlanders in the Year 1602, And, Afterwards from the Establishment of the 'Honourable Company' in the Same Island, till the Year 1757," 134–36.

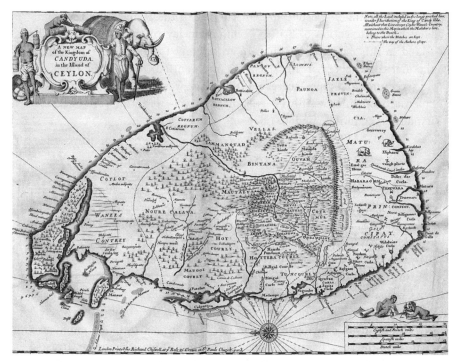

Fig. 2.1. "A New Map of the Kingdom of Candy Uda in the Island of Ceylon." From
Robert Knox, *An Historical Relation of the Island of Ceylon* (London, 1681) (detail).
Reproduced by kind permission of the Syndics of Cambridge University Library.

the seventeenth century: these were Kalpitiya and Puttalam on the west
coast and Trincomalee, Kottiyar, and Batticaloa on the east coast (fig. 2.1).[6]
Each of these ports was linked to a particular segment of the Kandyan
kingdom. The typical trade between Kandy and the outside world consisted of the export of areca nut and the import of cloth. In addition to
this, elephants were exported, paddy, or rice in husk, could come in or go
out of the island at various times, and salt and dried fish were brought into
the island. Kandyan articles of handicraft were also traded.

The trade at these ports was closely connected to the court, for the
king's produce from royal lands was sold to those who traded at the ports.
But this was also a network of trade which stretched by means of intermediaries to rural villages, where produce could be collected and transported

6. This paragraph relies on Sinnappah Arasaratnam, "The Kingdom of Kandy: Aspects of its External Relations and Commerce, 1658–1710," 110.

to the ports. The trade to and from the ports was conducted primarily by the Moors and Malabars, as noted in the previous chapter.

The role of these traders as "foreigners" is sometimes explained by the Sinhalese "abhorrence for commerce," and also the lack of a history of naval interests on the island.[7] Another explanation which plays up the ethnic factor in coming to terms with this trade asserts that the high symbolic significance of the external trade is connected to the clannish identity of the Nayakkar line. Accordingly, they are held responsible for loosening the "traditional moorings" of the kingdom of Kandy, "as a mountain girt fortress insular by nature and mistrustful of the outside world," by bringing it into the world of "international trade and diplomacy" by virtue of their experience as "astute traders with ambitions of dominating trade across the Palk Straits."[8] Yet all of these interpretations forget the fact that traders were heavily integrated into the kingdom in terms of the duties they performed and the taxes they paid. The Moor traders, for instance, had to deliver royal stores free of charge to the coasts.[9] They also had to pay a capitation tax, called *uliam*, in return for the right to act as traders. Anthony Bertolacci, writing at the start of the British period, wrote of this tax:

> When the Moormen, and Chetties or Malabars, first came to the island of Ceylon, previous to possession being taken of it by the Portuguese, they obtained the privilege of being, exclusively, authorized to keep shops in particular markets, for the retail of certain goods imported in seaport towns; and in consideration of such exclusive privilege, they bound themselves to work three months in the year for the Prince who governed the country, which they continued doing under the Portuguese and Dutch Governments.[10]

Through its maritime contact, Kandy sought to confirm its superiority to its European rivals, and to assert their vassal status. This is why the flags under which they sought to trade in the late seventeenth century are significant: in 1688, a vessel flying a "white flag charged with a red lion

---

7. See Ralph Pieris, *Sinhalese Social Organization: The Kandyan Period*, 111. For lack of naval interests, see Sinnappah Arasaratnam, "Sri Lanka's trade, Internal and External in the Seventeenth and Eighteenth Centuries," 406.

8. V.L.B. Mendis, *Foreign Relations of Sri Lanka: From Earliest Times to 1965*, 310.

9. Pieris, *Sinhalese Social Organization*, 112.

10. Anthony Bertolacci, *A View of the Agricultural, Commercial and Financial Interests of Ceylon* (1817), 385.

loaded with arecanuts" sought to evade Dutch scrutiny, and in 1689 one of
the vessels that sought to trade was said to fly "three white flags charged
with red lions."[11] The flags trumpeted the lion-ancestry of the kingly line
and its symbolic hold over the entire island.

The general pattern of Kandy's external trade relations in the late seventeenth and eighteenth century was dictated by the Dutch desire to take
greater control of these ports. However, this overriding factor was punctu-
ated at times from the Dutch side by a concern that regulation hampered
this trade by reducing the number of traders at the ports and unnecessarily
aggravating the king. In return, the Kandyans utilized the Dutch dependence on their lands for food and cinnamon as a strategic counterweight
in order to uphold their demand for access to the ports. On one occasion
in 1733 they told the Dutch that "none of the gravets [entry points to the
kingdom] should be opened until the port of Putulang [Puttalam] was declared free."[12] The Dutch on the coast and the Kandyans in the hills were
thus interdependent, for each to different degrees wished for access to the
land and sea in turn.

In the last decades of the seventeenth century, the Dutch and the Kandyans engaged in a "hot war" in disputes over these ports.[13] The Dutch occupied Kalpitiya in 1659 and then went on to occupy other Kandyan ports,
only to generate counter-attacks and a counter-blockade on the part of the
Kandyans, who decided to deny the Dutch the produce of the highland
kingdom. By 1675 the Dutch were forced to develop an alternative policy
and sought after friendship, and Kandy by this time was feeling the effect
of the blockades and was keen to negotiate. In 1687, the Dutch proposed a
treaty to the Kandyans, in which they suggested that all of the kingdom's
produce should be sold to the Dutch East India Company: this would isolate the traders and invalidate their role, which the Dutch argued was a
good idea. They even urged the political necessity of ending the influence
of foreign traders, from the exaggerated claim that the island would fall to
the Mughals, who would band together with Muslim merchants! In the
midst of debates about the way forward, the Kandyan trade was first freed
for some years, though by the first decade of the eighteenth century it was
finally taken into the control of the Dutch again. The position in 1708,

11. De Vos, ed. and trans., "A Short History of the Principal Events that Occurred in
the Island of Ceilon," 106, 109.

12. Ibid., 120.

13. This paragraph is based on Arasaratnam, "The Kingdom of Kandy."

whereby Dutch strategic concerns and worries about other European rivals dictated the closure of the ports to Kandy, held for some decades. Kandy became economically more isolated and dependent.

From the 1740s, however, the Kandyans became concerned yet again to have their ports back, and this interest coincided with a cultural resurgence in the interior kingdom. In the mid-eighteenth century, there was a particularly persistent bid on the part of Kandy to gain for the court a return of the elephant trade which had been a royal monopoly. The idea was to have elephants transported "at moderate expense by native vessels" rather than on Dutch ones, but the Dutch declined. This interest in elephants, as well as the wider concern with the sea in the mid-eighteenth century, is useful in making sense of Narenappa's course of behavior too.[14] War broke out between the two powers between 1760 and 1766, which tilted control more towards the Dutch again, necessitating a recognition on the part of the Kandyans of the Dutch as an equal power, rather than a vassal one. Van Angelbeek, later the Dutch governor of the island, even hoped to make the king pay tribute and to re-crown him under the VOC's authority, as a "perpetual remembrance of his vassalage," and to make him pay four elephants every year.[15] Before the terms of peace were made, one Dutch military man wrote: "*Candia* has been captured and plundered, and even the holiest temples and pagodas of the *Candians* and Sinhalese."[16] A treaty of 1766 secured formal Dutch access not just to the ports but to a strip of continuous coastal territory for the Dutch and also the right to peel cinnamon in the king's territories. It also required the Kandyans to trade in any "foreign commodities" through the Company.[17] According to the treaty:

> All the sea-borde round the Island of Lanka not held by the Company before the war, which is now drawing to a close, is to be entirely given over to the above-mentioned Honourable Company . . . putting the Company in absolute possession of the same.[18]

14. De Vos, "A Short History of the Principal Events," 143–44.

15. Ibid., 57.

16. Ibid., 56.

17. Alicia Schrikker, *Dutch and British Colonial Intervention in Sri Lanka, 1780–1815: Expansion and Reform*, 39; also A. E. Buultjens, trans., "Governor Van Eck's Expedition against the king of Kandy, 1765," 73.

18. Buultjens, "Governor Van Eck's Expedition," 71.

In turn the Dutch paid the Kandyans compensation for the loss of their ports. The status of the historic Kandyan ports was taken up in a new dimension, confirming Dutch fears of rivals, when first the French and then the British occupied the port of Trincomalee in 1782. Though the climate of rivalry between the Dutch and the British was accentuated in these years, the Kandyans were unable to turn this to their favor. Rather, it created even further paranoia about the ports and their security on the part of the Dutch, and exacerbated the isolation of the Kandyans in economic terms. As the most recent historian of this period notes, it was the issue of ports which defined the relationship between Kandy and the Dutch after 1782. In contrast to the end of the previous century, these decades might be seen as a cold war about ports.[19] Kandy wrote directly to the Netherlands about their claim to the ports; this met with a negative response, and the answer was delayed by the Dutch regime in Ceylon before reaching the Kandyan court.[20]

According to a classic work on the Kandyan kingdom, the dependence of the Kandyans and the increasing restriction of their external trade may be said to have had an impact on the circulation of specie in the kingdom.[21] By the eighteenth century currency was in very short supply. In fact more Europeans coins, such as the Dutch *stuvier*, were in circulation than was specie minted in Ceylon or India.[22] Robert Andrews, who was sent by the British on an embassy to Kandy, noted: "All sorts of Money is here very scarce: And they frequently buy and sell by exchanging Commodities. . . . Corn passeth instead of Money."[23] This lack of money is related by some scholars to the fact that Moors and "Malabars" were the prime money lenders in Kandyan territories; during the last years of the kingdom, the king's relatives who were "Malabars" charged an interest of 40 percent per annum, while the Moors were said to charge an interest of 20 percent.[24]

Yet the lack of currency, and the impact of Dutch and European control over the economy of the Kandyans, should not be read as supporting

19. Schrikker, *Dutch and British Colonial Intervention*, 114.

20. Ibid., 115.

21. This is Pieris, *Sinhalese Social Organization*.

22. For information about currency in Ceylon in the period leading into and following the British taking of the island, see B. W. Fernando, *Ceylon Currency: British Period, 1796–1936* (Colombo: Ceylon Government Press, 1939).

23. Cited in Pieris, *Sinhalese Social Organization*, 110.

24. John Davy, *An Account of the Interior of Ceylon and of its Inhabitants* (1821), 185.

a thesis of decline. There are several dangers in arriving at such an interpretation from the evidence presented thus far. First, it imagines the Kandyan trade to be distinct from the coastal trade with the outside world; in fact noncolonial trade from the island did not respect the political divisions between the Kandyan territory and territory under European control. Second, a thesis of decline that seeks to characterize Kandy's trade as restricted does so primarily from a view of the changing relations between the court and the Dutch; however, much of this trade was conducted informally. It was impossible for either the Dutch or the Kandyan court to claim absolute control over the commerce of the island, which was conducted out of almost every bay and inlet.

Further to this, the idea of Kandy's decline arose out of early British accounts. For instance, in 1816 Governor Brownrigg wrote to London, of Kandy's trade, "that it operated under undefined and oppressive dues and monopolies" which "threw the industry of the People and the produce of the soil under the paralizing [sic] influence of arbitrary Power, and that guided by the sole motive of Private Interest."[25] He added that the Kandyans were "wholly new to commercial ideas." Throughout this letter, Brownrigg's view of the commerce of Kandy was dictated by an exclusive interest in the functions of the court and also in a rhetorical confidence in British ideologies of free trade. These biases make his judgment on the trade history of the interior highlands difficult to take at face value. As late as 1833, Governor Horton could still describe the Kandyan territories as "hermetically sealed and not to be approached by a horse with a rider." He noted that "everything is carried in and out of it on men's shoulders or bullocks very lightly loaded."[26]

Taking a different perspective to the focus on the official encounter between the Europeans and Kandy, and looking from the littoral out to the sea rather than from Kandy to the sea, it is possible to discern several directions of trade. Muslim merchants sailed large vessels from Bengal and Golconda into Galle and Jaffna: the chief commodity that these vessels brought in was rice, but sugar, butter, and vegetable oils were also imported, in addition to silks and muslins. Right up to the early eighteenth century, the traders took back elephants, as Ceylon's elephants were deemed to be

25. Letter dated Colombo, 9 February 1826, from Robert Brownrigg to Earl Bathurst, CO/54/59, TNA.

26. Private letter dated Colombo, 25 October 1833, from Governor Horton to E. G. Stanley, CO54/130, TNA.

of superior quality to those in India and were useful for war.[27] In addition
to this rather long-distance trade to northern India, there was also a maritime trade which connected the coastal belt to western and southern India. To Surat, Wingurla, Cannanore, Calicut, and Cochin went areca nuts,
coir, and ropes and Kandyan lacquerware, in exchange for rice and textiles
again. To the coast from Travancore to Madras went areca nuts and betel
leaves, in exchange for coarsely woven cloth and other textiles. The Raja
of Travancore also had a monopoly of the tobacco trade from Jaffna. In
addition to Moors and "Malabar" traders, there were also others involved
in this trade from the littoral, including Chettie merchants and Roman
Catholics. It is difficult to provide a proper estimate of the significance of
this trade, as it was carried on informally in small vessels called dhonies,
which "unlike the European sailing vessels which had to go round the
island . . . touched the opposite coast in double quick time and were able
to offer substantially cheaper freight rates."[28] These dhonies also served to
connect points on the island with others along its own coast.[29]

In the late seventeenth and eighteenth centuries there are some signs
of a lull in this informal trade, for there were attempts on the part of the
Dutch to take it under their control, as they had the official trade of Kandy
through their ports. This partly arose out of a Dutch policy, evident after 1670, of seeking for monopolies in elephants and areca nuts, etc. The
Dutch also sought to take control of the supply of textiles into the island.
In particular they targeted the South Indian trade through restrictions.
There were, however, moments of liberalization, which coincided with
those noted above in relation to the Dutch-Kandyan relations over ports.

In summary, there was a concerted attempt on the part of the Dutch
to reorient the island's trades as part of a "well-knit system of the VOC's
eastern enterprises";[30] yet this was only partially successful as the informal trading networks continued to function despite the interference of
the need for passes and inspection.[31] At the same time the symbolism of
Kandy's reach and its right of access to the sea was asserted well into the
eighteenth century. More fundamentally, the economic history of the is-

27. For more on the British inheritance of these patterns of the elephant trade, see
Sujit Sivasundaram, "'Trading Knowledge': The East India Company's Elephants in
India and Britain."

28. Samaraweera, "Ceylon's Trade Relations," 14.

29. See Sinnappah Arasaratnam, "Ceylon in the Indian Ocean Trade: 1500–1800."

30. K. M. De Silva, *A History of Sri Lanka*, 174.

31. Arasaratnam, "Ceylon in the Indian Ocean Trade," 238.

land in this era was dictated not only by the real movement of commodities but also by the political tussle to accrue a ruling legitimacy by virtue of command of the seas and trade. For at the heart of this story is Kandy's desire to make the Europeans its vassals, while the Europeans sought for and succeeded in formalizing Kandy's vassal status.

TO ARRIVE AT AN ISLAND

With the arrival of the British there was an attempt again to intervene not only in the island's place in the economy of the Indian Ocean but also its symbolic place in the geography of this maritime region. Before turning to the trade history of early British colonialism, which recycled the Dutch policy of the late eighteenth century of making the island a node in an imperial system, it is pertinent to pay attention to how British colonists sought to bring a set of discursive meanings to bear on the place of the island in their expanding remit in this ocean.

Among travelers, there was a recurrent way of distinguishing Ceylon from India in aesthetic terms. Robert Percival, in his popular work on the island, put this oft-noted sentiment thus: "As you approach the island from the sea, it presents a fresher green to the eye, and has every way a more fertile appearance than most parts of the Malabar and Coromandel coasts." He continued by describing the typical aspect of the island. At the edge of the shore were groves of coconut trees, which led the eye into the interior, where there were plains of rice fields, which then drew the eye to the sight of mountains in the far distance covered with woods. "Such a prospect," noted Percival, "has the most pleasing effect on the eye, after being fatigued with the shores of barren white sand, which everywhere skirt the opposite continent."[32] This aspect of representation was implicitly a commercial form of imagining. For in playing up the islandedness of Ceylon, popular writers such as Percival were urging its importance as a trading emporium. On the second page of his book, Percival noted: "The vast importance of this island in both a commercial and political view" and linked it with its possession of "the only harbour either on the Coromandel or Malabar coasts, in which ships can moor in safety at all seasons of the year."[33] This was of course the harbor of Trincomalee, which was

32. Robert Percival, *An Account of the Island of Ceylon* (1803), 32. For another statement of this contrast, see Walter Hamilton, *A Geographical, Statistical and Historical Description of Hindostan* (1820), 485.

33. Percival, *An Account of the Island of Ceylon*, 2.

one of Kandy's ports in the eighteenth century. Charles Pridham, writing in 1849, noted the undeveloped state of the commerce of Trincomalee, and in particular how the "silver bosom" of the Mahavali river, which flowed into it, only brought "rafts of timber" from the interior.[34]

Speaking about Ceylon as an island node also brought in a comparison with other islands. At one level, the islandedness of Ceylon made it more British than the continent of India, in a period in which British identity itself was tied to the geography of the British isles. Maria Graham, the travel writer, noted: "I am so delighted with the place, and with the English society here, that if I could choose my place of residence for the rest of my time of absence from England, it should be Columbo."[35] Amelia Heber, the wife of the Bishop of Calcutta, in her journal of their visit to the island noted that the residents of Ceylon "make a distinction between the island and the continent, not allowing the former to be India"; and that excepting the "colour of the natives, everything wears a more English aspect than [they had been] accustomed to in India."[36]

The Bishop himself, in a letter to his mother published with the journal, took the idea of Ceylon's islandedness further by linking it to the emblematic island of this period:

Here I have been more than ever reminded of the prints and descriptions in Cook's "Voyages." The whole coast of the island is marked by the same features, a high white surf dashing against the coral rocks, which, by the way, though they sound very romantically, differ little in appearance from sand-stone: a thick grove of coco-trees, plantains, and bread-fruit, thrusting to their roots into the very shingles of the beach, and hanging their boughs over the spray; low thatched cottages scattered among the trees, and narrow canoes, each cut of the trunk of a single tree, with an out-rigger to keep it steady, and a sail exactly like that used in Otaheite. The people, too, who differ both in language and appearance from those of Hindostan, are still more like the South Sea islanders, having neither turban nor cap, but their long black hair fastened in a knot behind, with a large tortoise-shell comb, and seldom

34. Charles Pridham, *An Historical and Statistical Account of Ceylon and its Dependencies* (1849), 1:395.

35. Maria Graham, *Journal of a Residence in India* (1812), 100.

36. Reginald Heber and Amelia Heber, *Narrative of a Journey through the Upper Provinces of India from Calcutta to Bombay, 1824–1825, with Notes upon Ceylon Written by Mrs Heber* (1828), 2:152.

any clothing but a cotton cloth round their waist, to which the higher ranks add an old-fashioned blue coat, with gold or silver lace.[37]

In bringing Ceylon into the discourse that related to Tahiti, these words displaced it from the mainland and cast it into a different category of territories within the sway of the expanding British empire. Further, the comparison of Ceylon and Tahiti was a sexualized one. For instance, a tract printed in the *Asiatic Annual Register* noted how the women of Ceylon weren't kept in "slavish privacy and bondage" and that the "virtue of female chastity is not held to be a principle of honour. Such practices and sentiments are so diametrically opposite to the uniform and immemorial observances of all other Asiatic nations, and to the standing maxim of the Hindu law."[38] The author could not make up his mind of the veracity of these "extraordinary customs," which to his mind surpassed "in sensuality those of the luxurious savages of Otaheiti."[39]

SEPARATING THE SEA

Yet the impact of these words, which made Ceylon more luxurious, more British, and more exotic than the mainland, must not be overemphasized. These ideas also, like the categories that were used to classify peoples, emerged out of the way in which the British sought to make a state in Ceylon. They should be contextualized in relation to the physical mechanisms of navigation and travel.[40] Travelers' comments on how Ceylon was different from the mainland appeared just as both the Crown and the East India Company were coming to terms with the linkage of the two territories in sea routes. This observation takes on more force when read in the light of the conflict between the Crown and Company over the control that Indian vessels had over the island's trading practices.

When the time came for the Indian forces who had helped quell the rebellion of 1817–18 to return home, Governor Brownrigg asked the Ceylon government's agent in Bombay to arrange for vessels which were traveling between Calcutta and Bombay carrying rice to take back a battalion

37. Ibid., 2:244.

38. Anon., "An Historical Account of the Island of Ceylon" (1799), 111.

39. Ibid., 117

40. See Hamilton, A *Geographical, Statistical, and Historical Description of Hindostan*, 486. His comments on the contrast between the scenery of India and Ceylon was immediately followed by a description of the character of shipping.

on their return journey to Bengal. The agent, Messers. Rimington Crawford & Co., procured two ships, the *Dutchess of Argyle*, a 578 ton vessel,
and the *Dutchess of Cumbria*, a 720 ton vessel; the second of these belonged to the agent. When Rimington Crawford & Co. charged the fee of
15,000 rupees for the *Dutchess of Cumbria*'s role in transporting troops, a
dispute arose with the Ceylon government about overcharging. This dispute allows a snapshot into the character of this coast-bound pattern of
navigation between the mainland and the island. One fact which Rimington Crawford & Co. used to support their invoiced cost was the seasonal
nature of the system of navigation, depending on the monsoon. In October
the *Dutchess of Cumbria* might have been lying "on Colombo Roads un-
employed,"[41] but in April it was not normal for vessels to land at Ceylon
for trade, and indeed the ports of Colombo and Galle were shut to mercan-
tile shipping "when the increased risk and consequent higher rate of in-
surance, as well as finding a ship with additional ground tackle must enter
into the calculation."[42] This resulted both from the risky sea conditions
off the west coast of Ceylon and from the fact that a much faster passage
from Bombay to Calcutta was possible because of the winds, thus making
it unnecessary to stop on the island to load further trade.[43] In effect, the
company pointed to how the movement of cargoes out of the island was
predetermined by the shape of the mainland's trade, which had the first
shot at the best winds, and so that the charge was justifiable.[44]

The seasonal nature of shipping around the coast of Ceylon is also
borne out by Anthony Bertolacci, who wrote that to round the island is
"impracticable, except for waiting for changes of the regular monsoons":

> The south-west that blows from April till September, and is favourable
> to vessels going from Cape Comorin to Manar, or the coast of Ceylon
> near it, renders it impracticable to proceed thence to the point of Don-
> dera Head. The north-east, that prevails from October to the month of

41. Letter dated 31 May 1819, from Rimington Crawford & Co. to George Lusig-
nam, Secretary to the Kandyan Provinces, CO/54/74, TNA.

42. Letter dated 14 May 1819, from J. A. Rope to Rimington Crawford & Co.,
CO/54/74, TNA.

43. Letter dated 31 May 1819, from Rimington Crawford & Co. to George Lusig-
nam, Secretary to the Kandyan Provinces, CO/54/74, TNA.

44. For later commentary on the impact of the monsoons on the island's coastal
trade, see "Remarks on the paper published by Mr. Gordon," 1831, CO 54/113, TNA.
"A good ship cannot make the passage from Colombo to Trincomalee in less than six
weeks against the monsoon."

February, would facilitate the passage of these vessels from Manar to Dondera Head; but they must wait again for the south-west before they can proceed to Trincomalee, Point Pedro, and the coast of Coromandel. Even now that navigation is much improved, the Indian vessels that trade between Ceylon and the coast of Coromandel effect only one voyage in the year, and wait for the change of the regular monsoon to undertake their return.[45]

The Ceylon government's irritation at the informal pattern of navigation around its coasts, and also the hold of the East India Company over this pattern of navigation, led to an attempt to exercise more authority and to regulate the movements of vessels. In 1820, together with a tightening of custom duties, the Ceylon government announced that all dhonies and coasting vessels which originated from the ports of the island should possess a certificate on board.[46] Later that year, this regulation was broadened to encompass "every Ship, Vessel, or Doney sailing out of any port of [the] Island." It now became necessary for each of these boats to take out a Port Clearance with a "full and specific enumeration of the Cargo shipped on board the Ship, Vessel, or Doney, at such Port, and also whether the same has paid or is liable to pay duty or not."[47] Regulations were also brought in for shipping in the other direction from India to the island.

The context of the last decades of the Navigation Acts and the East India Company's charter is critical in providing the broader legislative framework for these interventions on the part of Ceylon.[48] In 1817 for instance the Advocate Fiscal of Ceylon noted the detrimental effect of the Navigation Acts on the island's trade with India. He wrote of the restrictions pertaining to sugar and tobacco grown in any plantation or territory under British Crown government, whereby it had to be sent to England or to some other plantation only, rather than having open recourse to trade.[49] He also noted how the Navigation Acts forbade the passage of European goods from Bombay, Madras, and Calcutta to the island. Soon after these complaints the meaning of the Navigation Acts for the Company's rela-

45. Bertolacci, *A View of the Agricultural, Commercial and Financial Interests of Ceylon*, 18.

46. Regulation of Government, 1820, No. 6, CO/54/ 77, TNA.

47. Regulation of Government, 1820, No. 19, CO/54/79, TNA.

48. For a old discussion of this period of the Navigation Acts, see J. H. Clapham, "The Last Years of the Navigation Acts."

49. Letter dated Colombo, 17 February 1817, from Hardinge Giffard, Advocate Fiscal and King's Advocate, Admiralty to the Governor, CO/54/65, TNA.

tions to the Crown was more carefully delineated, and certain exemptions were allowed. However this correspondence points to the mentality of surveillance, penalties, and bonding which characterized the Company and Crown's interaction in this period, which came from the ground of contestation between the Navigation Acts and the East India Company charter.[50]

One rather telling example of this, which is reminiscent of the discussion in the previous chapter of the use of Indians for military purposes in Ceylon, is the correspondence pertaining to how to regulate vessels owned by Indians sailing to Ceylon, which fell outside the terrain of the Crown government's regulations about registration noted above. In 1830 Governor Edward Barnes wrote to London that "conflicting opinions appear to exist with respect to the meaning of the term His Majesty's Subjects as regards natives of India."[51] He noted that the Navigation Act requiring the registration of vessels owned by British subjects was interpreted differently by different Company officials with regard to this class of vessels. He gave two examples: the vessel of "one James Roosmalecoey, a person of the class usually designated as European descendant," but which had no formal registration papers and sailed simply with a pass from the Government of Madras. And second, he noted a vessel owned by merchants at Negapatnam, "natives of India" and certainly not typically seen as British subjects, and yet possessing a registration certificate. Meanwhile the official line from India was rather vehement, in insisting that the Navigation Act requiring the registration of vessels owned by British subjects did not apply to vessels owned by Indians in Company territory. "'His Majesty's Subjects' may, in one sense be said to include all who owe allegiance to the Powers of England, but it has been the practice of the Supreme Council [in Calcutta] to regard the expression of 'British Subjects' in a more limited sense and so as not to include the Native Inhabitants of India." The Advocate General of Fort William also pointed to how no "Asiatic sailor, Lascar or Native" born in any territories within the charter of the East India Company could be deemed a "British sailor or mariner."[52]

50. See "An Act to exempt the Territories within the Limits of the East India Company's charter from certain of the Navigation Laws," 10 July 1817, CO 54/65, TNA. See also "An Act for the Encouragement of British shipping and Navigation" for the alignment of the Navigation laws with the laws pertaining to the registration of vessels in Ceylon, CO/54/93, TNA.

51. Dispatch dated Colombo, 24 November 1830, from Governor Barnes to Sir George Murray, Colonial Office, CO 54/107, TNA.

52. Letter dated Fort William, 14 June 1830, from John Pearson, Advocate General to H. J. Prinsep, Secretary of the Ceylon Government, CO/54/107, TNA. London simply

Moving to the East India Company's charter, the governors of Ceylon were concerned that the trade of the island did not have the favorable status enjoyed by the Company and that trade with Crown colonies was unfavorably affected by the charter.[53] Because of the Company's rights of trade, there was a ban in force on the export of coffee from Ceylon to Mauritius. In picking up complaints made by merchants about this ban, Barnes noted that there should be "no restriction" on "commercial intercourse" between "Ceylon, Mauritius and the Cape of Good Hope," all "being King's colonies."[54] A similar debate arose over the question of whether French vessels could import wine from Mauritius to Ceylon, and more generally with respect to the French trade into the island.[55] In 1837 the merchants of Ceylon saw the trade between Mauritius and Ceylon to be troubled by "vexations" and asked for it to be placed on the same footing as the trade between different Crown colonies in the Caribbean, which was the model that Ceylon aspired to in this period.[56] They hoped that the import of coffee from Ceylon could be met by the return of sugar from Mauritius.[57] The trade relationship with the Cape was another point of concern. It was justified as a historic one, dating to Dutch connections between the Cape and Ceylon.[58] The merchants of Ceylon hoped that coffee would find a ready market in the Cape, in addition to arrack, coconuts, coconut oil, coir, and planks.[59]

---

dodged this issue and referred it to the Commission of Eastern Inquiry, which was expected to arrive in Ceylon in the next few years. It rumbled on as an issue, even after the years of the Commission, when it was referred to the Company's Court of Directors, without a reply; see dispatch dated Colombo, 24 November 1830.

53. Dispatch dated Kandy, 4 January 1828, from Governor Barnes to Earl Bathurst, CO 54/101, TNA.

54. Dispatch dated Colombo, 24 July 1828, from Governor Barnes to Earl Bathurst, CO 54/101, TNA.

55. Dispatch dated Colombo, 3 December 1828, Governor Barnes to W. Huskisson, CO 54/101, TNA, and dispatch dated Colombo, 30 September 1835, from Governor Horton to Lord Glenelg, CO 54/141, TNA.

56. "Petition from merchants of Ceylon to the Honble Lord Commissioners of His Majesty's Treasury," dated Colombo, 20 March 1837, CO 54/154, TNA.

57. "Replies of the European merchants in Ceylon to the questions addressed to them by the Commissioners of Inquiry relative to the internal and external commerce of the island," Lot 19/60, SLNA. In particular, No. 17, "Trade with Mauritius."

58. See, for instance, "Letters from W. Huskisson to R. Dundas, wishing to know of his letters on the subject of trade between the Cape and Ceylon are being considered," Mss Afr. b.4.155; Mss Afr. b.4.177, Rhodes House Library, University of Oxford.

59. "Replies of the European merchants in Ceylon," Lot 19/60, SLNA. In particular No. 16, "Trade with the Cape Colony."

The investigations of the Commission of Eastern Inquiry sent to Ceylon, and discussed in later chapters, provide a detailed statistical overview of the navigation and commerce of Ceylon between the years 1820 and 1828. In Colombo, in these years, there were at most thirty-one British ships that discharged their cargoes in 1820 and at least thirteen in 1826 and 1827. In addition to this between two and sixteen pulled in for water, refreshment, and repairs annually. Over and above this between sixty-five and one hundred and seven smaller British brigs and vessels discharged their cargo in any one year, and between two and eleven of this category of vessel pulled in for supplies. As for Galle, there were approximately twenty ships and twenty brigs and smaller vessels discharging their cargo yearly, and a few in each category did the same each year in Trincomalee. These statistics reveal the relatively small number of vessels flying other flags calling at the island. In any one year, also calling at the island were at most four Portuguese ships and the odd smaller vessel under Portuguese control, a few French ships and brigs, and the very rare American, Dutch, and "Arabian" vessel.[60]

All of this bears out how Ceylon was becoming a node of British imperial commerce by this period. Though its relations with India continued, these were under strain, even as the island looked across the sea to other Crown colonies and the wider British empire.

The partitioning of the island from the mainland was particularly evident in relation to a set of surveys, pushed especially by the Madras Presidency, from 1822, to see whether the strip of sea which separated the two territories could be deepened in order to further partition the island and the mainland. [61] The Ceylon government also assisted with these surveys.

---

60. "Return of the number of ships arrived at and sailed from the ports of Colombo, Galle and Trincomalee in each year from 1820 to 1828," Lot 19/61, SLNA.

61. The Straits were first surveyed in September 1822 by Ensign Cotton, whose report went to the Marine Board of the Company and resulted in a further examination in 1823. By 1828 nothing had yet occurred, and a minute passed by the Governor of Madras urged for a trial in blasting and removing rocks that impeded the making of a passage. Another survey resulted from this new endeavor in 1830, and a report was written by Major Sim, Inspector General of Civil Estimates. Some blasting appears to have been carried out by Sim, but it was only by 1836 that a concentrated period of blasting commenced, and even this was not enough to make a passage for large vessels. This periodic enthusiasm may be explained to result in part from the clash between the Court of Directors in London, who did not wish money to be outlaid for this project without their express permission, and the eagerness for the venture on the part of the Madras Presidency. For the history of the attempts to survey and blast through a passage, see Anon.,

Throughout these surveys, each side was aware of their relative interests, and the correspondence of 1829 stated it to be "abundantly obvious" that a passage adjacent to Ceylon would benefit the island and one at the other end would benefit India.[62]

In this context of relative interest, the survey reports that resulted from various attempts at investigating the makeup of this passage reveal a desire to discover the "natural" break between the two territories. Two islands lay in the sea between Ceylon and India: the island of Mannar, on the side of Ceylon, and the island of Rameswaram, off the coast of South India (see fig. 0.2). Between these two islands was the fabled Adam's bridge, alternatively called Ram's bridge, which Engineer Sim, who wrote a report in 1830, called "a very extraordinary formation," which consisted "entirely of sand, partly above and below the water."[63] This bridge of sand was dismissed as a possible venue for the passage, as the wind and currents would ensure that any dent that was created in it would soon be refilled with sand, unless an inordinate expense was used to keep such a break open. This left the two channels between Mannar and Ceylon and between Rameswaram and South India. In the early survey of 1822, the latter was preferred.[64] In concluding the 1822 report, H. Fullerton, Civil Engineer of Trinchinopoly District wrote:

> In pointing out the part of the pass which I recommend to be deepened,
> I am disposed to adhere as much as possible to what is termed the river,
> both because the extent of work to be executed will in consequence

"Papers Regarding the Practicability of Forming a Navigable Passage between Ceylon and the Mainland of India" (1833). Also see the material in Lot 10/1, Lot 10/18, and Lot 10/204 in the SLNA and the material in files F/4/1250, F/4/1377, F/4/1523, F/4/1566 and F/5/1594, F/4/1599, F/4/1698, F/4/1719, F/4/1773, IOC, for the period in view here. For the period of blasting, see Augustus De Butts, *Rambles in Ceylon* (1841), 100–103.

62. "A Copy of the President's Minute sent to Ceylon Government," dated 25 November 1828, file F/4/1250, IOC.

63. Anon., "Papers Regarding the Practicability of Forming a Navigable Passage between Ceylon and the Mainland of India," 7.

64. Report dated 5 December 1822, from H. Fullerton, Esquire, Civil Engineer, Southern Division, Trinchinopoly District to Inspector of Civil Estimates, file F/4/1250, IOC. The term "river" here was taken to indicate how local people of the region referred to the pass; the sandstone that lay in this river was said to be "well adapted to blasting . . . [having] been undermined by the action of the water." In some places it even made "natural arches." Two specimens of stone accompanied this report to illustrate the action of water in making cavities.

be less, and likewise it may be presumed that this being as it were the natural opening formed by the action of currents, they will set more forcefully in this direction than any other.[65]

This tied into the argument that Rameswaram was at one point connected to South India. It was alleged that a breach had resulted from a storm of 1480, and was enlarged by other storms, and so the making of a deeper passage between Rameswaram and the mainland merely sped up the work already being undertaken by the forces of nature.[66] According to Pridham's account of this project, "Where Nature has in a sportive or capricious mood, barred or endangered the progress of man, it is to be observed, she has ever summoned forth increased energy, and resolution in her children for the encounter."[67]

Minor improvements that followed these surveys apparently increased the passage of vessels in the sea between the island and the mainland. At one stage, even samples of rocks were sent back to the Company to determine the natural action of the sea in eroding the rock, and so to determine the proper point for a passage.[68] In the late 1830s, a gang of convicts under the command of a Madras engineer, with the assistance of a diving bell provided by the Ceylon government and a series of catamarans, attempted to cut open a passage. "Great energy and perseverance was exhibited by all parties, the sappers and convicts working almost continuously."[69] In July 1837, the first English vessel with her sails set passed through the channel; and in the year that had just elapsed 738 vessels had used the passage conveying about 30,000 tons of cargo.[70] However, this was less than what had been anticipated by the Company. In 1839 Governor Stewart Mackenzie

65. Report dated 5 December 1822, from H. Fullerton.

66. "Major Sim's report," in "Papers Regarding the Practicability," 10–11. It is important to note that there were strict instructions to avoid biases in surveying the pass. Surveyors were sent by both the Crown and Company to survey their respective waters. See, for instance, "Instructions to Captain Dawson of the Royal Engineers and Chief Engineer in the island of Ceylon," dated 12 January 1829, F/4/1250, IOC.

67. Pridham, *An Historical and Statistical Account*, 2:507.

68. Report dated 5 December 1822, from H. Fullerton.

69. Pridham, *An Historical and Statistical Account*, 2:508.

70. See, for instance, "Extract of letter dated 19 July 1837 from Lieut. Lake, Adjutant of Engineers during duty at Paumban to Chief Secretary of Government," Lot 10/18, SLNA and "Extract from Minute of Consultation," dated 11 April 1837, Lot 10/204, SLNA. There are some signs of local appropriation and resistance to the program of surveying and blasting; see Lot 10/18.

of Ceylon wrote of the "partial deepening of the passage" in highly positive terms: "The increase of traffic from 26,000 to 50,000 tons in one year
proves the incalculable advantage to the Indian and Ceylon coasting trade
that Paumban pass where deepened will afford."[71]

In the second quarter of the nineteenth century, another variable entered debates about shipping around and between India and Ceylon, namely
steam power. Governor Mackenzie was anxious to procure a steamboat for
the colony, and also to ensure that the island would be a stopping point on
the steam lines.[72] In the same year that he had praised the partial deepening of the passage, he envisaged that Ceylon's steamboat would be used
for protecting the pearl banks, conveying "treasure" from one coast of the
island to another, transporting troops at times of emergency, and also carrying judges on their circuits. While it was thus envisaged that the island
must take its place on the map of steam lines, it was also hoped that steam
would make the island more governable by sea.[73]

The British attempt to intervene in the linking of the island and the
mainland reconfigured the evolving sacred and historical geography of the
region. But the colonial partitioning of these territories was incomplete.
For even if the legislative and representational aspects of the island's placement with respect to the mainland were tightened, Indian vessels carried
on circumnavigating the island.

ANTI-COMPANY POLICIES

Just as much as the connection of the physical spaces of the island and
the mainland was impacted by Crown and Company rule, both discursively but also structurally in terms of attempts to demarcate the maritime boundaries of Crown and Company, so were the commercial arrangements that connected the territories.[74]

The Crown government in Ceylon adopted a policy of economic disengagement from the mainland and attempted independence from the
East India Company in the period leading into the second decade of the

71. From Governor J. A. Stewart Mackenzie to Lord Glenelg, CO 54/170, TNA.

72. For Stewart Mackenzie's interest in the steamboat, see private letter dated 3 January 1838, from Governor Mackenzie to Lord Glenelg, CO 54/160, TNA; dispatch dated
Kandy, 11 May 1839, from Governor Mackenzie to Lord Glenelg, CO 54/170, TNA.

73. Dispatch dated 21 September 1839, from Governor J. A. Stewart Mackenzie to
the Marquis Normanby, CO 54/172, TNA.

74. The argument about anti-mainlandism in this section develops some of the
statements made by Samaraweera, "Economic and Social Developments."

nineteenth century. From the point of view of the Company and its supporters, the Crown government's actions and the results it achieved were
testimony to why India should not fall under Crown rule. A memorandum
dated 1813 in the mainland publication the *Asiatic Journal* noted that the
expenditure of Ceylon exceeded receipts by at least £70,000 annually, except in years when a pearl fishery could be held. The depreciation of the
currency, the adverse balance of trade, and the disposal of government bills
to the highest bidder had "almost annihilated commerce, and distressed
the inhabitants of all classes."[75] By 1828, the journal's critique of Ceylon's
financial condition had gone up a gear:

> We cannot help again adverting to the extraordinary facts which these
> official statements disclose, namely, that in Ceylon, which is the ex-
> perimental scene of those measures which are sought to be introduced
> into India, namely, colonization from Europe, the abolition of monopo-
> lies, freedom of trade, &c., where the hated dominion of the East-India
> Company is not felt, and where a liberal and constitutional government
> is established, which admits natives to public offices and to judicial
> functions as jurymen, the disbursements exceed the receipts, the rev-
> enues themselves are decreasing, the debt is augmenting, the sinking
> fund is swallowed up by the Government, and the trade, unshackled as
> it is, threatened to become wholly confined to imports, and these com-
> posed of a small and diminishing proportion of English goods![76]

Two years later, a letter to the editor carried this argument further by
pointing out that when the island was under the charge of the Company,
matters were handled much more equitably.[77] It ended with a challenge
to advocates of free trade: "What do our Rickards's, and Crawfurds, and
Whitmores say to this?" In these reports Ceylon serves as an experiment
in a very different kind of control, explicitly called "colonization," and set
against the type of governance practiced by the Company. In the eyes of
the *Asiatic Journal*, it was a failed experiment.

   This public debate between Ceylon and India can also be traced in the
official correspondence. In 1820, Governor Barnes imposed a system of
duties, which he hoped would shift the adverse balance of trade between
Ceylon and the mainland, and encourage trade between the island and Eu-

75. *The Asiatic Journal*, 1816, 448.
76. *The Asiatic Journal*, 1828, 313.
77. *The Asiatic Journal*, 1830, 211.

rope and other Crown colonies. He wrote of the need to "moderate" duties on British goods to the island so as to give them a "preference in competition in the market" and to "tax at higher rates than before such articles of import from India as are grown on the Island."[78] His recommendation had a particular bearing on the export of Indian cottons and rice to the island; the import duty on these articles was increased to 12 percent and 8 percent in 1820.[79] Two years later further changes were enacted: the import duty on Indian cottons rose to 20 percent and that on grain to 12 percent.[80] All of this was justified, on the part of the Crown government, with the argument of self-sufficiency: increasing the import duties on Indian products would "have the effect of stimulating the natives of the Island."[81] At the same time a greater tie would be forged with the "mother country" instead of the continental neighbor. In 1821 the Governor of Ceylon wrote to a number of British merchants on the island, asking for their views on the changes to the import and export duties.[82] Given that the cotton cloth supplied from Britain would be 20 percent cheaper than that which arrived from India and "of a superior quality," Beaufort and Huxham & Co. replied: "we are decidedly of the opinion that in a few years the whole quantity consumed on the Island will be imported from Great Britain."[83] In their reply, W. C. Gibson & Co. were enthusiastic in their support of the change in duties for cotton, but asked for a further reduction of the import duty on cottons from Great Britain, citing as one reason "the prejudices of the natives" and another "their [the inhabitants of the island's] own manufacture."[84]

The sale of British cotton cloth in Ceylon was tied to the feeling that this island was more "British" than India, and that the islanders were

78. Dispatch dated Colombo, 8 April 1820, from Governor Barnes to Earl Bathurst, CO 54/77, TNA.

79. See De Silva, *Ceylon under British Occupation*, 2:448 and chap. 15 for a general discussion of customs duties.

80. See De Silva, *Ceylon under British Occupation*, 2:449, and "Minute of the Governor to the Members of His Majesty's Council," 20 September 1822, CO 54/82, TNA.

81. Dispatch dated 8 April 1820, from Edward Barnes to Earl Bathurst, CO 54/77, TNA.

82. Letter dated 20 November 1821, from George Lusignan, Deputy Secretary of Government, to various merchant houses, CO/54/82, TNA.

83. Letter dated 22 November 1821, from Beaufort and Huxham & Co. to George Lusignan, Deputy Secretary of Government, CO//54/82, TNA.

84. Letter dated 28 November 1821, from W. C. Gibson & Co. to George Lusignan, Deputy Secretary of Government, CO/54/82, TNA.

more "civilizable" people. In the midst of an interview with the Commission of Eastern Inquiry, one merchant noted:

> The natives of Ceylon are more inclined to the use of British manufactures and the adoption of European manners than those of any country in India. In travelling through the island, few native houses are found without a table and chairs, they are frequently supplied with knives and forks, plates, glasses &c of British manufacture. The Modeliars or Headmen wear cloth coats, cotton waist-coats and neck cloths which is the nearest approach to European customs we have ever met with in India.[85]

Yet the increase of import duties on Indian goods left the feeling that the island's charges were too excessive, and that even where British goods were concerned the import duties on the island were greater than those for similar goods in India.[86] Policing the trade by duties in this way could thus have the counter-effect of making the island known as a colony of duties. In Ceylon itself, although the policy of economic disengagement with India arose partly with the encouragement of private merchants, such merchants also complained about the island's economic situation, and in particular of how a regime of excessive duties and a reduction in the circulation of currency made for difficult conditions of trade. In a collective letter to the Ceylon government in 1824, they noted: "If trade does not improve generally we fear there will be no alternative but for some of us to withdraw from the island."[87] Nonetheless, there were eleven firms established in trade in the early 1830s.[88] Some of these merchants were former civil servants, and others had worked in the army or navy; many of them kept up a close connection with England and made repeated journeys back and forth. Some of the traders were generalists, while others specialized in particular markets and areas. They often had a fulsome sense of themselves. Mr. Beaufort described the reach of his trade thus: "With Great Britain, France, Hapsburg and Denmark in Europe, to the Cape of Good

85. These are the views of Messrs. Ackland and Boyd, under No. 5, "Consumption of British Goods," "Replies of the European merchants," Lot 19/60, SLNA.

86. See complaints of Mr. Beaufort in "Replies of the European merchants," Lot 19/60, SLNA.

87. Letter dated Colombo, 31 July 1824, from the European merchants to George Lusignan Esq., Deputy Secretary of Government, in "Correspondence between the Governor and the merchants of Ceylon in 1824," Lot 19/63, SLNA.

88. Samaraweera, "Economic and Social Developments," 54.

Hope, Mauritius, Mozambique, Mombasa and Zanzibar in Africa. Bassora, Bashire, Muscat, Calcutta, Madras, Bombay, the Coromandel and Malabar coasts, Penang, Singapore, Sumatra, Java and China in Asia." Despite a desire to control all of the trade, what is striking is that European merchants naturally did better with the trade to Europe and to other colonies; "native" traders handled most of the trade to India. Imports, however, were often supplied to Ceylon via the ports of India, and this occasioned worry, as the East India Company's customs policy militated against its ports serving as entrepôts.[89] This division of the trade is an important context for understanding why official policy swung towards favoring trade across the Indian Ocean and to Europe rather than with India.

The Crown government and the Company did not get into a trade dispute over cotton cloth. Tobacco was different, and it can act as a case study of the Crown government's anti-Company trading policy. The British in India inherited from the Dutch their support of the South Indian Raja of Travancore's monopoly over the tobacco produced in the Jaffna peninsula, in the north of the island.[90] In 1798 the British government reimposed the export duty that the Dutch had levied on this commodity in the hope of adding to the coffers of the state. There was also a growing interest in the tobacco monopoly on the part of the Company; by guaranteeing its continuance the Company hoped to provide the Raja with the money that he required to pay for the subsidiary forces from the Company. However, in 1811, the Rani of Travancore considerably restricted her purchases and caused a blow to the Jaffna trade. In the context of the reduction of exports and the Company's vested interest, the Crown government decided to impose a counter-monopoly. All rights for the manufacture and export of tobacco now lay with the colonial state. This arrangement carried through with renewal until 1819.

But the entry of the Crown and Company into the tobacco trade shifted and redefined its character. Over these decades, the Crown's monopoly in Ceylon over the purchase and transport of tobacco excluded a range of merchants, brokers, and manufacturers and all their capital from the economy of Jaffna.[91] The Company's extended exchange with H. Powney, a tobacco merchant from Ceylon, reveals the impact of the involvement of Company

89. Ibid.

90. The discussion of tobacco here draws on the general contours provided by De Silva, *Ceylon under British Occupation*, 2:473ff. This paragraph summarizes De Silva. See also Bertolacci, *A View*, 165–77.

91. See Bertolacci, *A View*, 74.

and then Crown.[92] Powney shipped a consignment of tobacco to Tranvancore in 1803 following what he believed was a verbal agreement with the Raja, only to find that the port where it was to be received was shut against Jaffna tobacco because of new regulations which sought to combat smuggling.[93] Powney explained that traditionally the Travancore monopoly had never had a single incarnation, since the Travancore government had three separate routes for supplying itself with tobacco. First, it formed contacts with traders in Travancore or Ceylon who purchased tobacco from Jaffna or from other traders; second, it bought tobacco from foreign importers; and third, it invited imports of tobacco and gave permission for its sale in Travancore subject to a tribute to the Raja. Powney accused the Company, and most particularly the Resident of Travancore, of "introducing extraordinary changes in the interior councils and management of the Travancore Durbar."[94] He quoted the Raja's words to him as a way of backing up his claim that British structures had interfered with traditional patterns of trade. The Raja was alleged to have told him that the new contracts were "drawn up contrary to the usages of his country and after the Mahomedan manner & style, under the direction and control of the Resident. . . . They were tendered not to his own subjects, but to foreign Europeans & strangers of whom he would have but little security or tie."[95] Powney must have been at least a little biased, but his comments are suggestive of the way in which the tussle between Crown and Colony formalized and redefined the tobacco trade even at this early date.

The Crown's involvement in the tobacco trade also prompted complaints on the part of the Raja of Travancore, which were communicated through the Company. In this correspondence, it is difficult to differentiate between the Raja's interests and those of the Company. The complaints related to smuggling and to the quality, price, and means of payment for the tobacco. In response to the first of these worries, in 1821, Governor Barnes bonded those who shipped tobacco out of the island to guarantee the place of its destination and then to provide proof in Ceylon that the tobacco had reached that destination.[96] This regulation arose directly out of

92. For the Company's correspondence with H. Powney, see files F/4/245, F/4/354, and F/4/339, IOC.

93. Letter dated 15 September 1807, from H. Powney to the Chief Secretary of Government, F/4/245, IOC.

94. Undated letter from H. Powney to N. B. Edmonstone, Secretary of Government, Calcutta, F/4/339, IOC.

95. Undated letter, ibid., F/4/339, IOC.

96. Dispatch dated 9 April 1821, from Edward Barnes to London, CO/54/79, TNA.

the curious dispute about the brig *Anna Laura*, which left Jaffna with two hundred bales of tobacco alleged to be of the Acheen or Ache Assortment. The brig was declared to be bound for Negapatnam, Colombo, and Bombay, but it was discovered plying the Travancore coast, keeping out at sea during the day and approaching the coast during the night, and then dispatching tobacco to small boats. The arrival of this tobacco in Travancore was alleged to have reduced the price of tobacco to one sixteenth of its original value.[97] Yet Barnes's regulations about bonding with regard to destination did not take account of the flexible patterns of trade, and this is apparent even in the journey for which the *Anna Laura* originally gained permission. In setting off from Jaffna, the *Anna Laura*'s owner Mr. James asked for permission for a complicated journey, which did not merely amount to the taking of tobacco from one destination to another. He applied to dispose tobacco at Negapatnam to the vessels used by the Nagur merchants bound for Ache, and if this did not work to load rice at Negapatnam for Colombo and after landing such rice at Colombo to go to Calpentyn and load arrack, and then go to Bombay and sell both the tobacco and the arrack.[98]

By 1824, the Raja's contract for tobacco had been canceled, and this resulted less from the issue of smuggling than other sticking points such as price, means of payment, and quality. Previously, in 1819, J. Munro, the Company's Resident in Travancore, wrote of the Raja's great dissatisfaction, having had to "procure Star Pagodas in the adjoining Provinces at a very great loss," and also to draw a large sum of pagodas from Bengal to Madras in order to pay for tobacco in the coin specified by the Ceylon government. He continued by noting of the Raja's ministers: "While they had laboured to execute the contract as far as was possible, the Ceylon government had progressively increased the price of the tobacco." He also noted how Ceylon "had supplied tobacco for three years, but especially in the Malabar year 992 of a quality so inferior and bad as to occasion an immense additional loss to the Circar."[99] The Raja thus laid these difficulties directly at the feet of the Crown government.

The Crown government insisted on being paid in pagodas because of

97. Letter dated 18 December 1821, from "Vencatron," the Raja's representative to Mr. J. Rodgers, Agent for Jaffna tobacco, CO/54/79, TNA.

98. Letter dated 3 January 1821, from W. H. Hooper, Collector, CO/54/79, TNA.

99. Letter dated 25 May 1818, from J. Munro, Resident to the Chief Secretary of Government, Ceylon, CO/54/74, TNA. The Star Pagoda was an Indian gold currency, and in 1796 a Star Pagoda was valued at 45 fanams, the Ceylon copper currency.

a steady drain of specie away from the island.[100] This dispute about currency was settled at first when the Crown agreed to be paid for tobacco in Madras rupees.[101] However, when the contract fell apart in 1824, Governor Barnes was quick to punish Travancore by fixing a "heavy duty on exports to that Territory."[102] His logic was that this heavy duty would encourage new markets for tobacco elsewhere rather than in India.[103] The net effect of this duty was the slow decline of the Jaffna trade in tobacco.

A similar story can be told for other commodities in this period.[104] The Crown government also attempted to wrestle cinnamon, the product for which the island was mythologized in the early modern period, out of the Company's grasp (fig. 2.2).[105]

Before the arrival of the British, the Dutch had imposed draconian controls over the trade in cinnamon.[106] The British at first sent the island's cinnamon to the mainland. By 1802, the Crown government had agreed that 400,000 lbs of cinnamon of good quality would be sold at Colombo to the Company. In fact when Ceylon came under Crown rule, the Company was anxious to secure this monopolistic right of purchase over the island's cinnamon trade.[107] Governor North initially dreamed of being able to produce twice the quantity of cinnamon then available for consumption around the world by restricting the growth of cinnamon to government plantations alone, therefore following the restrictive policies of the Dutch. Yet this was but a dream, for Ceylon struggled even to produce the quantity stipulated by the Company's monopoly. When the kingdom of Kandy fell in 1815, however, more cinnamon could be supplied. But the Company was now less enthusiastic: the increase in supply meant that it

---

100. For the financial context and for more information about currency, see De Silva, *Ceylon under British Occupation*, vol. 2, chap. 17.

101. Ibid., 2:447.

102. Dispatch dated 26 July 1825, from Edward Barnes, CO/54/89, TNA.

103. Regulation of Government, 1824, no. 3, CO/54/86, TNA.

104. For instance, arrack, another staple of the export trade, was also hit hard by an excise duty imposed by the Company in 1812. See De Silva, *Ceylon under British Occupation*, 2:451.

105. For a summary of the cinnamon trade monopoly in this period, see De Silva, *Ceylon under British Occupation*, vol. 2, chap. 14. The material in the paragraph that follows summarizes De Silva and also Samaraweera, "The Cinnamon Trade of Ceylon." Before the arrival of the British, the Dutch had imposed draconian controls over the trade in cinnamon. See Bertolacci, *A View*, 241.

106. Bertolacci, *A View*, 241.

107. See Samaraweera, "The Cinnamon Trade of Ceylon," 420.

Fig. 2.2. *"Laurus cinnamonum."* From James Cordiner, *A Description of Ceylon* (London, 1807), vol. 1, facing page 405. Reproduced by kind permission of the Syndics of Cambridge University Library.

had a surplus of stock, and the Company claimed that the Kandyan cinnamon was of an inferior quality. Further, the price of cinnamon was falling, partly because of the way Ceylon was supplying cinnamon to the east of the Cape of Good Hope, which then made its way to Europe and interfered with the Company's monopoly, and also because of its trade in the cinnamon rejected by the Company. In this context, the Company insisted in 1821 that, if the contract were to be renewed, Ceylon had to restrict its production of cinnamon to 5,000 bales, and to restrict its sales to the east; in return, the Company said it was willing to buy up to 4,500 bales.

Ceylon and London found these terms unacceptable, and the contract was not renewed. Instead the Ceylon government decided to control the cinnamon supply itself and to sell the commodity in Colombo. Perhaps it was this continuation of monopolistic practices which was in the mind of one pamphlet writer who noted: "The finest spice the world produces is kept from the world; to make the sacrifice complete we ought to avow that we burn it as our most acceptable offering to the shrine of the demon of monopoly; let us acknowledge whom we serve."[108] This scheme was an utter failure; the uptake on cinnamon was slow and patchy and prices continued to fall. The Company was offended by the Crown's noncompliance with its conditions and did not enter the market. In the context of a rising tide of criticism of the monopolistic practices of the cinnamon trade, especially with the emergence of a group of private traders on the island, and the recommendations of the Commission of Eastern Inquiry, which sought to reform the governance of the island away from monopolies, the trade was finally declared free in 1833.[109] Ironically, the cinnamon trade was freed just as it went into eclipse. European consumers had begun to prefer cassia bark, an inferior type of cinnamon, and Java overtook Ceylon as the main source of this commodity.

The conflict between the two regimes with respect to cinnamon amounted to the failing of two separate attempts to monopolize the market, first by the Company and then by the Crown. Governor Barnes in particular, soon after his arrival in the island, was critical of the Company's monopoly of cinnamon and suggested at first that it should be thrown open. In a private letter to London, he wrote that the Company's profits from cinnamon bore no relation to the price they paid Ceylon; furthermore, he criticized the way in which the Company rejected a good amount of Ceylon cinnamon as being too coarse, leaving the Crown government to

108. Letter to the editor of the Madras Courier, dated 27 November 1827, reprinted in Peter Gordon, *India; or Notes on the Administration of the Establishments in India*, 68.

109. Letter dated Nuwera Ellia, 9 April 1833, from James Stewart, Master Attendant at Colombo, to Governor Horton, CO 54/128, TNA, on the continuation of an illicit trade. For more on the concern with smuggling which coincided with the imposition of a high export duty, see "Extract from minutes of Council held at the King's House on 9th July 1833," CO 54/129, TNA.

cope with the loss that this entailed.[110] The Ceylon government sold this rejected cinnamon in Colombo to purchasers who were bonded to the assurance that they would not transport this inferior cinnamon to the west of the Cape of Good Hope, so as to threaten the Company's monopoly in Europe. In another critical letter from 1821, Barnes wrote again suggesting that the Company itself was complicit in transporting this rejected cinnamon to Europe; he cited the report in a Calcutta newspaper of the exports and imports out of that port.[111] It was in the context of irritation and resentment of the Company's control over the market, the price of cinnamon, and the channels of trade in cinnamon that the contract was withdrawn. Later, when the Crown took on itself a monopoly of cinnamon, the Company, with its command of the market, that saw to it that the new system failed.

It is important to contrast this anti-mainland sentiment with that expressed by the island's early Governors in their desire to cooperate commercially with the Company, or to intervene in the trade of India. For instance, in 1816 Governor Brownrigg proposed that the harbor of Trincomalee could serve as an "entrepot" between the "Indian settlements" and Europe and between "one part of India and another,"[112] and this rather echoes Robert Percival's comment in 1803 about this harbor, quoted above. The onset of British Crown rule did not see an automatic swing against the Company; rather, differences arose slowly over the course of the first decades of the century and were then consolidated into a concerted policy of disengagement. The economic agreements, like the discourses surrounding the human and geographical linkage of the island with the mainland, saw at first an attempt to solidify bonds. But when this policy of accommodation did not work, it yielded a hard line of detachment.

CONCLUSION

The colonial transition at the end of the eighteenth and the start of the nineteenth century in South Asia is often seen in terms of landed interests, pertaining to subsidiary alliances, regimes of taxation, and the role

110. Private letter dated 31 July 1819, from Governor Barnes to London, CO/54/74, TNA.

111. Dispatch dated 27 December 1821, from Governor Barnes to London, CO/54/80, TNA.

112. Dispatch dated 1 November 1816 from Robert Brownrigg to London, CO/54/61, TNA.

of indigenous financiers, middlemen, and private traders. If the ocean comes into view it is often through a focus on strategic considerations and imperial rivalries. Yet considering this moment of transition to British control from a particularly oceanic place, such as Sri Lanka, provides a different vantage point. First, it illustrates that the British were interested in monopolizing the seas around Ceylon in order to stabilize their control over the land. Such control became linked to the structural form of colonialism, which meant that Crown versus Company made a definite impact in shaping the maritime policy of the colony of Ceylon. In the spirit of the argument about movement and recycling, this chapter highlights how creating units in maritime economies was central to political and colonial legitimacy, from the Kandyans to the Dutch and the British, though the means through which such acts of territorialization were conducted changed significantly by the nineteenth century. The emergence of the bureaucratic regime of Crown governance set in motion a different form of territorial state, while connecting that state to a longstanding concept of the specific place of this island in the sea. Sri Lanka had been a center of trade and overlapping systems of exchange, rather like South India had been; the British continued to exploit it as such while bounding it in new ways.

The decline of the economy of the subcontinent used to be a mainstay of explanations of how the British became supreme in South Asia; but this has been significantly revised of late. Systems of trade, merchant groups, and forms of accountancy have all been described as more robust than originally thought. The analysis of Kandy's relation to the seas fits in here: we ought also to be circumspect about thinking that this interior kingdom was in terminal decline. The Kandyan mode of trade was advanced, and it encompassed trade out to sea as part of political legitimacy, just as the Dutch and British did. At the same time it is important not to reduce precolonial patterns of trade to that in the official control of the Kandyan court. The legislative regime of the British did orchestrate a change of gear, but it needs to be understood in the light of the legacies of an ideal of kingly sovereignty. It is important not to exaggerate the shift from South Asian polities to European empires in relation to changing conceptions of the sea.

In some ways the British takeover of Sri Lanka in this period can be summarized under the term fixity. This is in keeping with the argument of the last chapter that the British sought to fix the inhabitants of the island around categories of indigeneity and foreignness. In commercial terms, the British sought fixity, by a precise definition of boundaries, by surveying the passage of vessels, by bringing in a system of duties and passes, and by shifting the trade of the island to conform to the pattern of British imperialism, and this was more far-reaching in bureaucratic terms than that undertaken by the Dutch before them. In all, this militated against a more informal system of trading practices, carried on out of every nook and cranny along the coast, as well as a defined pattern of trade conducted by the kingdom of Kandy. Such attention to fixity had an important foundation in ideas of territoriality as well as in more global systems of capitalist exchange. Indeed Prasannan Parthasarathi has identified such fixity to be an important aspect of the moment of colonial transition, in a critical work of counter-revisionism which pertains to the status of laborers in South India in this period. Parthasarathi argues that weavers' power decreased as the East India Company fixed bounds and terms with respect to mobility and work, which were contrary to South Indian norms. This led to a great "settling" in nineteenth-century South India.[113] It is critical to distinguish between fixity and a sense of place: in Sri Lanka fixity includes but does not equate with locality. For a sense of this island's place in the Indian Ocean was prevalent in the thinking of the Dutch and the Kandyans. Yet the British had the power to fix and to demarcate the bounds of this island in a way that neither of their predecessors had. This sense of fixity does not mean that Ceylon became isolated under British colonialism, though it does mean that it was subject to a mechanism of partitioning from its earlier relations to the mainland. Following Vijaya Samaraweera: "There was much truth in the contemporary colonial view that the two territories of Ceylon and India were governed as 'rival possessions.'"[114]

The next chapter takes this analysis of colonial transition in a different direction by examining the intellectual history of the kingdom of Kandy and British orientalism. In doing this it brings into central view another key set of agents in Lankan history, the Buddhist monks. Yet even then it seeks to make sense of the permutations of scholarly work on religion through attention to processes of imperial state-making. It argues that though there were entanglements between the Kandyan kingdom and British structures, the placement of ideas about the island's religion and past were reconfigured, in global webs of the passage of texts, people, and ideas, by the arrival of the British empire. This counted as a recycling and redefinition of the "indigenous."

---

113. Prasannan Parthasarathi, *The Transition to a Colonial Economy: Weavers, Merchants and Kings in South India, 1720–1800*, 145, for idea of "great settling."

114. Samaraweera, "Economic and Social Developments," 53.

CHAPTER THREE

Scholars

British Ceylon must take a central place among those that gave birth to the modern idea of Buddhism as a world religion. Indeed, one of the first books in English to use the word Buddhism arose from material from Lanka.[1] The British state's takeover of the island explains this, for the British aimed to govern in accord with Buddhist tradition, as was evident in the constitution marking the taking of Kandy, and they used the same principles of colonization as in mainland India, where the study of Hinduism and Hindu history was seen to be integral to governance. However, despite the wealth of work on the history of orientalism in India, the story from Ceylon is virtually unknown, except perhaps to scholars of Buddhist studies.[2] This chapter reconsiders the intellectual history of the colonial transition from an island space that saw texts and other forms of religious culture arrive, depart, and sustain themselves across the sea. The argument does not emphasize the "invention" or even the "Protestantization" of Buddhism; instead it explains how British colonial policy in Ceylon shifted in line with changing notions of Buddhism, consistent with the global expansion of the British empire.[3]

1. Probably the first book to use the word "Buddhism" in English is Edward Upham, *The History and Doctrine of Buddhism* (1829). The word was also used by Brian Hodgson, "Sketch of Buddhism, Derived from the Bauddha Scriptures of Nipal" (1829).

2. For recent work on orientalism in India, see Tony Ballantyne, *Orientalism and Race: Aryanism in the British Empire*; Michael Dodson, *Orientalism, Empire and National Culture: India, 1770–1880*; Javed Majeed, *Ungoverned Imaginings: James Mill's "The History of British India" and Orientalism*; Robert Travers, *Ideology and Empire in Eighteenth-Century India*; and Thomas Trautmann, *Aryans and British India*.

3. The thesis of the "invention" of Buddhism follows work such as Philip C. Almond, *The British Discovery of Buddhism*. For general accounts of Buddhism in Sri

After Christopher Bayly's seminal work on the war over intelligence at the heart of the East India Company's expansion, it is not surprising to find that extant networks for the circulation of information cast a long shadow on the powers of British colonialism.[4] Late-eighteenth-century Kandy witnessed a comprehensive reorganization of Buddhism, which saw the restoration of scholarly monks and the reintroduction of higher ordination for the Buddhist clergy from Siam. At the center of this program of reorganization was a monk by the name of Valivita Saranamkara. Among the first generation of British orientalists, Alexander Johnston, the island's chief justice, and John D'Oyly, who became the first Resident of Kandy upon its fall to the British, were arguably the leading scholars, as far as influence was concerned. The evidence that is presented here shows how Saranamkara is linked to both Johnston and D'Oyly. The intermediary who links the reformer to the orientalists was an intriguing Buddhist monk by the name of Karatota Dhammarama (1737–1827), who became chief priest of the southern territories which were under colonial control, and who was seen as a friend of foreigners in the kingdom of Kandy.

Dhammarama traced his pupillary descent, an important marker of status, to Saranamkara; he then tutored D'Oyly in Sinhala and provided the texts and contacts that were necessary for some of Johnston's most important translation projects, including the first translation into English of a commentary of the now canonical Buddhist chronicle *Mahavamsa*, which describes the history of the island for over twenty-five centuries.

In working through this complicated scholarly lineage, this chapter argues that linguistic skills and ways of conceiving of patronage passed from the late-eighteenth-century Buddhist program of reformation to British orientalism.[5] An excellent study of the first publications of the *Mahavamsa* has identified the agency of Buddhist monks as an essential element in early orientalism.[6] It locates that agency in the fact that the texts

---

Lanka, see K. Malalgoda, *Buddhism in Sinhalese Society*, and Gombrich and Obeyesekere, *Buddhism Transformed*.

4. This follows the discussion of Bayly's *Empire and Information* in the introduction.

5. For a comparative argument regarding Burma, see Michael W. Charney, *Powerful Learning: Buddhist Literati and the Throne in Burma's Last Dynasty, 1752–1885*.

6. Jonathan S. Walters and Matthew B. Colley, "Making History: George Turnour, Edward Upham and the 'Discovery' of the *Mahavamsa*." For another account of the agency of Buddhist monks in the orientalist works, covering the later nineteenth and the early twentieth centuries, see A. Guruge, *From the Living Fountains of Buddhism: Sri Lankan Support to Pioneering Western Orientalists*.

presented to the orientalists were commentaries rather than the original *Mahavamsa*; this is in keeping with the manner in which monks read the Buddhist chronicles. The discussion here extends the line of argument: colonial recycling is discernible not only in the character of the texts but also in the social and intellectual relations that linked the orientalists to the monks. In addition to this, the claims presented here take up Anne Blackburn's detailed study of the mid-nineteenth-century monk Hik-kaduve Sumangala. Blackburn shows how a Buddhist in Lanka could inhabit plural spaces and times, responding to colonialism as well as recasting traditions of older Buddhist knowledge and regional affiliation.[7]

It is important to clarify that this is not an argument about the passage of the authentic precolonial into the modern colonial. Andrew Sartori writes: "The relative indigeneity of any particular practice or concept would seem to matter less than the transformation of its significance within the changing historical contexts of capitalist modernity."[8] Following this sentiment, continuity does not capture the story when what appears from this narrative is how the local is contextualized in relation to wider networks of global transformation. At the same time, an emphasis on dialogic exchange between colonizer and colonized should be set aside. Instead what is important is the question of how ideas from elsewhere were replanted, circumvented, or rejected in local contexts, and how ideas were tied to global concerns.[9] Accordingly, in the nineteenth century, scholarship on the island was constantly evolving in line with wider imperial networks. The late-eighteenth-century Kandyan reorganization of Buddhism, which drew on Southeast Asia, led to the passage of texts, priests, and images to Britain, and then set in context a new consolidation of island-based British colonialism by the 1830s, which sought a different relation with monks and other assistants. Yet such a line of claims must not minimize the fact that priestly agency existed in the first place.

This chapter and the next are concerned with early-nineteenth-century British understandings of Buddhism in Lanka. While this chapter focuses on scholarly activities, the next will consider sacred religious sites. As a pair they urge the need to combine an understanding of the high texts

7. Anne Blackburn, *Locations of Buddhism: Colonialism and Modernity in Sri Lanka*.

8. Andrew Sartori, *Bengal in Global Concept History: Culturalism in the Age of Capital*, 21.

9. See the special issue of *Modern Intellectual History* 4 (2007), esp. "Preface" by Shruti Kapila, 3–6, and "Afterword" by C. A. Bayly, 163–69.

of orientalism with the practices of visual, artistic, and archeological description, linking programs of translation to pilgrimage. While the first part of this chapter focuses on intellectual networks and the passage of skills, artifacts, and texts between eighteenth-century Buddhism and the British, and between Ceylon and London, the second part shows how these scholarly traditions were molded by colonialism in a global age of empire.

TEACHERS AND PUPILS

A poem composed in southern Sri Lanka at the start of the nineteenth century has attracted attention ever since. The poet was a middle-aged woman known as Gajaman Nona and also as Dona Isabella Cornelia Perumal (1747?–1814). She had been left without support for herself and her children by the deaths of her father and of both her first and second husbands.[10] Her poem, written in Sinhala around 1805, was a petition for assistance addressed to John D'Oyly, the newly appointed British agent at Matara. Its sensual prose is striking:

> You are certainly like the full moon that illuminates the city with soft
>   rays,
> Your hands are like the boughs of the wish-tree that grants the people's
>   wishes,
> In relation to wisdom you are like Brahaspati, as you understand the
>   meanings of poetic compositions,
> By your figure, Lord of this district, you are none other than the God of
>   love.
>
> Goddess Lakshmi dwells on your shoulders and bedecks the ladies of
>   directions [common poetic convention of comparing the east, west,
>   north, south etc. to women] with the garlands of glory,
> Gem-rays of your compassion have reached the shrine of the guardian
>   deity of the city.

10. The sources for the information on Perumal and D'Oyly's correspondence are D. A. Tillekeratne, "Gajaman Nona"; Brendon and Yasmine Gooneratne, *This Inscrutable Englishman, Sir John D'Oyly (1774–1824)*, esp. chap. 5; and P.B.J. Hevavasam, *Matara Yugaye Sahityadharayan ha Sahitya Nibandhana* [The Literary Works and Writers of the Matara Era], 65ff.

> To this Lord, the Government Agent, whose lotus like mouth is always
>   full of nectar-like words,
> I appeal for support, worshipping his feet, together with my children.[11]

Perumal was unusual: she had been educated by men, one of whom was Karatota Dhammarama. The few verses of her compositions that have survived show that she was well versed in flattering her patrons and getting what she wanted.

John D'Oyly, in responding to Perumal's petition, took the unexpected step of interviewing her in person, and he conferred on her and her descendants the ownership of the village of Nonagama, near Hambantota in the south of the island, so that they could accrue revenue from the tenants on the land. It is no surprise that this encounter has become a well-known national myth in the history of Sinhala literature. The story is still told of how Perumal had a romance with D'Oyly and was paid for her services. Yet a duller and wiser verdict is more likely. D'Oyly was trying to play the part of a classic royal patron, and Perumal's poetry was reminiscent of that written in honor of a king. Perumal wrote poetry of this kind, combining classical Sinhala idiom with folk language, for all of her patrons.

D'Oyly himself was a fantastic character. How could a recently arrived young Briton be so well versed in Sinhala to read the poetry of someone who was part of the *literati*? Indeed, Perumal may now be identified as one of the leading lights of a particularly fruitful period of Sinhala literature, termed the Matara era for its home in the south of Sri Lanka in the eighteenth century. Perumal's own surprise about D'Oyly's competence in Sinhala verse is evident in the stanza just cited. To understand how and why D'Oyly learned Sinhala, and why Perumal could enroll him as a patron, it is necessary to rewind the clock to about a half a century prior to D'Oyly's arrival in the island.

As discussed in chapter 1, by the middle of the eighteenth century there were no ordained monks in the island, which meant that the Buddhist *sasana* had become dominated by members of the aristocratic class, called *ganinnanses*, who had little interest in scholarship.[12] The *ganinnanses*

---

11. This is a translation undertaken by Prof. Udaya Meddegama of the University of Peradeniya, Sri Lanka, from H. H. Gamage, ed., *Andare Saha Gajaman Nona* [Andare and Gajaman Nona].

12. This paragraph draws on Malalgoda, *Buddhism in Sinhalese Society*, 49–58, and Anne Blackburn, *Buddhist Learning and Textual Practice in Eighteenth-century*

ignored monastic traditions regarding dress and naming. Their pupillary succession followed biological descent rather than intellectual lineage. *Ganinnanses* spent more time acquainting themselves with the rituals of the temple than the exegesis of texts, and this meant that scholarly endeavor was at a low ebb. One historian has described *ganinnanses* as priests and magicians rather than Buddhist monks, because of their interest in worldly concerns and the lowly sciences of astrology and divination.[13]

It was into this context that there emerged a monk, who according to one of his devoted followers, "illuminated this Lankan isle," so that "what had been like a darkened home, [came to seem] like a tree full of lights." His students "descended into the ocean of advanced learning, investigating with the eye of wisdom, [until] they saw grammar and logic, [related to] Sanskrit, Pali and Sinhala."[14] This monk was Valivita Saranamkara (1698–1778) who presided over a comprehensive reorganization of Buddhism in Kandy in the mid-eighteenth century.[15]

The closing chapters of the *Mahavamsa* pay sustained and unusual attention to texts, and this emphasis may well reflect the interventions made by Saranamkara and the fraternity that he initiated. The performative role of texts for patronage and kingly beneficence comes across very clearly. For instance, the chronicler writes of how king Kirti Sri displayed his piety by commissioning "a magnificent golden book." "On its golden leaves he had many Suttantas inscribed such as the Dhammakka Sutta and others and had these recited by preachers of the true doctrine the whole night long."[16] This attention to texts is also evident in the historical chronicle *Narendra*

Lankan Monastic Culture, 35–38. See also Lorna Dewaraja, *The Kandyan Kingdom of Sri Lanka*, chap. 9.

13. Malalgoda, *Buddhism*, 58.

14. Cited in Blackburn, *Buddhist Learning and Textual Practice*, 1, from *Samgharajavata*, probably composed shortly after 1778–79, upon the death of Saranamkara.

15. Rather than an organizer, Saranamkara has often been called a revivalist. It is but a short step from that term to the claim that he set the stage for a renaissance of Buddhist scholarship and literary attainment, which then spread to the other arts and to other parts of the island such as Matara in the south, where Perumal lived. Yet the category of revivalism must be treated with caution, for Saranamkara was first cast as a revivalist by his own followers. It cannot be denied that Buddhism changed in this period, but an exaggerated sense of opposition between early- and late-eighteenth-century Buddhism around the cyclic terms of decline and revival is problematic. This is the argument of Blackburn, *Buddhist Learning and Textual Practice*.

16. W. Geiger, trans., *Culavamsa, Being the Most Recent Part of the Mahavamsa*, 2:99, lines 28ff.

Caritavalokana Pradipikava, a lesser-known text with independent cover-
age of this period of the Kandyan kingdom.[17] Here Kirti Sri is said to have
"learnt from sermons" the importance of Buddha's teaching, and caused
halls for preaching to be built at many places. He is also said, in keeping
with the *Mahavamsa*, to have made "gold leaves" from "nine thousand
gold coins," on which he ordered sermons to be inscribed. The writer notes
how he updated a history of the island, bringing it up to his own reign. In
these texts the narrative of kingly patronage encompasses the copying of
texts, the act of preaching on them, the process of institutionalizing monastic education, the celebration of learning in generous fashion, and the
gaining of merit from all of this.

Saranamkara himself appears in these works. In the *Mahavamsa* description of the reign of King Narendrasimha, the reader is told of "a poet,
one learned in the scriptures, ready of speech, teacher of a host of disci-
ples, renowned, who devoted his life to his own and to others' weal, who
shone like the moon in the heaven in the Order in Lanka."[18] Narendrasimha helped to institutionalize Saranamkara's new fraternity by ordering
the formation of an educational center at Niyamakanda for his monks and
by providing the land for it; by Kirti Sri's time this had led to the establishment of a network of temple centers for Saranamkara's followers.[19]

The Silvat Samagama, or "Pious Ones" as Saranamkara's order was
first named, was increasingly scorned and ostracized by the aristocratic
monks, the *ganinnanses*. The "Pious Ones" did not accept money, and
dedication to poverty was one of their ideals. They also shed the insistence
on caste and family connections, common among *ganinnanses*.[20] The tensions between the two led to a royal decree by Narendrasimha against the
Silvat Samagama, ordering the latter to show greater respect to the *ganinnanses* and to keep away from Kandy.

Texts were critical to the various stages of the Pious Ones' evolution
into the Siyam Nikaya, the name they took after Siamese monks reintro-
duced higher ordination in the reign of Kirti Sri. Saranamkara himself received the rite from these monks.[21] Before Saranamkara's rise to popular-

17. This relies on a translation completed by Prof. Udaya Meddegama of University of Peradeniya, Sri Lanka, from P. A. Hewavitarana, ed., *Narendra Caritavalokana
Pradipikava* by Yatanvala Mahathera of Asgiri Vihara.

18. Geiger, *Culavamsa*, 2:97, lines 48–49.

19. Blackburn, *Buddhist Learning and Textual Practice*, 49–51.

20. Malalgoda, *Buddhism in Sinhalese Society*, 59.

21. Higher ordination has to be conferred with the consent of five fully ordained
monks, who question the novice monk.

ity at the court, his followers are said to have traveled widely for study and for the instruction of the laity: "They taught reading and writing; from the *Hodiya* [an alphabet text for Elu Sinhala script] to youngsters who didn't know their letters, and from the *Anavum* and *Sakaskadaya* [two texts, the latter a biography of Buddha] to virtuous people who knew the alphabet already, and from the sataka pot [collections of Sanskrit poems] to some people."[22] After Narendrasimha helped institutionalize the Silvat Samagama at Niyamakanda, students spent long hours with their texts, and received particular training in how to preach. Niyamakanda was effectively a node for the collection, correction, and composition of Buddhist texts. When Siamese monks arrived in 1753, they too brought a long list of texts with them, and this led to the consolidation of a set curriculum for study. The curriculum ascended through various stages: new entrants learned the *Hodiya* alphabet and then graduated on to various texts, including one on the life of Buddha. Crucial to the curriculum was recitation and memorization, and students routinely consulted commentaries and handbooks in their studies.[23]

European colonization was crucial to the rise of the Siyam Nikaya, as Dutch vessels gave Saranamkara the opportunity to establish links with Siam. Monks who resided in Dutch territory along the coasts were thus beholden to two patrons, the Dutch and the king of Kandy, and these dual loyalties were difficult to manage. On their part, the Dutch took steps to patronize Buddhist monks, aware of the power and status these monks held in the island. It was in this context that the Dutch Governor van de Graff (1785–93) recommended that the Karatota Dhammarama, who had by this time become one of the leading monks in the South, receive a monthly stipend of twenty five rix-dollars.[24] Dhammarama is an important figure, for he had as his patrons the Dutch, the British, and the Kandyans, and among his students were both of the unusual characters with which this discussion began, Perumal and D'Oyly.[25]

Among Buddhist monks pupillary descent is a critical badge of honor and is carefully recorded, and Saranamkara was the teacher par excellence. Indeed the Siyam Nikaya's success arose in part because of the importance

---

22. From the *Samgharajasadhucariyava*, which dates from 1779, cited in Blackburn, *Buddhist Learning and Textual Practice*, 48.

23. Blackburn, *Buddhist Learning and Textual Practice*, 55ff.

24. Malalgoda, *Buddhism in Sinhalese Society*, 83.

25. John F. Tillekeratne, "The Life of Karatota Kirti Sri Dhammarama, High Priest of Matara in the Southern Province of the Island of Ceylon."

of the curriculum devised by Saranamkara. D'Oyly is linked by pupillary descent to Saranamkara through Dhammarama. For Dhammarama was taught by another monk, Attaragama Rajaguru Bandara, and Rajaguru Bandara was taught directly by Saranamkara. Rajaguru Bandara was the author of some texts in the standard curriculum for the Siyam Nikaya.[26]

This pupillary lineage reveals the changing nature of the patronage of Buddhist scholarship. This ties into the critical claim of this chapter that global changes came to mold the connections and evolution of that scholarship. Dhammarama's career is illustrative: throughout his life the powers in Kandy saw him as a friend of foreigners. In 1807 the British Governor Thomas Maitland proposed that a committee of monks in the southern provinces should be established, which could adjudicate on Buddhist legal matters, and Dhammarama was appointed to the head of this committee. The chief priest of Kandy informed the last king of Kandy of Dhammarama's alliances, and this led to Dhammarama losing revenues from a village in Sabaragamuwa, which had been given to him in recognition of his talents as a poet.[27]

Yet Dhammarama's status as a scholar must not be disputed. For he undertook a reformation of the Sinhala alphabet and so utilized his education in Kandy with the Siyam Nikaya to good effect. The poetry which won him his revenue in Sabaragamuwa was particularly impressive: it was a panegyric of the Buddha, a *Baranamagabasaka*, "A Diagram of Twelve Stanzas," where the syllables are arranged in diagrammatic form to compose the stanzas, which are read along each line, and where certain lines combine to spell the name of the poem.[28] It is said to have been composed in the Saka year 1708 (around 1786) when he was imprisoned by King Rajadhi Rajasimha, for heady ambition, for making the mistake of writing to the king asking to be appointed chief priest. When the poem was received, the king is said to have asked various individuals to recite it, but to no avail. Dhammarama was summoned out of prison, "looking more like a veddah [aboriginal] from the wilds of Bintenna, than an educated pupil of Attaragama, and was ordered to read the verse."[29] The king was suitably impressed and urged his release. This story is apocryphal and suggests that Dhammarama and his supporters were also versed in how scholars and kings should relate, and in the power of literary compositions.

26. Blackburn, *Buddhist Learning and Textual Practice*, 55.
27. Malalgoda, *Buddhism in Sinhalese Society*, 85.
28. C. E. Godakumbura, *Sinhalese Literature*, 249–50.
29. Tillekeratne, "The Life of Karatota," 205.

Even Dhammarama's pragmatic cultivation of the Dutch and British does not lessen the surprising fact that he agreed, together with another monk of the Siyam Nikaya named Bovala Dhammananda (d. 1835), to provide his expertise in Pali in order to help with a translation of the Gospel of Matthew into Sinhala by the Colombo branch of the British and Foreign Bible Society in 1812.[30] The Bible translations were under the supervision of the British civil servant William Tolfrey (d. 1817), and the monks' willingness to lend their expertise is even more striking in that the committee of translators included two apostate monks, Petrus Panditta Sekara and George Nadoris De Silva.[31] In addition to this, it appears that Dhammarama also helped translate William Paley's *Evidences of Christianity* into Sinhala.[32] As late as 1847, this tradition of assisting in Bible and other translations continued; by this time the monk Lamkagoda Dhirananda was advising the Bible Society.[33]

Does this mean that Christianization in early British Ceylon was accelerated by a reorganization of Buddhist scholarly endeavor? Indeed, it seems likely that in the first three decades of the nineteenth century Buddhist monks saw Christian ministers less as opponents and more as compatriots, who like them were religious men of the word. The Wesleyan missionary Robert Spence Hardy, the author of several critical texts on Buddhism which were important in the emergence of the later-nineteenth-century public debates and polemics between Christian missionaries and Buddhist monks, held the view that "there is an unnatural, sinful, and pernicious connexion between the British government of Ceylon and idolatry."[34] Yet he documented the lack of opposition on the part of the Buddhist clergy to Christian missionaries.[35] He noted elsewhere: "In travelling through unfrequented parts of the interior, as was once my wont and my delight, I usually took up my abode at the pansal [Buddhist temple], and seldom was I refused a night's lodging or a temporary shelter during the heat of

---

30. R. F. Young and G.P.V. Somaratna, *Vain Debates: The Buddhist-Christian Controversies of Nineteenth-Century Ceylon,* 52.

31. *The First Report of the Colombo Auxiliary Bible Society* (1813).

32. See "Translations of Some Slokas in the Sanskrit Language addressed to Sir Alexander Johnston," in Lot 25/25/48, SLNA.

33. Young and Somaratna, *Vain Debates,* 52.

34. Robert Spence Hardy, *The British Government and the Idolatry of Ceylon* (1839), 10. For more on Hardy's views of Buddhism, see Elizabeth Harris, *Theravada Buddhism and the British Encounter: Religious, Missionary and Colonial Experience in Nineteenth-Century Sri Lanka,* chap. 6.

35. Hardy, *The British Government and the Idolatry of Ceylon,* 43.

the day."[36] He noted that the priests would bring out the alms bowl, to tempt him to partake of the offerings that had been made to them, "or they would bring tobacco or some other luxury, to express their satisfaction at my visit." The wonderful picture of a missionary and priest sharing an alms bowl and tobacco was a surprising one for Britons to come to terms with. Even more surprising was the fact that Buddhist monks asked for the use of missionary schools for Buddhist preaching, a request which the missionaries refused.[37]

The connection between Saranamkara's reformulation of Buddhism and Christianization therefore fits perfectly within the wider argument about the recontextualization of eighteenth-century Buddhism in the colonial era. The new system of Buddhist pedagogy and scholarship which was institutionalized in the mid-eighteenth century, and which radiated outwards from Kandy, had unexpected legacies. In addition to mission translations, this new system came to set the context for the emergence of early British orientalism, and indeed was essential for the education of the most important early orientalists, such as D'Oyly, known as the first Briton to learn Sinhala, William Tolfrey, probably the first Briton to learn Pali and who worked under D'Oyly as his chief translator in Kandy, and even Sir Alexander Johnston, the island's Chief Justice, as will become clear shortly.

This argument might also be reinforced by the suggestion that the form of British orientalism was also guided by Saranamkara's scheme of teaching, with its preference for high Sinhala and Pali. For instance, the Sinhala translation of the Bible undertaken by Tolfrey and his committee set out to correct "the Dutch idiom" of the previous Sinhala rendition undertaken in 1783 and reissued from Serampore outside Calcutta in 1813. A handwritten note at the front of Tolfrey's version says this:

> The learned natives pronounced this version greatly superior to the old, and in clearness, purity, and perspicuity of style; but soon after its publication, it was discovered that it was not sufficiently vernacular, and that if the old version was so mean and low as to disgust any Cingalese of moderate literary knowledge, the new one contained many words too high for common readers.[38]

36. Robert Spence Hardy, *Eastern Monachism* (1850), 312–13.

37. Ibid., 313.

38. Handwritten note in Cambridge University Library copy of *The Singhalese Translation of the New Testament of Our Lord and Saviour Jesus Christ*, ed. W. Tolfrey and A. Armour (1817).

At the same time the important early commentaries on Pali and Sinhala
published by the missionary Rev. Benjamin Clough display a marked preference for Pali over Sinhala. Clough's *Compendious Pali Grammar with a
Copious Vocabulary in the Same Language* (1824) arose out of the unpub-
lished papers of William Tolfrey, which were left after the latter's death.
Clough had in view as readers Indian orientalists, such as Thomas Henry
Colebrooke, who was mentioned by name in the preface. He bemoaned
the lack of attention to Pali by Europeans and stressed that Pali had gone
into decline when Buddhists were expelled from India by the brahmins. In
conclusion he claimed that Pali was "one of the most ancient and perfect
scions of the Sanskrit": the vocabularies of its nouns and verbal roots were
nearly the same and the grammar of Pali was on almost the same model
as that of Sanskrit. There was also detailed discussion of the relationship
between Pali in Burma and in Ceylon.[39]

This enthusiasm for Pali might be compared with Clough's preface to
the Sinhala dictionary he published later and dedicated to Governor Edward Barnes. Clough bemoaned the imprecision of Sinhala: "At present
the generality of the *natives* write and speak according to fancy, and this
subjects none to any inconvenience, an appeal to their own books is of no
importance, as they generally look upon them as belonging only to ages
that are past." Clough hoped that his dictionary of Sinhala would follow
the "purest standard of Singhalese," which was evident in the translation
of the Sinhala Bible.[40]

Beyond these passages of linguistic expertise, it is possible to see how
the idea of patronage also linked Kandy and the British state. Given how
the last kings of Kandy valued texts, it is no wonder that D'Oyly attempted
to follow suit, even in Matara, in responding to his female petitioner. A
keenness to be a patron and to find a patron was central to early British
orientalism, as is evident from Clough's bold dedication pages. Britons
sought after lost and pure languages and abstract scholarly norms, and this
fitted with traditions of kingly patronage and monastic endeavor that they
inherited from the Kandyan kingdom.

39. Benjamin Clough, *Compendious Pali Grammar with a Copious Vocabulary in
the Same Language* (1824), "Preface"; quotation at v.

40. Clough, *A Dictionary of the English and Singhalese and Singhalese and
English Languages* (1821), "Preface." However, this attitude was not followed by less
scholarly Wesleyan missionaries, such as John Callaway, who praised the beauty of Sinhala in his *A Vocabulary in Cingalese and English* (1820). For more on the controversy
surrounding these Bible translations, see Malalgoda, *Buddhism in Sinhalese Society*,
198–200.

GLOBALIZING BUDDHISM

If there was one event which accelerated the encounter between metropolitan British orientalists and the island's Buddhist clergy, it was Sir Alexander Johnston's taking of a collection of Buddhist manuscripts and other materials, together with two Buddhist priests, to London. Johnston's patronage of scholarship on Ceylon widened the network of scholars writing on Buddhism and therefore brought a global intellectual politics to bear on studies of Buddhism on the island. This global intellectual politics eventually led to an intense debate about veracity and methodology in relation to the first translations of the Buddhist chronicle, *The Mahavamsa*, or in fact what turned out to be commentaries on this text.

Johnston was instrumental in the establishment of the Royal Asiatic Society in Britain in 1823 and was its first vice-president. It is known that he was impressed with the function of the Buddhist committee appointed by Governor Maitland and headed by Dhammarama to advise on Buddhist legal matters in Ceylon. He had a portrait of Dhammarama commissioned for himself.[41] In addition to this, Johnston was involved in conferring on Dhammarama the title of chief priest of the southern provinces.[42]

In Alexander Johnston's farewell papers, relating to his departure from Ceylon in 1817, lies an intriguing exchange.[43] There is a document titled "Translation of some slokas in the Sanskrit Language addressed to Sir Alexander Johnston on his departure from Ceylon and presented to him in the name of all the Buddhoo Priests of the Southern Provinces by their high priest Keertisene Darma Rama Nayake Oonancey" and a response to it by Johnston. With this address in hand, the author of this document called in person on Johnston prior to the judge's departure. Johnston thanked his visitor for this honor, which was bestowed "notwithstanding the very advanced period of [his] life" and the "great distance" that the priest had had to travel. Karatota was eighty years old, and he had traveled from Matara to Colombo. Though the name Karatota does not appear in the exchange, Karatota also had the name of Kirti Sri, which could have possibly been corrupted as Kirti Sena. Indeed, another letter addressed to Johnston, about

41. Malalgoda, *Buddhism in Sinhalese Society*, 84n.

42. Letter titled "The humble petition of Karototte Damme Ramanayeke Teroonnase, Chief Priest," undated, to Alexander Johnston, in Folder 6, Alexander Johnston box, New York Public Library Mss. Col. 1578.

43. See "Translations of Some Slokas Addressed to Sir Alexander Johnston."

a census of Buddhist priests begins: "I Damma Ramanayeke Terunnancy of Caretotte."[44]

The stanzas are full of praise for Johnston's intelligence and knowledge in the "arts and sciences." He is said to have an "accurate knowledge of several languages" and to have had "intercourse with all the wise and skilful persons of this country." His commitment is said to extend not just to the people of Ceylon but to its "priests, cattle, paddy fields, villages, gardens, forests, temples, images and religion":

> The aforementioned Chief Justice is as the sun in dispelling the dark designs of unrighteous men. The Goddess named Sriyahantawe dwells near his arm. His Lordship is successful in pleasing the inhabitants and is extremely active and temperate and righteous and therefore may His Lordship be protected by the King of Gods called Sekkradhi Diwiyo.

Yet the closeness of Johnston's dealings with this priest become clearer in the judge's response. Johnston noted his "frequent communications" with the priest, and the "alacrity with which [he had] at all times afforded [him] with the information required." Johnston had been permitted to consult books in the high priest's temples. The judge also gives us the information necessary to solve one small mystery in Sri Lankan history about how the first translation of a commentary of the now canonical Buddhist chronicles was completed:

> The translations into English which you have enabled me to procure of the three most celebrated histories of your country and your religion, the Mahavamsa, Rajavalli, and Rajaratnakari, and the numerous extracts which you have made for me from your other Sanscrit, Pali and Cyngalese books . . . form a valuable collection of materials. . . .

Johnston thanked the priest for his assistance with his experiment in trial by jury, and noted that his visitor had headed the committee to which Buddhist judicial matters were referred. This last piece of evidence seems to clinch the case that Karatota Dhammarama and Johnston's visitor were one. Johnston concluded by praising his visitor's "liberal sentiments" on religion, and noted the zeal with which two priests from the southern provinces were insisting on accompanying him to England. Karatota

---

44. See letter to Alexander Johnston, undated, in Folder 6, Alexander Johnston Box, New York Public Library Mss. Col. 1578.

Dhammarama, and the historical legacy of Buddhist scholarship that he represented, had come to play a defining role in the major achievements of early orientalism in Lanka.

Back in England in 1828, Johnston presented the Royal Asiatic Society of Great Britain with an illustration of the two Buddhist priests who had accompanied him home from Ceylon, pictured together with Rev. Adam Clarke (1762–1832), the Wesleyan minister and scholar (fig. 3.1). This image, which follows the traditional genre of paintings of noble savages brought home in the late eighteenth and early nineteenth centuries, shows the three in Clarke's library. The impression created by the rendition of the priestly robes is reminiscent for instance of the iconic images of the robed Mai, brought back by Captain Cook from the Pacific. Clarke appears to be instructing the priests, and one of them holds his hand towards an open volume. Clarke wrote on taking charge of the priests on Johnston's invitation:

Fig. 3.1. "Dr. Adam Clarke and the Priests of Buddha." An engraving presented by Alexander Johnston to the Royal Asiatic Society in 1828. Photograph: © National Portrait Gallery, London.

[They] are about five feet six inches, and quite black: they have fine
eyes, particularly the eldest, regular features, and the younger has a
remarkably fine nose: there is a gentleness, and an intelligence in their
faces which has greatly impressed me in their favour; in short, they
are lovely youths, for whom I feel already deeply interested; their hair,
which is beginning to grow, (for as priests, they are always shaven,) is
jet black; their clothing is imposing in appearance: it consists of three
parts. . . .[45]

From Clarke's papers it appears that they were instructed in science and
Christianity; he tells of the experiments he performed with them, and the
glee with which they walked on an icy pond and melted snow. The presence of the Buddhist priests in England did not completely escape the British romantic and literary imagination; one of the priests translated a poem
by the philanthropist Hannah More into Sinhala.[46] They also provided Sir
Joseph Banks, who accompanied Cook to the Pacific, with a translation of
a Sinhala manuscript.[47] Eventually the two priests were baptized in front
of "hundreds of deeply interested and attentive persons" in a Wesleyan
chapel in Liverpool; they took the names "Adam Sri Munni Ratna and
Alexander Dharmma Rama."[48] It is likely that they took the first names of
their two benefactors in England: Adam Clarke and Alexander Johnston,
for it seems to have been common for priests who converted in this period
to take the names of their sponsors.[49] The ex-priests, Adam and Alexander,
were then sent back to Ceylon, and Alexander's letter to Clarke included
this line: "Sir I will try to be Englishman long as I live; and if any try to

45. J.B.B. Clarke, ed., *An Account of the Religious and Literary Life of Adam
Clarke by a Member of his Family* (1833), 2:351–52.

46. This may have been the poem that More sent to Johnston in 1818, with the subject of the judge's measures to abolish slavery in Ceylon. More wrote: "I consider it no
small distinction to have become Poet Laureate to the Cingalese Slaves." Letter dated
18 November 1818 from Hannah More to Alexander Johnston, Lot 25/25/1, SLNA.

47. Clarke, *An Account of the Religious and Literary Life of Adam Clarke*,
2:357–58.

48. Ibid., 2:370; *Ceylon Literary Register*, 1887, 160.

49. See J. W. Bennett, *Ceylon and its Capabilities: An Account of its Natural
Resources, Indigenous Productions and Commercial Facilities* (1843), 340–41, for an
account of a monk who went to Amarapura in Burma in 1808, and who returned to be
converted by Christian missionaries, and to take the name George, after Rev. George
Bisset.

make me Sinhagalese man, that I not like."[50] By 1827, both Adam and Alexander were in the government service of Ceylon: Adam was a proponent in the Church of England and Alexander was Maha Vidana Mohandiram in Moratuwa and Galkissa.[51]

While Karatota's friendship with Johnston appears to have been central to the passage of these priests, that relationship also bore fruit in relation to the manuscripts which gave rise to the writer and orientalist Edward Upham's books on Ceylon. Edward Upham (1776–1834), was a retired bookseller and mayor of Exeter, who devoted his later life to orientalist works. The year 1829 saw the publication of his *The History and Doctrine of Buddhism, Popularly Illustrated*, and the year before his death, 1833, saw the publication of *The Mahavansi, the Raja-ratnacari and the Raja-vali, forming the Sacred and Historical Books of Ceylon*. While more attention has been directed to the second of these, it is important to start with the first, which despite its main title is actually an account of Buddhism in Ceylon; it is dedicated to Sir Alexander Johnston, from whose "manuscripts and drawings" it is said to be chiefly derived.[52]

The importance of this work in the printing of the text of the *Mahavamsa* commentary can be read from the preface. Upham noted how little information had been gathered from the "priesthood of Ceylon": "No source, however, can, it is presumed, be less exceptionable than original pictorial representations, combined with explanatory precepts."[53] This was why *The History and Doctrine of Buddhism, Popularly Illustrated* was what in our terms would be a coffee table work, replete with forty-three plates drawn from Johnston's collection of Buddhist drawings (figs. 3.2 and 3.3). They depict *jataka* stories, or the lives of the Buddha, other deities such as Pattini, signs of the zodiac, and devils. It would appear that little assistance was given in the production of this volume by anyone with expertise in the Sinhala language: the Sinhala letters appear to have been

50. Clarke, *An Account*, 2:377.

51. Paul E. Pieris, "Two Bhikkus in England, 1818–1820." Adam and Alexander were just two of a group of Buddhist priests who converted to Christianity. For more biographical information, see Young and Somaratna, *Vain Debates*, chap. 2. For another account of a conversion of a Buddhist monk who had traveled to Burma, see Bennett, *Ceylon and its Capabilities*, 340–41.

52. Upham's earlier work on Buddhism has recently been considered by E. Harris, *Theravada Buddhism and the British Encounter: Religious, Missionary and Colonial Experience in Nineteenth-Century Sri Lanka.*, chap. 2.

53. All quotations from Upham, *The History and Doctrine of Buddhism*, vi.

Figs. 3.2 and 3.3. Plates from Edward Upham, *The History and Doctrine of Buddhism, Popularly Illustrated with Notices of the Kappooism, or Demon Worship, and of the Bali, or Planetary Incantations of Ceylon* (London, 1829), showing the god Saman and four devils, facing pages 52 and 130. Reproduced by kind permission of the Syndics of Cambridge University Library.

copied stroke for stroke and are sometimes unreadable. This is unsurprising given that Upham personified the type of orientalist who presumed to write about Ceylon or the Indian subcontinent without ever having traveled there. These plates also indicate the fascination with the visual culture of Buddhism, and show how Johnston collected not only texts from Buddhist temples but images as well. Indeed Johnston was given these images by Buddhist priests who were aware that Johnston could not read the languages of the island.[54]

Johnston therefore carried on the traditions of patronage that we encountered in the late Kandyan court, which were central to scholarship. This orientalism was about the gathering of a variety of objects for his collection, rather like the kings of Kandy who surrounded their persons with works which reflected their learning. In this enterprise Johnston followed the example of his teachers and mentors who were renowned orientalists, who he encountered during his time in Madura, while his father, Samuel Johnston, had the post of paymaster. It is striking that he counted among his associates some of the main names in South Indian orientalism. Sir Thomas Munro, the Governor of Madras and architect of the *ryotwari* system, taught the younger Johnston Latin, Colin Mackenzie, the Surveyor General of India, taught him mathematics, and the missionary Christian Schwartz of Tanjore taught him Christianity.[55] Upham's commentary, in the meanwhile, bore many marks of the scholarly contentions of the Indian scholar Francis Buchanan, who had also written on Buddhism.[56] It is plausible to imagine that with these examples of Indian orientalism in mind, and with the legacy of Kandyan kingship as a framework, Johnston transported the texts that would give rise to the printing of the *Mahavamsa*.

Upham's publication of what he called the "sacred and historical books of Ceylon," which included a wide range of texts including the *Mahavamsa* commentary, again trumpeted Johnston's role. Upham wrote in the introduction how this was "the first specimen of an original and genuine Buddhist history that has been offered to the [British] public," and included verbatim a letter from Johnston to the Court of Directors of the East India

54. Some of these images are still housed at the Royal Asiatic Society in London; see Raymond Head, *Catalogue of Paintings, Drawings, Engravings and Busts in the Royal Asiatic Society Collections*, 201. A letter which accompanied Johnston's gift of thirty-three drawings to the RAS noted: "While I was holding a session in the southern provinces of the island, and enquiring into the state of Buddhism, Astrology, and the worship of Demons in that part of the island, that some of the most intelligent of the Buddhist Jurymen, knowing that I could obtain little or no useful information from any books . . . laid before me the greatest part of these drawings."

55. William Jerdan, "Alexander Johnston," *National Portrait Gallery of Illustrious and Eminent Personages* (1832), 2. James T. Rutnam, *The Early life of Sir Alexander Johnston, Third Chief Justice of Ceylon.*

56. See Harris, *Theravada Buddhism and the British Encounter*, chap. 2.

Company about how Johnston had procured the documents.[57] In this letter, Johnston noted how he wished to frame a code of law for the island, and had sought the best information relative to the customs, manners, and religion of the people. After consultation, the Buddhist monks had presented Johnston with the texts in view, and Johnston then compared them with "all the best copies of the same works in the different temples of Buddha on Ceylon." "Two of the ablest priests of Buddha" then carefully revised and corrected the copy, and an English translation was made by Johnston's official translator, under the superintendence of "the late chief of the cinnamon department, who was himself the best native Pali and Singhalese scholar in the country." This translation was revised by "Revd. Mr. Fox, who resided on Ceylon for many years as a Wesleyan missionary," who Johnston called the best Pali and Sinhala scholar in Europe. The long list of hands through which this manuscript passed establishes very clearly the fact that it was a collaborative effort, linking priests to civil servants and missionaries, and tying orientalism based in the island to that in Europe.

Yet the ease with which knowledge passed across this network linking islanders to Britons must not be exaggerated. Three years later, a paper war erupted about the publication of these chronicles. George Turnour (1779–1843), a civil servant in Ceylon who had learned Pali, published another account of the *Mahavamsa*, this time with text in English and Pali, rendered into Roman script. This was hailed as the first publication in the Pali language and consisted of the first twenty chapters of the *Mahavamsa* and an extended introduction; more chapters followed the next year. In fact this translation was also based on a commentary. It was undertaken with the patronage of the Crown government.[58] Turnour intended it to supplement the advances made in the study of Pali by Tolfrey and Clough.[59] Yet he denounced Upham's version of the *Mahavamsa* as "one of the most

---

57. Edward Upham, ed. and trans., *The Mahavansi, the Raja-ratnacari and the Raja-vali* (1833); all quotations from the introduction. See also Upham, *Proposals for Publishing by Subscription the Sacred and Historical Books of Ceylon.*

58. See letter dated Kandy, 26 April 1836, from George Turnour to the Secretary of State for the Colonies, CO 54/154A, TNA.

59. For Turnour's request to the Royal Asiatic Society of Bengal to patronize the work, see *Journal of the Royal Asiatic Society of Bengal* (1835), 407–8. For his dispatch of two copies of a pamphlet describing this work to the government, see letter dated 26 April 1836 from George Turnour to the Secretary of State to the Colonies, CO 54/154A, TNA.

extraordinary delusions, perhaps, ever practised on the literary world."[60] Johnston's Buddhist priests were accused of making a compilation of their own rather than translating the original. The head of the cinnamon department was said to have as much Pali as a modern European. Fox himself was said to have no Pali.

Turnour wrote that, having taken up his appointments in the Kandyan provinces, he had been led to study both Sinhala and Pali. "I have possessed the advantage, from my official position of almost daily intercourse with the heads of the buddhistical church, of access to their libraries, and of their assistance. . . ."[61] It is said that Turnour's time in Ratnapura as Government Agent of Saffragam, and his friendship with the chief priest of Saffragam, was central to his acquisition of a commentary of the *Mahavamsa*, from Mulagirigala temple in South Lanka, and to his study of Pali.[62] Turnour's translation of the *Mahavamsa* was certainly more accurate, given that he himself had more competence in Pali. One contemporary reviewer got it right in urging that since it came out of Ceylon rather than England, it was bound to have fewer errors. Nevertheless, Turnour was accused of being "a little too keen in his critique of the rival work."[63] The disconnection between the two works also resulted from the fact that Johnston's priests were from the lowlands while Turnour's closest priestly acquaintances were from Kandyan territories. This is perhaps why Turnour wrote that he had not been able to ascertain the identity of the priests who had collaborated with Johnston.[64]

This contest is not merely an arcane matter of priority and method: it reveals how orientalism in Ceylon began to mutate. For with the increasing acquaintance with Sinhala, Pali, and Sanskrit that Britons were able to draw upon, there came higher standards and more critical reliance on collaborators. Together with this came a bifurcation of local orientalism and metropolitan popular writing, which was another means through which Lanka was islanded. These points are important in moving beyond an em-

---

60. George Turnour, *Epitome of the History of Ceylon* (1836), v–vi.

61. Ibid., cxi.

62. Anon., "George Turnour (1799–1843)." See also Anon., "Upham's Sacred and Historical Books of Ceylon." Walters and Colley note the irony that if Turnour's commentary came from Mulgirigala, then the monks from whom it originated would have been closely associated with those that supplied Johnston his texts. See Walters and Colley, "Making History," 142 n. 19.

63. Review at end of Turnour, *Epitome*. Turnour's work was recommended by Augustus De Butts, *Rambles in Ceylon* (1841), 133.

64. Turnour, *Epitome*, viii.

phasis on the continuities between indigenous and colonial traditions in the history of orientalism.

However this is not to imply that indigenous agency wholly disappeared in orientalist writings by the middle of the century, or that there were not some similarities between the work of Upham and Turnour. Turnour needed collaborators too, but he had less of a need for collaborators, when compared with the way Johnston worked. In spite of their differences, both Upham and Turnour deployed a consciously articulated sense of reason in the act of translation. Upham was keen to produce what he felt was a proper chronology and derided the indefinite way in which numbers were used in the *Mahavamsa*. He noted: "What appears in these histories as fabulous, because literally impossible, is merely the highly figurative language employed which is quite familiar to the Asiatics."[65] Turnour, in the meanwhile, appended a table of the monarchs of the island, starting with the legendary arrival of "King Wijeya" in 543 BC. The story, as retold in the *Mahavamsa*, of how Vijaya had a lion as grandfather finds no place in Turnour's appendix; neither is Kuweni, his first wife, identified as a demon as she was in the original manuscript. Kuweni is said to be the "daughter of a native prince" in Turnour's objective genealogy.[66] The imposition of the norms of European historiography is also evident in the several papers which appeared under Turnour's name in the *Journal of the Asiatic Society of Bengal*, which sought to iron out discrepancies of Buddhist chronology and align the discoveries from Pali texts in Ceylon with those from Buddhist texts elsewhere.[67] Turnour critically synthesized and analyzed his data in a way that is even now seen as "pioneering" by scholars of Buddhism. These papers also show the reception of his work in India.

By the later 1830s, British orientalism was reconceptualizing what counted as the indigenous and who could serve as a collaborator. British scholars were becoming less dependent on indigenous traditions of scholarly endeavor, even as British power on the island was on the increase. This sat together with a greater awareness of the global network of intellectual production on Buddhism, tying work in Ceylon to India and to London,

---

65. Upham, *The Mahavansi*, xiii.

66. See "An Epitome of the History of Ceylon and Translations of Historical Inscriptions, Originally Published in the Ceylon Almanacs, 1833 and 1834," in Turnour, *Epitome*, 13.

67. See George Turnour, "Examination of Some Points of Buddhist Chronology" (1836); Turnour, "Examination of the Pali Buddhistical Annals" (1837); Turnour, "Account of the Tooth Relic of Ceylon" (1837); Turnour, "Examination of the Pali Buddhistical Annals," nos. 2, 3, and 4 (1838).

and the distinction between work done in the field and in the metropolis. Moving from the first to the second translation of the *Mahavamsa*—or in fact its commentaries—indicates neither simple continuity nor radical change, but a renegotiation of the links and contexts of indigenous and European historical traditions, in line with widening networks of intellectual communication. For the island to be made, the British needed to fix this conception of the indigenous.

GOVERNING ORIENTALISM

Early British orientalism in Ceylon was not merely a scholarly endeavor; it was also deeply connected with the colonial state. By using the official lives of Johnston and D'Oyly, it is possible to see how their intellectual work fitted into the narrative of the consolidation of the British hold of the island. Yet these two lives also demonstrate how at rare times the relation between intellectual production and governance could be overturned, for this was a restless union of interests.

The second British Governor of Ceylon, who took over following North's disastrous defeat at the hands of Kandyan troops, was Sir Thomas Maitland, who was of good Scottish nobility. His residence in Mount Lavinia, now a grand hotel, became a regular meeting place between Maitland and Johnston. The judge's papers show that the governor and chief justice got on very well.[68] In 1806, when Maitland was detained on the coast while awaiting the arrival of the new Governor General of India, he asked Johnston to deputize for him and to undertake a tour of the island. In Johnston's own words he was ordered to "collect all the information I could get, respecting the customs & laws which are peculiar to each of the respective provinces in which these territories are divided and to submit my opinion to [Maitland] as to the practicability of adopting or replacing either a part or the whole of them."[69] Central to the unanimity of governor and judge was a commitment to liberalism and friendship with local peoples. Maitland encouraged the study of Sinhala and offered rewards for

68. There was a constant flurry of close correspondence between Maitland and Johnston from 1805 to 1809. These letters form the basis of the comments about Maitland's relations with Johnston and Johnston's deputising for the Governor. See Lot 25/1/34, Sri Lanka National Archives. For Maitland's instruction that Johnston represent the views of Ceylon in London, see Anon., "Sir Alexander Johnston's Memorandum."

69. Undated letter from Sir Alexander Johnston, possibly from 20 July 1807, Lot 25/1/34, SLNA.

doing so. Maitland wrote: "The sole object of Government is . . . to ensure the prosperity of the island solely through the medium of generally increasing the prosperity and happiness of the natives."[70]

It was after Johnston's tour of the island that he emerged at the head of a network of information on all things related to the island, especially after his return to London. From the northern provinces around Jaffna, he started to collect information about "Malabar" laws and customs. Rev. Christian David, who was also connected to the Tanjore mission and to Schwartz, was one of his informants from that quarter. David translated various palm leaves from Tamil into English for Johnston, and procured a copy of the legal work, the *Tesavalamai*.[71] Another informant from Jaffna had the name of B. Rodrigo.[72] Johnston also had a close connection with the area around Galle, where he had a house.[73] Johnston orchestrated the surveys undertaken by Captain G. Schneider, from 1806 onwards, by suggesting to Maitland that they should be undertaken and by providing precise instructions. Later, the reports from these surveys were deposited at the Royal Asiatic Society in London, and Johnston wrote in 1831 to the Colonial Office, offering copies to it.[74] At this point he noted that his 1806 tour had resulted in "a very large collection of manuscript maps and charts, of every district of that island, of every harbour, and of every part of the sea coasts and of every pearl and chank [sea shells used for ornaments] banks in the Gulf of Mannar." In addition to all of this he had collected "a complete set of working models of every instrument, tool, or machine employed in any description of art, manufacture or agriculture on that island." Johnston had a definite idea of the communities which inhabited the island; in dispatches to London, he presented the "customary laws" of Malabars, low-country Sinhalese, Kandyans, "Mohammedans," and "Parsees" separately.[75]

Johnston collected information with the aim of reforming the gover-

70. From H. M. Chichester, "Maitland, Sir Thomas (1760–1824)."

71. For David's letters to Johnston, see Lot 25/25/55, SLNA.

72. Letter dated Jaffnapatam, 8 December 1814, from B. Rodrigo to Alexander Johnston, in Alexander Johnston Box, New York Public Library.

73. See Johnston's reply to the address directed to him by "the European descendants of Europeans and Burghers residing in the District of Galle," dated 20 September 1817, prior to Johnston's departure. Lot 25/25/47, SLNA.

74. Letter dated 23 April 1834 from Alexander Johnston to the Right Honble E. Stanley, CO/54/126, TNA.

75. For a reading of Johnston's dispatches, see H. W. Tambiah, "The Alexander Johnston Papers."

nance and laws of Ceylon, which he envisaged as a test case for the governance of the whole of British India. In this scheme Ceylon would be an island laboratory. In 1832, when Johnston was interviewed by the committee on India in the House of Commons, it was noted that he gave it "as his decided opinion, that every class of inhabitants in India, natives as well as Europeans, without distinction of religion, color or descent, should be considered eligible to every office, civil and military, under Government."[76] It was this liberalism which led to the achievements for which Johnston is best remembered, which were beautifully illustrated by the Russian émigré painter, James Stephanoff.[77] Stephanoff portrays the scene of the trial of five high caste Sinhalese for murder, in order to commemorate the introduction of a new charter of justice to Ceylon by Johnston, including trial by jury for local peoples and the abolition of slavery, which led to the declaration that children of slaves born after 1816 were free. The original drawing upon which this was based was drawn up by a Sinhala member of the panel of jurors depicted in the image. A copy was reprinted in the *Saturday Magazine* of 1835 and was accompanied by a detailed article by James Cordiner, which helps identify all the people in the image (fig. 3.4):

> The picture accordingly exhibits with great accuracy, the costume of the priests of the Hindoo, Boddhoo, and Monhammedan religions, as well as that of the Protestant and Catholic Missionaries, and that of the Malabar inhabitants of the North, and likewise that of the Cingalese of the South, that of the Malay princes and their attendants from the Eastern islands, that of the people of the Laccadive and Maldive Islands, that of the Hindoo population of the coasts of Malabar and Coromandle, and lastly that of the Moguls, Arabs and Parsees.[78]

The judicial charter introduced by Johnston arose in the context of a series of unpleasant wrangles between the judiciary and the executive arms of government in Ceylon. As early as 1806, Johnston had sent the

---

76. Pamphlet titled "Representation of the Dumfries District of Burghs," 7, in Lot 25/25/99, SLNA.

77. For an account of this print and Stephanoff, see Ismeth Raheem, *A Catalogue of an Exhibition of Paintings, Engravings and Drawings of Ceylon by Nineteenth-Century Artists.* The Stephanoff painting faces the frontispiece, and see 43ff.

78. *Saturday Magazine* (1835), 106.

Fig. 3.4. "The Jury Court of Ceylon." From *Saturday Magazine*, 21 March 1835. Reproduced by kind permission of the Syndics of Cambridge University Library.

Whig politician and then foreign minister Charles James Fox (1749–1806) a paper on how he hoped the governance of India might be reformed, which included thirty-one recommendations.[79] He proposed that India's inhabitants should have a general system of education, that they be eligible for all judicial, revenue, and civil offices, that domestic slavery be abolished, that forced unpaid labor be annulled, that all religions, including Catholicism, be accorded the same rights, that "half-castes" be placed on the same footing as others, that restrictions on European settlement be taken off and that the press be made free, among much else.[80] The charter was an attempt to put into practice these very ideas in Ceylon, and these ideas continued to guide his enthusiasm for reforming Indian governance after his final return to England in 1817.[81] Yet Maitland felt that Johnston's reforms

79. For Johnston's close friendship with Fox, see the pamphlet titled *Alexander Johnston, Third Chief Justice of Ceylon*, in Lot 25/25/43, SLNA.

80. See "Paper sent by Sir Alexander Johnston, in 1806, from the Island of Ceylon to the late Charles Fox" in Lot 25/25/43, SLNA.

81. For the ways in which Johnston pushed forward a reforming agenda upon his return to England in 1817, see "A Statement," in Lot 25/25/56, SLNA.

were too sweeping and made representations against them to Lord Liver-pool. Some changes were enacted. While trial by jury was retained, it was reformed so that Europeans could only be tried by Europeans.[82]

Given Johnston's enthusiasm for the participation of Ceylonese in the governance of the island, it might be imagined that he had a close friendship with the period's other orientalist in high standing in the government, John D'Oyly. There is a relative lack of correspondence be-tween the two, which may be explained by the fact that D'Oyly became a recluse in the eyes of the British after he became the first Resident in Kandy. In a letter merely dated 29 September, which was probably written prior to Johnston's final departure, D'Oyly penned these terse lines to the judge:

> I return with many thanks your paper on the Buddha religion & hope that I have not detained it too long. It appears to me in general to con-tain a tolerably correct account, but I assure you, I have as yet had op-portunity of taking no more than a cursory view of it.[83]

After his time in Matara, where we encountered him receiving sensuous verses, D'Oyly was appointed Chief Translator to the government in 1805. In that position he was invaluable to Maitland's successor, Robert Brown-rigg, in his war with Kandy in 1815. Throughout the years leading up to 1815, palm leaves in Sinhala went back and forth between the British and the kingdom of Kandy, under D'Oyly's superintendence and authorship. In taking responsibility for translation, D'Oyly took on himself the du-ties that had until this date fallen to the Sinhala elite who collaborated with the British.[84] In this official correspondence, which was directed to the Buddhist priest Moratota Dhammakkanda and also to ministers of the Kandyan court, the British sought to secure the release of the English cap-tives taken by the Kandyans in the disastrous war of 1803, most particu-larly Major Davie.[85]

D'Oyly slowly learnt the social etiquette that was necessary for com-munications with the Kandyan court. For instance, upon the earlier ar-rival of the new governor Thomas Maitland, Moratota complained: "In

---

82. See Colvin R. De Silva, *Ceylon under British Occupation*, vol. 1, chap. 10.

83. Letter dated 29 September from J. D'Oyly to Alexander Johnston, Lot 25/1/35, SLNA.

84. See Paul E. Pieris, *Tri Sinhala: The Last Phase, 1796–1815*, 87.

85. Rambukwelle Siddhartha Thera, ed. and trans., "Letters of J. D'Oyly."

former times when a Governor arrived at Colombo, a suitable *Keydapana* [letter addressed to or by the king] is dispatched in a respectful manner to render that event acceptable to the happy lotus-like mind of the Divine Supremely Great King, the Ruler of Tri Sinhala."[86] In responding to this critique, Maitland wrote of the "stupidity of the Candians"; yet, D'Oyly did his homework by seeking to acquaint himself further with the norms of Kandy.

D'Oyly's growing competence with Sinhala correspondence on palm leaves, and its connection to colonial takeover, appears not only in the official letters. His diary, running with some gaps from 1805 to 1815, is a very curious document. It is written almost in code and points to how D'Oyly was a master of subterfuge. It presents a record of the receipt and dispatch of numerous secret "olas" or palm leaves.[87] D'Oyly cultivated a wide network of spies, who gathered information on the inner workings of the kingdom of Kandy. He was given £1,500 in order to pay his spies, in secret.[88] Some of these spies were low-country Sinhala nobility, who lived in British territory immediately adjacent to Kandyan land, others were Buddhist priests within Kandy or Muslim traders, while yet others posed as priests and traders.[89] In addition to this D'Oyly successfully bought the friendship of the leading ministers in the court of the king of Kandy. The first of these was the chief minister Pilima Talauve, to whom he sent gifts of cloth, salt, fish, and plates. This minister asserted that only fifty British soldiers were necessary to take over Kandy.[90]

After Pilima Talauve's assassination by the Kandyan king, Ahalepola, who we encountered in chapter 1, succeeded him as chief minister. Ahalepola kept up a steady correspondence with D'Oyly too.[91] In 1813 Ahalepola confirmed that the captive Major Davie was dead. In response, D'Oyly prodded him to determine where his loyalties lay: "Ourselves also being anxious to promote in the Happy Island of Lanka the Prosperity of the World and Religion, I shall rejoice to receive from you an explicit communication by what means you propose to accomplish that beneficial ob-

86. Pieris, *Tri Sinhala*, 89.

87. H. W. Codrington, ed., *Diary of Mr. John D'Oyly.*

88. Ramesh Somasunderam, "British Infiltration of Ceylon (Sri Lanka) in the Nineteenth Century: A Study of the D'Oyly Papers between 1805 and 1818."

89. For a full list of D'Oyly's informants, see Somasunderam, "British Infiltration," appendix 2.

90. See Somasunderam, "British Infiltration," but also Pieris, *Tri Sinhala*, 98–99, and De Silva, *Ceylon under British Occupation*, 1:134–35.

91. Gooneratne and Gooneratne, *This Inscrutable Englishman*, 108–9.

ject."[92] In reply Ahalepola stressed the discontent in Kandy: "Now women, men and all other persons residing on this side of the limits are disaffected to the Great Gate who governs our country. . . . If you have a desire for our Country it is good that anything which is done be done without delaying."[93] D'Oyly pulled away somewhat, perhaps suspecting that the correspondence was being undertaken with the knowledge of the king, with a view of trapping the British, as in 1803. In the end when Ahalepola revolted against the King in 1814, he found little assistance from D'Oyly, despite his pleas for British soldiers or even twenty Malays.[94] Ahalepola was forced to flee into D'Oyly's protection, and the civil servant arranged for his removal from territory close to Kandy.[95]

The time was now right for the taking of Kandy, and all that was needed was an excuse for war: Ahalepola supplied information about the king's military and financial resources, his treasure, and his lines of retreat in case of invasion. He described the king as "living as a fish encompassed by a net."[96] The pretext for war came when the king of Kandy's officers detained some traders, who they suspected to be British spies. These men had their noses and one arm and ear each cut off. They were sent back by different routes to Colombo, with their severed limbs hanging round their necks.[97] The British propaganda machine had what it wanted. Indeed the colonists made further use of tales of tyranny, in suggesting that great pits had been formed in Kandy for the reception of the dead bodies of rebellious people.[98] A further excuse for war followed when the king's troops crossed into British territory and burnt a house in Ruvanvalla.[99] D'Oyly accompanied the 1815 troops as Commissioner and gathered information from his informants as they neared Kandy. He saw to it that Sri Vickrama Rajasimha, the last king, was taken captive.

The only image we have of D'Oyly in adulthood shows him deep in conversation with the ministers Ahalepola, Molligoda, and Kapuwatte, after the taking of the kingdom, his top hat put to one side, with a secretary in the background, dressed only from the waist down, taking notes

92. Pieris, *Tri Sinhala*, 126–27.

93. Ibid., 127.

94. Ibid., 134.

95. Ibid., 140.

96. Ibid., 145, and De Silva, *Ceylon under British Occupation*, 1:152.

97. Pieris, *Tri Sinhala*, 147, and De Silva, *Ceylon under British Occupation*, 1:153.

98. Pieris, *Tri Sinhala*, 152. See also the introduction for the slaughter of Ahalepola's family.

99. Ibid., 155.

Fig. 3.5. "John D'Oyly in conversation with the chiefs of Kandy,
March 1815." From Paul E. Pieris, *Tri Sinhala: The Last Phase, 1796–
1815* (Cambridge, 1939), frontispiece. Author's photograph.

(fig. 3.5). It is more a scene of Kandyan diplomacy than British colonialism:
the men are seated on white linen, a custom for those in high office, and
the secretary is writing with a stylus on palm leaf. Entitled "Conference
in the Hall of Audience of the Kings of Kandy, 19th March 1815," it was
sketched by Captain William King of the Royal Staff Corps.[100] The drawing shows D'Oyly discussing the new appointments in the government
of Kandy with the ministers. Ahalepola, who had hoped for the throne,
was deeply disappointed and rejected any offer of ministerial appointment
in the British government; Molligoda was appointed chief minister. All
of the structures of Kandy were kept intact, and a British board of governance was superimposed on them in place of the king. D'Oyly as Resident
presided over this board, and had two assistants, the Revenue and Judicial
Commissioners.[101] The Convention that announced the taking of Kandy
was drafted by D'Oyly in two days and was effectively an agreement between the British and the ministers.[102] It promised to rule according to tradition and custom, in accord with the famous clause, which was cited in

100. Gooneratne and Gooneratne, *This Inscrutable Englishman*, 157.
101. De Silva, *Ceylon under British Occupation*, 1:165.
102. Gooneratne and Gooneratne, *This Inscrutable Englishman*, 152.

the introduction above, that Buddhist religion was "inviolable." D'Oyly's friendship and respect for the Buddhist clergy had thus come to have a decided impact on the settlement of Kandy. The British had become like kings.

There was a steady stream of criticism of the constitution. Robert Spence Hardy, in *The British Government and the Idolatry of Ceylon* (1839), capitalized the word "inviolable" from the clause of the Convention and wrote that this word was "strange, unwarrantable" and "an error."[103] He noted:

> The British Government of Ceylon appoints the principal Buddhist priests of the interior province—confirms in their appointments the priests of the palace at Kandy—appoints the lay chiefs of the principal temples dedicated to the worship of Hindoo deities—incurs the annual expense of the Perahara at Kandy—pays the expenses of other heathen festivals—and sanctions devil-dances, invocation to evil spirits, (a ceremony banished from his palace by a heathen monarch,) in the name of our Sovereign Lady the Queen. Can it seriously be maintained that these are trifles, by those who hold the divine authority of the word of God?[104]

In London, the British policy came to the attention of William Wilberforce, who criticized it.[105] Brownrigg was forced to write defensively that the security of the British possession of Kandy "hinged upon this point."[106] D'Oyly took on himself the customary duties of the king of Kandy, by joining ritual processions and carrying the sacred insignia, and by presenting the offerings customarily sent by the king to the four temples.[107]

D'Oyly's last years in Kandy have attracted much less attention than those that led to the taking of Kandy, and among unread letters lies further evidence of how he "went native." The severest challenge to D'Oyly's governance of Kandy came in the form of the rebellion of 1817–18, which he did not fully envisage, in his belief that the chiefs, whom he had worked with and whose friendship he had cultivated, would stay loyal. The rebellion began when Brownrigg, who had benefited from the assistance of Mus-

---

103. Hardy, *The British Government and the Idolatry of Ceylon*, 37.
104. Ibid., 35.
105. De Silva, *Ceylon under British Occupation*, 1:197.
106. Ibid., 1:164.
107. Gooneratne and Gooneratne, *This Inscrutable Englishman*, 158.

lim traders, in recompense appointed a Muslim chief, for the first time, in the remote provinces of Uva and Vellassa. A "Malabar," who the British claimed to be a pretender to the crown but who may have been the brother-in-law of the King Rajadhi Rajasimha, was found and proclaimed himself king. From here rebellion spread to other provinces. Even D'Oyly's confidant Ahalepola was found to have worked with the rebels and was eventually sent as an untried state prisoner to Mauritius.[108] A case has been made that Brownrigg held D'Oyly to be responsible for this rebellion, since the civil servant had taken the course of conciliating the chiefs too far.[109] Even in the context of the rebellion, D'Oyly found it difficult to adopt a more punitive policy with respect to the ministers and Buddhist clergy. To one of his subordinates in Pasbage, D'Oyly wrote asking whether any rebels had been spotted, but continued that "no trespass of any kind be committed" upon two temples in the area.[110] To another of his civil servants, he wrote complaining:

> 2 Priests have just arrived with a grievous complaint that the soldiers have notwithstanding their earnest Remonstrances shot their cow whilst tied up in the court of their temple with its calf near, and they naturally say, if such violence take place, whilst I am here, having promised them security for their property to whom can they look for protection? & how can they remain?[111]

As the rebellion progressed D'Oyly's tone with his agents grew sterner. To H. Wright at Badulla, he wrote calling to attention "the principles of British Govt and British justice," when various depredations at temples in the area came to his notice.[112]

The last decade of D'Oyly's life shows that his extensive knowledge of Sinhala Buddhism led him to intervene in very intricate matters. He sought to restore its ancient structures by bringing lowland temples within the tributary remit of Kandy, as they had been prior to British colonialism.[113] He guarded the rights of temples with respect to land and

---

108. De Silva, *Ceylon under British Occupation*, 1:185.

109. Gooneratne and Gooneratne, *This Inscrutable Englishman*, 184.

110. Letter dated 19 April 1818 from J. D'Oyly to Lieut. McKenzie Lot 21/51B, SLNA.

111. Letter dated 2 May 1818 from J. D'Oyly to Lieut. Elmslie, Lot 21/51B, SLNA.

112. Letter dated 3 May 1819 from J. D'Oyly to H. Wright, Lot 21/51B, SLNA.

113. Letter dated 20 June 1819 from J. D'Oyly to James Sutherland, Lot 21/51B, SLNA.

service, especially in view of new requirements for labor on the roads.[114] He sought to keep a watchful eye on the badge of Kandyan kingship, the Relic of the Tooth, and recommended that his *artachy* or writer and translator, Don Simon Perera, be placed in its charge.[115] He was able to advise in detail on matters of succession to Buddhist temples.[116] He recommended that the government provide an allowance to forty-two Buddhist priests in the Kandyan territories.[117] In 1817, he set out this list of cloths that were required as annual presents for the temples and priests in Kandy:

> 1 Silk Katchy–yellow or red, 20 or 22 cubits long.
>
> 15 Sallu Kawani Katchies—each about 22 cubits long
>
> 55 Sal Katchies, Pullu Katchies or Kalu Nul Katchies—each 38 or 40 cts long
>
> 8 Katta Surattu Tuppotties—each 13 or 14 cts long
>
> 32 Cubits of fine chintz or silk for priests fans
>
> 10 ordinary cloths called chita each about 7 cts long value about 4 Rix Drs each[118]

What were D'Oyly's motives by this stage? Though his scholarly study of Sinhala and Buddhism, and his contacts with priests and ministers, were vital to the taking of Kandy, those who watched his doings increasingly questioned their value to the British empire. As early as 1810, Sir James Mackintosh, who had known D'Oyly while he was an undergraduate in Cambridge, recorded that he "live[d] on a plantain" and "invites nobody to his house": "I was struck by the change of a Cambridge boy into a Cingalese hermit."[119] In the aftermath of the rebellion, Brownrigg complained that D'Oyly was a "private Gentleman of very retired and unostentious habits."[120] A missionary could, in the meantime, write of how the Resi-

---

114. Letter dated 15 November 1819 from J. D'Oyly to Major Martin, Lot 21/56, SLNA.

115. Letter dated 1 January 1819 from J. D'Oyly to G. Lusignan, Lot 21/51B, SLNA.

116. Letter dated 13 July 1819 from J. D'Oyly to G. Lusignan, Lot 21/52, SLNA.

117. Letter dated 30 January 1817 from J. D'Oyly to James Sutherland, Lot 21/51A, SLNA.

118. Letter dated 2 October 1817 from John D'Oyly to James Sutherland, Lot 21/51B, SLNA.

119. Quoted in Paul E. Pieris, ed., *Letters to Ceylon, 1814–1824, Being Correspondence Addressed to Sir John D'Oyly*, 12.

120. Quoted in Gooneratne and Gooneratne, *This Inscrutable Englishman*, 184.

dent at Kandy "has for a long time been a worshipper of Budu. . . . He takes off his Shoes & offers flowers &c. &c. to Budhu."[121]

Alexander Johnston and John D'Oyly epitomize two very different trajectories of oriental learning. One quickly rose to the head of a network of assistants and informants and returned to Britain, while the other "turned native." Both of them sought to use their studies to intervene in the judicial and political format of colonization in Ceylon; their orientalism was tied to a liberal commitment to protecting the lot of local peoples. Johnston was instrumental in bringing out more publications than D'Oyly. Yet after the latter's death, the judge played a role in seeing that D'Oyly's only publication found a readership.

"A Sketch of the Constitution of the Kandyan Kingdom," an unfinished summary of the laws and customs of Kandy, was published by the Royal Asiatic Society in 1834. D'Oyly had by this time been dead for ten years: he had succumbed to a fever and was buried in the Garrison Cemetery, behind the temple of the Tooth Relic, in land that is still contested and claimed by the temple. At the head of the Society's publication of D'Oyly's précis of the legal code appears a letter from Johnston. "The late Sir John D'Oyly and I were engaged, the whole time we were together on Ceylon, in inquiring, amongst other objects of literary and antiquarian curiosity, into the numerous remains which are still to be found in every part of the island of ancient Hindu history, laws, customs, manners, science and literature." Johnston praised D'Oyly and held up his *Sketch* as accurate, detailed, and authentic. The *Sketch* began with this sentence: "The power of the King is supreme and absolute. The ministers advise but cannot controul his will."[122] It ranged over a variety of matters, the classes of inhabitants, the jurisdiction of ministers, the punishment for crimes, and the ownership of property.[123] Further editions of D'Oyly's digest saw additions made to its content by others devoted to the recovery of the island's traditions and who played a role in the governance of Kandy, such as Simon Sawers, who was Revenue Commissioner of Kandy from 1816.[124]

In 1828 George Turnour, the translator of the *Mahavamsa*, was ap-

121. Quoted in ibid., 160.

122. From J. D'Oyly, "A Sketch of the Constitution of the Kandyan Kingdom, By late Sir John D'Oyly, Communicated by Sir A. Johnston," 191–93.

123. See L.J.B. Turner, ed., "A Constitution of the Kandyan Kingdom by John D'Oyly."

124. For Sawers' additions, see Simon Sawers, "Memoranda and Notes on the Kandyan Law of Inheritance, Marriage, Slavery etc. by Simon Sawers."

pointed Revenue Commissioner of Kandy. His career saw the style of
governance, which tied the study of local languages to an insistence on
precedence, carried on into 1840. Yet Turnour's papers show that by the
later 1820s and early 1830s orientalist governance was opposed in some
quarters, and in particular by Governor Edward Barnes. Barnes sought to
intervene in force and so establish British commercial, military, and social superiority. For instance, he resented the fact that Turnour proposed
that the system of compulsory labor on the roads should be reformed. By
1832–33, when the Commission of Eastern Inquiry arrived in Ceylon, their
interviews revealed the place of oriental learning in the civil service of the
island.

In his interview with the Commission, Turnour claimed that the "natives are still very decidedly attached to their religious institutions, so
much so that the local Government has hitherto considered it politic to
give them support."[125] "Any interference with a view of change would
certainly be regarded by the chiefs and people as well as by the priests
as subversive of those institutions." Turnour noted how his knowledge of
language had given "confidence to the natives in [his] official proceedings
and decisions" and lamented the lack of linguistic skills on the part of his
contemporaries in the civil service, which he put down to the fact that
civil servants were obliged to change their stations too swiftly, and because promotion was tied to seniority rather than linguistic capability. He
added, in contradiction to Johnston's earlier views, that "the standard of
the native character in point of view of intellect and morals is still too low
to admit of their being enlisted with responsible duties, which are not under the immediate controul of European functionaries." At the end of the
transcript of the interview appears a memorandum from Turnour, recom-
mending a scheme for encouraging the study of language in Ceylon's civil
service. He proposed that returning missionaries from Ceylon could be
employed at a sum, and asked to reside close to London in order to afford
instruction in Sinhala and Tamil.

Even more striking evidence for how the consolidation of the civil service in Ceylon did not see an advance of orientalism appears from the Commission's interviews with other civil servants.[126] Most of the Agents of Gov-

125. All quotations from "Evidence of George Turnour Esq. Revenue Commissioner
of the Kandyan Provinces, Kandy 22nd September 1829," Lot 19/106, SLNA.
126. All quotations from "Replies of the several Agents of Government in the Kandyan Provinces to the questions of the Commissioners of Inquiry relative to the laws
and mode of conducting judicial proceedings," Lot 19/103, SLNA.

ernment who were interviewed held that a knowledge of local languages was not indispensable for their duties, and they asserted that most of the laws of the land were unwritten and oral. When asked to comment on D'Oyly and Sawers' digest of laws, most of the Agents of Government seemed blissfully unaware of its existence. The Agent of Saffragam noted: "No digest has been formed, the laws prevailing among the Candians are oral and traditional." The Agent of Matale observed: "I have heard that a digest of the native laws and customs have [sic] been formed since the establishment of British Government. I have never seen it, nor has it been sent to court."

Oriental learning did not die out, for in many ways the Colebrooke and Cameron reforms that resulted from this inquiry encouraged the entry of the island's inhabitants into the civil service and to playing a greater role in governance. But the norms of orientalism shifted in this period, and so also did what counted as the indigenous tradition of the island. Johnston had forged an extended network and relied very heavily on his collaborators, who included priests; D'Oyly had close friendships with priests and devoted time to language study; and Turnour took the study of language even further by devoting himself to Pali, and hoped to rid himself of the need of interpreters and assistants, at least in public service. Turnour told the Commission: "It is very important that public servants should be able to communicate with the people in their own language, and without the assistance of Interpreters."[127] The evolution of the styles of learning mirrors a shift in colonial policy, from a need to govern according to local precedent to a new form of liberalism, which sought to reform local customs and traditions. There was also a separation of the spheres of scholarship represented by Buddhist priests and British civil servants. When the Ceylon branch of the Royal Asiatic Society was formed in 1845, it had no non-European members.[128]

The issues of the Society's journal for the duration of the first decade contained a steady stream of papers written by non-European authors, such as Simon Casie Chetty (1807–60), who on Johnston's recommendation became a corresponding member of the Asiatic Society of Great Britain, and James De Alwis (1823–78), who had Pali and Sanskrit in addition to English and Sinhala and who was a member of the Legislative Council of Ceylon.[129] However, their biographies show how British colonialism in

127. "Evidence of George Turnour Esq," Lot 19/106, SLNA.

128. See *Journal of the Royal Asiatic Society Ceylon Branch* 1 (1845): i–iii.

129. See Mervyn Casie Chetty, "Introduction," in Simon Casie Chetty, *Ceylon Gazetteer*, i–vi; Michael Roberts, Ismeth Raheem, and Percy Colin-Thome, *People in*

Ceylon slowly engendered a new sort of scholar among the local population, who spoke and wrote in English and who belonged to the urban social elite and was part of the civil service. These were hybrid individuals, who were critical of indigenous traditions and committed to critiquing colonial governance. Some of them in turn cultivated their own connections with Buddhist monks long into the nineteenth century. Ceylon was now heading for economic laissez faire reform, which gave rise to a new colonized elite. This was a decidedly different context to that in which the first British oriental scholars had worked, with the assistance of Buddhist priests who had first served as collaborators and then as teachers of language. Further, a new distance emerged between the Buddhist priests and British colonial officers, in the midst of public debates about religion and missionary pamphlets.[130] This does not mean that Buddhist monks lost their status as scholars or that they did not initiate new educational endeavors. With new patriotic sentiment and religious reformism, connected with a theosophical interest in Buddhism from the West, the Buddhist monk-scholar reemerged in the later nineteenth century.[131]

CONCLUSION

Johnston and D'Oyly's scholarship, in turn, can be equated with the consolidation of British knowledge of the lowlands and highlands of Ceylon and with the advance of the British state in the place of the Kandyan kingdom. Yet the irony is that both Johnston and D'Oyly benefited from the afterlife of the reorganization of eighteenth-century Buddhism in Kandy.

The study of the history of orientalism needs to be historicized in relation to colonial change and policy-making. Such an enterprise of historicism will need also to pay attention to the global and transnational dimensions of scholarly endeavor, in addition to shifts in specific places like the island of Lanka. Lanka was given a place on a global stage of religious thought, was classed as a separable unit, and was seen as Buddhist. Work

---

Between: The Burghers and the Middle Class in the Transformations within Sri Lanka,
1790s-1960s, 79–81; and Yasmine Gooneratne, *Relative Merits: A Personal Memoir of*
the Bandaranaike Family of Sri Lanka.

130. See Young and Somaratna, *Vain Debates.*

131. For more on this see Blackburn, *Locations of Buddhism,* and Guruge, *From the Living Fountains of Buddhism.*

on the island became distinct from that done elsewhere, such as in India, because of the permutations of Crown rule.

In following the networks through which artifacts, information, and texts moved, it is possible to see the radical changes that were wrought to the study of Buddhist traditions. The change from a traditional orientalism, dependent on intimate connections with Buddhist priests, to a newly confident orientalism, where scholars hoped to acquire linguistic skills themselves while allowing indigenous elites to act as scholars who could in turn cooperate with monks, is part of a broader story of the impact of liberalism on colonialism across the imperial realms. Yet such a global picture of the colonial power dynamics of scholarship should not eclipse the individual—witness D'Oyly's sympathy for Buddhist monks or indeed Karatota's unusually productive relationship with Johnston.[132]

This chapter has a thread of argument that is more fully elaborated in the next. Throughout I have suggested how the study of Buddhist tradition was not only about the content of texts. The project of textual exegesis was performative and linked to kingly patronage. For British orientalists, as for the Kandyan kings, it was a process of collection, which enhanced political credentials. Oriental collection encompassed both images and texts, and sometimes the monks themselves, as is evident from the two prelates who accompanied Johnston to London. The next chapter widens the analysis of the character of orientalism to other objects, and in particular to the realm of ancient sites and their associated material traces. Orientalism was an activity carried out not merely in the study but in the field. It was closely connected to topography and to the imagination of the territory of the island; orientalists' sites operated concomitantly with the making of the local and the governable.

Lineage has been a central idea in this discussion of early British orientalism. The identity and intellectual lineage of Dhammarama, and his relations to the Kandyans, the Dutch, and British, and to Alexander Johnston and John D'Oyly, is important. If lineage was critically important to Buddhist monks, what was at stake in the British becoming students and patrons of this eighteenth-century lineage of scholarship? Scholarship for both the Kandyan kings and the British orientalists served a wide purpose in the generation of a public presence. This would mean, in contradiction to Donald Lopez, that the study of Buddhism was indeed guided by the

132. This claim also returns to Blackburn, *Locations of Buddhism*, which insists on the agency of the individual.

threat which the religion posed to the advance of colonialism in Ceylon.[133] As the next chapter shows, the performance of orientalism by the British was an act of good governance that made it possible for the British to be seen as inheritors of Kandyan kingship and the rightful governors of the island.

133. See Donald S. Lopez, "Introduction" in Donald S. Lopez, ed., *Curators of the Buddha: The Study of Buddhism under Colonialism*, 11.

CHAPTER FOUR

Sites

Perhaps the most cherished register of travelogues is the rhetoric of discovery—claiming to have seen a site, or even a sight, for the first time.[1] In Ceylon, the language of discovery is discernible in how travelers mapped, described, and explored historic sites in the first decades of the nineteenth century. Yet it is important to critique the language of discovery used by British orientalists in Ceylon. For every claim of discovery recycled extant traditions in the midst of colonial expansion.

In Lanka, the British were not exceptional in their interest in travel. Colonial travel operated both intentionally and unwittingly, within notions of pilgrimage which were an important element of Kandyan kingship as well as popular religious culture. Indeed, the anthropologist Gananath Obeyesekere argues that, in the age before mass communication, pilgrimage served as a vital means of forming an identity greater than a regional one.[2] In their journeys, the British and the Kandyans shared modes of engagement with Buddhist priests, a turn to visual description in order to authenticate travel, and a commitment to the performative role of travel to historic places for political purchase. Because of these symmetries, the British may have been cast into the role of pilgrims by Lankans. Even the power of their scientific instruments, which symbolized the modernity of British travel in the nineteenth century, could either be objected to or reinterpreted by onlookers into a religious culture of travel.

---

1. For a recent account of the relationship between travel and knowledge-making, see David Arnold, *The Tropics and the Travelling Gaze: India, Landscape and Science, 1800–1856.*

2. Gananath Obeyesekere, "Buddhism, Nationhood and Cultural Identity: A Question of Fundamentals," 237.

Two sites of pilgrimage are at focus in this chapter: the ancient capital
city of Anuradhapura in the north-central provinces, which served as a seat
of kingship until 1017 AD, and the fabled mountain of Sri Pada, Samantakuta, Sumanakuta, or Adam's Peak, in Sabaragamuwa in the southwest,
believed to hold the footprint of the Buddha, Siva, or Adam. Both these sites
have one point in common: Buddhists considered them to be among the sixteen most hallowed places of the island; in fact, several sites within Anuradhapura and Sri Pada counted among the sixteen. While colonists cast
Anuradhapura in a discourse of ruin and abandon, Sri Pada was a mountain
that was heavily frequented by pilgrims. Discovery in the case of Anuradhapura proved to be a ready mode of self-presentation—for Britons sought
to portray themselves as bringing a lost city to life from the very jungles
that were taking it over. Anuradhapura was thought of as an elusive site. In
contrast, Sri Pada was visible even from the shore. Travelers to the island
commented upon it, seeing it to be the highest mountain of the island; the
peak's visibility was also key to islanding discourses associated with it.

The connection between the local and the global was an important element in the shifting traditions of both the Buddhist *sasana* and early British orientalism. This chapter expands the argument of the last by opening
up the local history of knowledge connected with Anuradhapura and Sri
Pada and so takes the local to a different level of specificity, by moving
from England, Colombo, and Kandy, which were the prime points of attention in the previous chapter, to two of the peripheries of the Kandyan kingdom. Working with the broader claim of how the island was being made
into a unit during this period of colonial transition, this chapter breaks
the island space within itself. The two peripheral sites sustained rival and
hybrid religious, political, and cultural identities which undercut those
emanating from the centers of Kandy and Colombo. Though the British
islanded Ceylon, the speed with which Anglicist culture reached into its
deepest realms in the decades following 1815 must not be assumed. Resistance to the islanding of Ceylon arose from localities such as those studied here.

There has been a tradition of writing about the historicization of Anuradhapura from a different perspective. Pradeep Jeganathan and Elizabeth
Nissan in their articles on how Anuradhapura became a sacred symbol
agree that Britons rediscovered the old city in the nineteenth century, after they conquered the kingdom of Kandy.[3] For Nissan, local texts such

---

3. Elizabeth Nissan, "History in the Making: Anuradhapura and the Sinhala Buddhist nation"; Pradeep Jeganathan, "Authorizing History, Ordering Land: The Conquest

as the *Mahavamsa* do not gesture to any precolonial historical tradition that fed into British historiography; instead, she points to the accidental congruence between these chronicles and European historicism.[4] Jeganathan is more explicit about how Britons set in motion a new history of the settlement; he speaks of a "radical rupture in the nineteenth century."[5] He does not subscribe to Nissan's thesis of accidental congruence because it limits the constructive power of colonial makings of history and archaeology. Despite their differences, both Jeganathan's and Nissan's accounts are similar in the lack of attention they pay to the eighteenth-century history of Sri Lanka. For both of them the decisive moment in the modern history of Anuradhapura is 1833, when the British established an Assistant Agency at the town, which led to its restitution. In one brief paragraph Jeganathan dismisses the importance of Anuradhapura to the Kandyan kings of the eighteenth century; the ancient capital is also said to have had no significance for the anti-British rebellion in 1817–18.[6] The present discussion counts as an attempt to understand the history of this site, and of Sri Pada as well, without overemphasizing the discursive powers of British orientalism.

In keeping with the perspective on colonial transition in this work, this is an argument not for continuity or change but rather for the consistency of continuity and also for the ever-changeability of change. Both of these facts make for a dynamic context and meaning for indigenous precolonial tradition in the cultural history of travel. This may be illustrated by thinking of Anuradhapura and Sri Pada in the 1830s, the end point of this chapter. Once a road had been built to Anuradhapura and once Britons had climbed Sri Pada many times over, the regime of colonial knowledge was able to assert its power over these places in radical ways. Yet this did not preclude the fact that regional cultural forms, connected to myths, forests, and popular practice, continued to exist alongside the newly reified colonial order of control, even in the middle of the nineteenth century. The way in which these regional loyalties asserted themselves was via the colonial state's courts and through its government agents.

---

of Anuradhapura." See also E. Valentine Daniel, "Afterword: Sacred Places, Violent Spaces," 234.

4. Nissan, "History in the Making," 68. This is also the argument of her doctoral dissertation, Elizabeth Nissan, "The Sacred City of Anuradhapura: Aspects of Sinhalese Buddhism and Nationalism," 11.

5. Jeganathan, "Authorizing History, Ordering Land," 107.

6. Ibid., 117, 128.

Anuradhapura receives more attention than Sri Pada in this chapter. The first section illustrates how the Kandyan kings came to terms with the heritage of the ancient capital, even while an eclectic and independent tradition of cultural identity flourished in the region in which Anuradhapura lay. The discussion then moves to how early British understandings of Anuradhapura and of archaeological explorations were hemmed in by earlier kingly and popular renditions of the site. The last section presents a similar argument about Sri Pada, while illustrating also how this early phase of slippage passed into a scientific tradition of description of what was called Adam's Peak by the 1830s, which in turn had a place for the "indigenous."

MULTIVALENT HISTORIES

In the late Kandyan kingdom, kingship was in part a territorialized discourse. The *Mahavamsa*, which Turnour caused a furor over, demonstrates how the Nayakkar kings hoped to bring the provinces directly under the spiritual superintendence of the capital in Kandy, and so to unify the island as a political entity. We are told of the spectacular ceremony that resulted from instructions given by Sri Viyaya Rajasimha. Lamps were burnt in every town and in the temples of every province on the same night. While this ritual was in progress around the land, the king gathered a crowd and celebrated a sacrifice of lamps, with seven-hundred-and-ninety-thousand six-hundred lamps. "Thus with the burning lamps the Ruler of Lanka made the land of Lanka like to the star-strewn firmament."[7] In a similar vein, Kirti Sri Rajasimha brought "all the inhabitants of Lanka" to the capital, and had "people from the individual provinces separated and made them dwell in different places, provided with standards." Together with this throng, and accompanied with different symbols and banners, Kirti Sri marched around Kandy "like the Prince of the gods, with great (and) royal splendour."[8] In the words of one historian, this ceremony, which is now held annually and is called the *asala perahara*, established horizontally and vertically synchronized relationships between Kandy and the provinces, and between the human king and the gods.[9]

7. W. Geiger, trans., *Culavamsa, Being the Most Recent Part of the Mahavamsa*, 2:98, lines 63–64.

8. Ibid., 2:99, lines 43–52.

9. John Holt, *The Religious World of Kirti Sri: Buddhism, Art, and Politics in Late Medieval Sri Lanka*, 32.

Kingly pilgrimages to isolated sites might in this context have resulted from the desire to proclaim the spiritual right to rule outlying districts. Nayakkar pilgrimages to Anuradhapura are of particular interest here. Narendrasimha (r. 1707–39) is the first to have accrued merit by journeying to the ancient land of the kings. "At the head of a great retinue he left the great city, went forth to the great [city] Anuradhapura and celebrated a great sacrificial festival."[10] Sri Viyaya Rajasimha, his successor, also visited the historic city. The king is recorded as having eased the passage of pilgrims by making stone bridges across waterways lying on the route to sacred places, such as Anuradhapura.[11] Later in the century, Kirti Sri Rajasimha also became a pilgrim to the old capital. His party is described in lavish terms: "Yearning for merit the Lord of men betook himself with his retinue to superb Anuradhapura. Here the King sacrificed to the Bodhi tree and the sacred cetiyas with elephants, and horses, with gold, silver and the like."[12] Anuradhapura survived in official memory as a place of spiritual significance. It was certainly not forgotten.

In Kandy, Anuradhapura might have also been identified with the province of Nuvarakalaviya, in which it was found. Each of the provinces or *disavanies* was represented by an official called the *disapati* who resided in Kandy. He traveled to the province under his control only when instructed by the king, for instance to collect greater taxes from the people.[13] The various *disavanies* were ranked according to importance. In 1769, a Dutch governor was told by Buddhist priests that there were seventeen *disavanies*; Nuvarakalaviya was ranked sixteenth.[14] Symbolism was used to publicize the hierarchy of regions in Kandy. Each of the *disavanies* had a banner. Nuvarakalaviya was represented by a beast called *gajasinha*, compounded of an elephant and a lion. The *disapati* of Nuvarakalaviya alone had the right to have this banner carried before him, to the accompaniment of music.[15] On occasions such as the *asala perahara*, each *disapati* was put in order of importance. Even though Nuvarakalaviya may have been insignificant in purely political terms, there is evidence that control was exerted from the center. For instance, communication was made by

10. Geiger, *Culavamsa*, 2.97, lines 32–33.

11. Ibid., 2:98, line 86; see also R. W. Ievers, *Manual of the North-Central Province of Ceylon* (1899), 203, for an account of the road to Anuradhapura from Dambulla.

12. Geiger, *Culavamsa*, 2:99, line 36.

13. Lorna Dewaraja, *A Study of the Political, Administrative, and Social Structure of the Kandyan Kingdom*, 169–70.

14. Ievers, *Manual of the North-Central Province*, 59.

15. Ibid., 60.

means of *bola*, a letter carried in a silk handkerchief, sent by hand from village to village, along with a staff.[16]

The spiritual and symbolic significance of Anuradhapura was keenly disseminated not only in Kandy but in other areas of the island as well, so that the meaning of Anuradhapura was not simply created by the center. In reviving monastic communities in various provinces, Kirti Sri Rajasimha proclaimed his granting of land in citations called *sannasa*. A *sannasa* dated 1752 from Urelawatte, for instance, speaks of how Kirti Sri revived Buddhist practice: "Hundreds of ruined temples were repaired in different places, including the Relic Temple, offerings of gardens, fields, and flower gardens were made to Anuradhapura, Samanatakutaya, Mahiyanganavehera, the Dalada Maligawa, and other vihara."[17] These and other royal land grants carried the news of the revival of religious sensibility and of the restitution of sacred Anuradhapura to monastic communities in the regions.[18]

Kirti Sri was also a great patron of temple paintings; the themes and narratives of these illustrations were consciously chosen to enhance the king's religio-political standing and to provide aids for meditation by monks and lay people alike. It has been argued that these paintings constituted a visual liturgy that required the active participation of the observer. They are said to have been especially effective with illiterate agrarian and service-caste village Buddhists who made simple offerings to the Buddha at their village temples.[19] The paintings commissioned by the king were never identical in all of the provincial temples, yet there are some important continuities. For our purposes, the recurrent portrayal of the *solosmasthana*, or sixteen most sacred places of the island, is noteworthy. The *solosmasthana* brought together the eleven sites visited by the Buddha according to the *Mahavamsa* and added various stupas that had been built by pious kings in the past. The geographical distribution of these sites allowed the linkage of the various regions with a narrative of religiosity. Their illustration, by the patronage of the king, supported the righteousness of Kirti Sri's reign. The ancient settlement of Anuradhapura was represented by the stupas of Ruvanvalisaya, Thuparama, Jetavana, Abhyagiriya, and Mirisavati, and by the Sri Mahabodhi, the sacred Bo-tree.

---

16. Ibid., 67–68.

17. A. C. Lawrie, *A Gazetteer of the Central Provinces of Ceylon* (1896), 887; for another *sannasa* from 1751, see x.

18. For more on *sannasa*, see Holt, *The Religious World of Kirti Sri*, 35–39.

19. Ibid., 47.

Fig. 4.1. An eighteenth-century mural on the back of a boulder in the southwest corner cave at Dambulla, displaying the ploughing of the sima or monastic boundaries of Anuradhapura. From John Clifford Holt, *The Religious World of Kirti Sri: Buddhism, Art, and Politics in Late Medieval Sri Lanka*, plate 20. Photograph courtesy John Clifford Holt, Bowdoin College.

The accessibility of these images of Anuradhapura in well-visited regional temples encouraged information about the old capital to spread outside Kandy. Indeed one function of the *solosmasthana* was to identify sites that were worthy of pilgrimage. The later years of Kirti Sri's reign saw this genre of paintings extended in the large caves at Dambulla.[20] Several striking paintings from Dambulla publicize the early history of Buddhist practice in Anuradhapura. One of these shows the plowing of the *sima* or monastic boundaries of the new temple complex in Anuradhapura (fig. 4.1), another the arrival of the sacred sapling in the settlement, and a third the placement of the Buddha's relics in what the *Mahavamsa* identifies as the first stupa to be built in the country, called the Thuparama. These paintings suggest that Anuradhapura had come, by the time of Kirti Sri's reign, to be connected with the origins of Buddhist practice in the island.

The *Mahabhinikman Jataka*, a ballad dating in its earliest form to the late sixteenth century, also provides evidence for this. This poem circulated widely, and additions were made as it passed from mouth to mouth. An extant manuscript from 1747 presents some additions made by Walmoruwe Kivindu, who resided in Matale as Secretary of the Gate, which lay on the route to Nuvarakalaviya from Kandy. While the earlier version pertained to the life of the Buddha, this later version appended verses relating to the arrival of Buddhism in Sri Lanka and to the erection of the

20. I am relying here on ibid.

chief stupa at Anuradhapura.[21] Therefore, in the late Kandyan kingdom, the biography of the Buddha and the early history of the island merged; the distant past was appropriated and connected with Anuradhapura.

Though this evidence suggests that Anuradhapura took on a defined identity that came from outside the region in which it lay, the situation looks very different when viewed from Nuvarakalaviya. With the break-down of the kingdoms of Anuradhapura and Polonnaruva in Nuvarakala-viya, the area came under the control of a class of chieftains known as the Vanniyas. The Vanniyas of Nuvarakalaviya wielded great power, especially the Maha Vanni Unnahe or Vanni Bandara, who controlled all the tracks away from the king's territory to Dutch land. Robert Knox, a sailor in the captivity of the king of Kandy in the late seventeenth century, noted that a watch was constantly kept at Anuradhapura so as to monitor the move-ment of people across the borders of the king's domain.[22] A tax of 10 per-cent was levied on all merchandise, as was one and a quarter pice on all individuals who passed through the gates.[23] This money was credited to the Vanniya, and he in turn paid the king of Kandy a tribute.[24] The pow-ers of the Vanniya extended also to judiciary and police duties.[25] It is clear then that the region of Nuvarakalaviya enjoyed a surprising degree of in-dependence. In this context, it is plausible that the inhabitants established an identity which was separate from that created for them in Kandy.

The author of a paper delivered to the Royal Asiatic Society of Ceylon in the mid-nineteenth century noted that the people of Nuvarakalaviya held an "intense dislike of strangers." He continued: "So strongly does this feeling still exist, that we have even now to take the greatest care not to bring roads too near to villages."[26] This comment may point either to a response to British practice or to the prejudices of the author. Yet it might also be interpreted as an indicator of a regional identity which was in flux. There is evidence for instance that the everyday practices of fish-

21. See Hugh Nevill, *Sinhala Verse (Kavi)*, vol. 1, no. 121, and vol. 3, no. 714.

22. Robert Knox, *An Historical Relation of the Island of Ceylon* (1681), 2:30. For Knox's description of the ruins of Anuradhapura, see 2:421–22.

23. The pice was a copper coin, equivalent in value to a quarter of a fanam or half a rix dollar.

24. Ievers, *Manual of the North-Central Province*, 112, 45–46.

25. For a helpful summary, see Lorna Dewaraja, *A Study of the Political, Adminis-trative, and Social Structure of the Kandyan Kingdom*, 190–92.

26. A. Oswald Brodie, "Topographical and Statistical Account of the District of Noowerakalawiya," 151.

ing and cultivation served to unite the villages of the region. These were associated closely with the ruined reservoirs that dotted the region, which were built by the kings when Anuradhapura and Polonnaruva were in turn capitals of the island. In the wet season, for example, the reservoirs were filled with fish. At agreed times, the villagers would assemble and fish in these reservoirs. The catch was then divided according to a prescribed formula. A type of lotus that grew in these reservoirs supplemented the diet of the region; the stalks were boiled and curried, the seeds were eaten raw and cooked, and the kernel was used as a herb, which could be boiled and roasted.[27] Game also formed an item of food; in the dry season every village put up a platform at every available water-hole to facilitate the slaughter of animals.[28]

Everyday practices such as these might be traced in any eighteenth-century village community in Sri Lanka, yet the role of historic monuments such as reservoirs in the organization of the community is noteworthy. For example, *Hen habe,* a late-nineteenth-century song from this region, tells of the misery experienced by a group of villagers at having to drink well water while in jail. Being accustomed to drinking reservoir water alone, this punishment weighed heavily.[29] The reservoirs also played a part in the division of the paddy fields. After the ground was cleared, a line was stretched and marks were put in according to measurements taken from the banks of the reservoirs, which were used in dividing land. The owners of every plot of land had to pay a share toward repairs of the reservoir, irrespective of the harvest from their plots. The first and last plots of land were less fertile, since the channel from the reservoir entered and left the cultivable land at these points. These plots were therefore larger than the rest.[30]

These rules and customs both united and created divisions among the population. A British official named R. W. Ievers, in his late-nineteenth-century manual to the north-central province, noted that the custom of fishing with baskets was adopted only by a group that he identified as the Sinhalese of Nuvarakalaviya. The rest of the population used a rod and line.[31] The

---

27. Ievers, *Manual of the North-Central Province,* 193.

28. Ibid., 195.

29. Nevill, *Sinhala Verse,* vol. 3, no. 650.

30. Brodie, "Topographical and Statistical Account," 157.

31. Ievers, *Manual of the North-Central Province,* 197; for fishing with baskets, see also the plate in Knox, *An Historical Relation,* vol. 2, image facing 98.

headmen of the villages received a double portion in the allocation of paddy fields.[32] Ievers added that there was a special caste in the region constituted by a group who claimed that their ancestors had come to Sri Lanka with the Bo-tree.[33] He reported the attendant oral tradition: this caste claimed their ancestors had been given the task of protecting the Bo-tree from monkeys by using arrows. The historian must be cautious in assigning value to comments on social organization and caste drawn from later colonial sources.[34] In particular, Ievers' account reverts to a timeless cultural description of Nuvarakalaviya even as it bears evidence of how the region was in flux. It is indisputable that the people of Nuvarakalaviya shared a vibrant set of traditions, but these were forged in relation both to the historical artifacts in the region and to their contextualization in relation to state structures and norms of governance in the historic present. In keeping with the methodology of this argument, colonial sources need to be interwoven with songs and narrations on palm leaf, to recover the entangled and inextricable paths taken by the colonial and precolonial.

The large collection of palm-leaf manuscripts in London, first put together by Hugh Nevill, demonstrate that the region surrounding Anuradhapura sustained an exceptionally large number of legends. For instance, there is a class of poems called *vandana kavi* which were supposed to be chanted by pilgrims on their way to sacred sites. The *Solosmasthana vandanawa* provides an example of this genre. This poem is said to be at least three centuries old: at Anuradhapura the pilgrim is directed to worship the Bo-tree on Sunday, the Mirisavati stupa on Monday, the Ruvanvalisaya stupa on Tuesday, the Thuparama stupa on Wednesday, the Abhayagiriya stupa on Thursday, the Silacetiya on Friday, and the Jetavana on Saturday.[35] The palm-leaf collection of the Colombo Museum also houses copies of *vandana kavi*. The *Ruvanvali Dagab Varnava*, for instance, describes the city of Anuradhapura as consisting of "nine lakhs of double storied buildings and ninety lakhs of one storey buildings."[36] The *Ruvanvali Vistaraya* presents an image of Anuradhapura as a city of luxury in the reign of Dutthagamani; at its four doors pilgrims find clothes to be worn and rich

32. Brodie, "Topographical and Statistical Account," 157.

33. Ievers, *Manual of the North-Central Province*, 92.

34. See Susan Bayly, *Caste, Society and Politics from the Eighteenth Century to the Modern Age.*

35. Nevill, *Sinhala Verse*, vol. 1, no. 147.

36. *Ruvanvali Dagab Varnava*, palm-leaf manuscript in the Colombo Museum Library, AR 9.

foods to be eaten prior to worship. There are also sixty-four different kinds of jewelry for different parts of the body.[37]

Historically, Sinhalese Buddhism was comfortable with a pantheon of gods. The guardian gods of Sri Lanka were subject to the Buddha, and under them were the regional gods and devils of the people.[38] The oral records of seventeenth- and eighteenth-century Sri Lanka suggest that the vibrant traditions connected with Nuvarakalaviya did not only pertain to the relics of the Buddha. There were also ballads connected with regional gods and devils. For instance there is the tradition of the Vanni Bandara. His love of the forest and his enthusiasm for catching elephants were retold in verse.[39] There is also the rich tradition of Ratna-valli, who was worshiped by a telambu tree that occupied the site of the Ruvanvalisaya stupa in Anuradhapura. This tree had to be cut down by King Dutthagamani when the site was chosen for the stupa. The king appeased Ratna-valli with blood sacrifices and named the stupa after her.[40] By the late nineteenth century this tree was still partially sacred; the name of the Ruvanvalisaya stupa reminded inhabitants of the tradition of Ratna-valli. A verse from one of the poems about Ratna-valli has been translated as follows; it presents the goddess's words to the king:

> Think not covetously of my excellent telambu tree,
> You fall into the toils, Maha raja, think not wrongly,
> Like water in the river, making blood flow propitiously,
> Build in my name the Gem dagoba.[41]

The ballads relating to the region in which Anuradhapura lies were inspired by the circumstance of the heavy vegetation that covered the area. Robert Knox commented on the dense forests of Nuvarakalaviya. The sailor's escape route lay through the province, which he described as the northernmost piece of land in the dominion of the king of Kandy: "As you traviell through this country there is nothing to be seene but woodes, the

37. *Runvanvali Vistaraya*, palm-leaf manuscript in the Colombo Museum Library, 1899/7/7.

38. This hierarchy is well explained in Gananath Obeyesekere, "The Great Tradition and the Little in the Perspective of Sinhalese Buddhism."

39. Nevill, *Sinhala Verse*, vol. 1, no. 202.

40. L. D. Barnett, "Alphabetical Guide to Sinhalese Folklore from Ballad Sources," 86–87; also Nevill, *Sinhala Verse*, vol. 2, no. 355, vol. 3, nos. 765, 767, 877.

41. Ibid., vol. 3, no. 877.

trees growing over the rodes."[42] There is some speculation that the etymology of the word Vanni lies in a derivation from *vana*, meaning forested area.[43] The connection between ballads and forests is exemplified for instance by *Kadawara puwata*, which provides the account of the origin of the Kalavava reservoir.[44] According to this poem, a man who was disgraced by the conduct of his wife decided to abandon civilization and to keep company with the deer for twelve years. The king, who heard of this man, assembled his court and captured the outcast. Upon questioning, the man told the king about a sea of water that he had seen on his travels. When the king inspected this site, he decided to build the bund of a reservoir there and entrusted the keeping of the reservoir to this man. But when the reservoir burst because of the man's inefficiency, the man was overcome with grief and threw himself into the breach, after which he was reborn as Kadawara Deva.

There are other ballads in the Nevill Collection which relate to the Minneri reservoir, which was built in the Pollonnaruva period, the Pattini goddess, and various traditions from this historic region. Taken together these poems point to a genre of popular historical knowledge relating to the remains of Anuradhapura, and more generally to Nuvarakalaviya and the Vanni. Anuradhapura had a place in the spiritual cartography of the island, which was at once both elite and popular, and cultivated from the metropolis and within the regions, in the period prior to the British advent. The historic artifacts of Nuvarakalaviya mediated in the formation of social communities and practices in the immediate locality, even as they symbolized the region far afield. These meanings were attributed to Anuradhapura simultaneously, and could compete, diverge, and coalesce. These multivalent narrations urge a healthy skepticism of a reductionism which looks out from Kandy or Colombo alone.

THE BRITISH STRUGGLING TO CONTROL THE PAST

British orientalism is often said to have created new structures of knowledge. Yet the colonizers did not have this degree of creative freedom in the early years of their presence in Kandy. This is exemplified in how the

---

42. Knox, *An Historical Relation of the Island of Ceylon*, 2:5.

43. Michael Roberts, *Sinhala Consciousness in the Kandyan Period, 1590s to 1815*, 72.

44. Nevill, *Sinhala Verse*, vol. 2, no. 454.

British recontextualized the historical memory of Anuradhapura that pre-dated their arrival.

For instance, the notion that Nuvarakalaviya was a densely forested region led by independent chiefs comes to the fore in early British accounts. Soon after the British taking of Kandy, the Assistant Government Agent at Mannar on the northwest coast of the island wrote of how Suriakula Kumarasinha, the Maha Vanni Unnahe, arrived at Mannar seeking marks of respect and honor from the British government. The Agent wrote: "To this application I readily assented, considering that the restoration of a person of his rank, who in the hour of distress and persecution by the late ruler of Kandy had thrown himself upon us for protection would be approved."[45] Kumarasinha was therefore sent back on a palanquin, accompanied by servants, local people in the employment of the British, and twenty Sinhala copies of the British governor's proclamation after the taking of Kandy for distribution in his district. Yet just as much as the Kandyan kings found the Vanniyas resistant to simple rule, three months after this episode the Collector of Mannar reported how Kumarasinha had refused to allow a traveler with a passport issued by the Resident at Kandy to travel through his district. Kumarasinha allegedly ill treated, seized, and detained the traveler along with all his cattle and merchandise.[46] British state-making, despite its muscularity in military terms, still worked with Vanniyas.

The ambiguous nature of the relationship between the Vanniyas of Nu-varakalaviya and the British government is also illustrated by the events following the rebellion of 1817–18, which we came across in the previous chapter in relation to John D'Oyly's dealings with priests.[47] When the rebellion reached Nuvarakalaviya, the Maha Vanni Unnahe threw in his lot with the rebels. Kappitipola, one of the disaffected chiefs and leaders of the rebellion, issued acts of appointment to the Vanniyas of Nuvarakalaviya.[48] When the British propaganda machine pronounced that Doraisami, who had proclaimed himself to be of royal blood, was an impostor, the leadership of the rebellion fragmented. In Nuvarakalaviya, however, the Maha

45. Letter dated Mannar, 15 March 1815, from the Records of the Mannar Kachcheri, reproduced in Ievers, *Manual of the North-Central Province*, 44.

46. Letter dated 26 June 1815, from the Collector of Mannar to the Hon. John D'Oyley, Resident at Kandy, reproduced in ibid., 45.

47. For more on the rebellion, see Paul E. Pieris, *Sinhale and the Patriots, 1815–1818*, and Kumari Jayawardena, *Perpetual Ferment: Popular Revolts in Sri Lanka in the Eighteenth and Nineteenth Centuries.*

48. Ievers, *Manual of the North-Central Province*, 50.

Vanni Unnahe found a new pretender to the throne; he sent to a devale or Hindu temple in Arippu on the west coast and brought back a man he pretended was one of the royal family.[49] There is some evidence that this new plot was initiated by Pilima Talauve, the first Adigar or chief minister of Kandy at this time.[50] Pilima Talauve was alleged to have harbored the plan of exhibiting to the people this "new Phantom King" at the Dambulla vihare, which we have encountered before as one of the sites of the paintings undertaken by Kirti Sri.[51]

The symbolic value of Anuradhapura was crucial to the rebels' new plan. For instance, Governor Robert Brownrigg wrote to Earl Bathurst that he had sent a detachment to Nuvarakalaviya under the command of Captain Fraser: "My object in directing Captain Fraser's march was, by obtaining possession of that venerated spot, the ancient capital of the kingdom, to prevent delusion being spread from thence by the still obstinate chiefs who had rumoured among the people that a true scion of the royal stock had been brought thither from the Coast of Coromandel."[52] This statement speaks of the British attempt to take control of the meanings assigned by an existent tradition of historical memory. The British military marches to Anuradhapura, prompted by the rebellion, might also be cast alongside eighteenth-century attempts by the Nayakkar kings to bring Nuvarakalaviya under their power. According to a proclamation made by Governor Brownrigg in response to the rebellion, the British officers in the kingdom of Kandy would consciously seek to appropriate Nayakkar forms of ceremony and dress to impose their authority.[53]

In the years following the rebellion, the British fought hard to take control of the meanings the rebels had assigned to Anuradhapura and Nuvarakalaviya. Fraser took into his custody the two rebels, Pilima Talauve and Kappitipola, in October 1818, and the account of their taking was evidently retold many times.[54] Just before their taking the rebels were alleged to have been near Anuradhapura, and it became legendary that "when

49. Ibid.

50. Letter from Brownrigg to Bathurst, dated Colombo, 24 April 1819, CO/54/74, TNA.

51. Letter from Brownrigg to Bathurst, dated Kandy, 9 October 1818, CO/54/76, TNA.

52. Letter from Brownrigg to Bathurst, dated Kandy, 9 October 1818, CO/54/76, TNA.

53. "Proclamation by His Excellency General Sir Robert Brownrigg, given at Kandy on the twenty first day of November, 1818," CO/54/73, TNA.

54. Brownrigg to Bathurst, dated Kandy, 31 October 1818, CO/54/76, TNA.

passing the so-called tomb of Elara, Pilame Talawa, though weary and ill, alighted from his palanquin and walked until the ancient memorial of the Tamil king was passed and customary reverence shown."[55] Brownrigg also wrote to Bathurst about the capture of the Maha Vanni Unnahe, who he called "the Nuwerewewe Moodianse or Chief of the Northern part of the remote and Jungly District of Nuwere Kalawiye."[56] Doraisami went into hiding in Nuvarakalaviya and was only found twelve years later in November 1830.[57] But despite these captures, which in effect were an attempt to rid this densely forested region of outlaws, the British found Nuvarakalaviya difficult to bring under their control. In 1823, Governor Colin Campbell, who served in the island from 1841 to 1847, reported the rumor that a priest in Nuvarakalaviya had set himself up as a royal.[58] Fraser was promptly dispatched again and complained of the lack of connection between the British government and its subjects in Nuvarakalaviya.[59] Governor Campbell at first proposed placing a permanent Government Agent at Anuradhapura, but in the end he recommended that the Agent at Matale and the Agent of the Seven Korles might make periodical visits to different parts of Nuvarakalaviya.[60] John D'Oyly wrote to the Agent of Matale asking him to visit the province "at such season as may be deemed most healthy."[61]

The claim that the rebellion in Nuvarakalaviya might be seen as a contest between rebels and colonists to take control of historical symbolism is supported by their shared modes of engagement with Anuradhapura. Both sides sought to appoint representatives and to kindle and utilize loyalties to the traditions of Nayakkar kingship. In the end, the British won this contest, partly because of the extent of power they could wield over formulations of history. This context is illuminating, in coming to terms with British archaeology in the region. The physical remnants of the ancient civilization of Anuradhapura were encountered quite by chance, and in the context of military operations by the British. When T. R. Backhouse, the Collector of Mannar, was in pursuit of the rebels in 1818, he noted the ruins of Anuradhapura. He wrote in his diary where the ruins lay and

55. Ievers, *Manual of the North-Central Province*, 51.

56. Brownrigg to Bathurst, Colombo, 24 April 1819, CO/54/74, TNA.

57. See K. M. De Silva, *A History of Sri Lanka*, 193.

58. Campbell to Bathurst, dated Colombo, 5 March 1823, CO/54/84, TNA.

59. Campbell to Bathurst, dated Colombo, 16 March 1823, CO/54/84, TNA.

60. Campbell to Bathurst, dated Colombo, 30 June 1823, CO/54/84, TNA.

61. Letter dated 3 April 1823, from John D'Oyly to Lieut. Col. H. Stacpoole, Lot 21/56, SLNA.

added the comment that the name Anuradhapura denotes "the precise spot where the temple is situated."[62] The seminal paper that created public interest in Anuradhapura came much later. It is striking that it was authored by a military man. Captain I. J. Chapman of the Royal Artillery published "Some Remarks upon the Ancient City of Anarajapura," which appeared in the *Transactions of the Royal Asiatic Society* in 1832. Chapman began his paper with the sentence: "In December 1828, when quartered in Colombo (Ceylon), I joined a party on a shooting excursion, in the course of which we visited Anarajapura."[63] Military affairs, shooting, and the exploration of antiquities were perfectly congruous occupations. This was also the case for Major Jonathan Forbes, the Assistant Government Agent and district judge of Matale, another important early explorer of the ancient capital. His first visit to Anuradhapura was around March 1828.[64] His account of the ruins, including detailed descriptions of the Maha Vihare or great-temple and the sacred Bo-tree, was interrupted by this admission:

> The quantity of game in the immediate neighbourhood of the ruins was astonishing. . . . No native would have ventured to transgress the first commandment of the Budda, viz. from the meanest insect to man, thou shalt not kill. . . . While employed in examining the ruins in the presence and with the assistance of the priests, I deemed it advisable to commit no murder . . . but on the last day of our stay we left the gentlemen of the long yellow robe behind.[65]

It is intriguing to explore how the inhabitants of Anuradhapura may have responded to this combination of interests. In the Kandyan kingdom all elephants were the property of the crown. In the province of Nuvarakalaviya, the chiefs were allowed to hunt for themselves and to take possession of tusks and tuskers as their perquisites.[66] By shooting in this province, the British unintentionally projected themselves as rulers in line with a discourse of governance in Nuvarakalaviya.

British archaeology might also be cast into a longer history of attempts

62. Ievers, *Manual of the North-Central Province*, 47.

63. I. J. Chapman, "Some Remarks upon the Ancient City of Anarajapura" (1832), 463.

64. Jonathan Forbes, *Eleven Years in Ceylon: Comprising Sketches of the Field Sports and Natural History of that Colony and an Account of its History and Antiquities* (1840), 1:187.

65. Ibid., 1:218.

66. Ievers, *Manual of the North-Central Province*, 262.

to restore ancient knowledge, which follow from the Nayakkar pilgrimages to this region. While it is tempting to see British archaeology as quite unlike anything that came before, this would be an exaggeration. In fact Buddhist priests who inhabited the region also sought to restore ruined structures in the early colonial period. Forbes notes for instance that in 1828–29 the Abhayagiri dagaba was cleared of jungle by a priest "whose zeal in the difficult and dangerous task had been nearly recompensed with martyrdom, a fragment of the spire having fallen on, and severely injured, this pious desecrator of the picturesque."[67] In 1841, a Buddhist priest collected a large sum of money and restored the Thuparama dagaba; the British criticized this act of restoration on the grounds of style.[68] But regardless of this criticism, it is important to keep in mind that Britons were not alone in demonstrating an interest in the archaeology of the ruined cities in the early nineteenth century. Supposedly precolonial and colonial practices cannot be cleanly separated. It is intriguing to reflect whether local inhabitants made sense of the British interest in Anuradhapura by seeing it as an example of "merit-seeking," just as they had of Nayakkar interest in the region.[69]

However, the intensity of British interest in ruins was at a level unseen before their arrival; archaeological enquiry was accelerated with the formation of the Ceylon branch of the Royal Asiatic Society in 1845. As in India, Britons displayed an obsessive concern with numbers in their study of monuments, and they used their calculations as a means of suggesting the decay of ancient structures. George Turnour compared Major Thomas Skinner's measurements at Anuradhapura with those of the author of the Buddhist chronicle, the *Mahavamsa*:

Lieutenant SKINNER makes the Jaitawanaaraamaya dagoba in its present dilapidated state, 260 feet high; the historical account makes it 140 cubits, equal to 315 feet. The Abhayagiri dagoba, deprived of its spire and pinnacle, he makes 230 feet; according to the native account, it was the highest of all the dagobas, and measured 180 cubits, equal to 405 feet. The remains of the Ruwanwellis dagoba now measure 189 feet. By the native account it was 120 cubits or 270 feet high.[70]

67. Forbes, *Eleven Years in Ceylon*, 1:240.

68. C. E. Godakumbura, "History of Archaeology in Ceylon," 10.

69. See Nissan, *The Sacred City*, 263ff. Nissan urges that while merit-seeking motivated Buddhist priestly restoration, this contrasted with overly Christian archaeology.

70. George Turnour, *Epitome of the History of Ceylon* (1836), Appendix, 59.

Mathematical calculations took on a performative role. Chapman's paper classified the dagabas into three types according to size and included copious measurements.[71] He also mentioned how Mr. E. Layard, a civil servant, had opened a dagaba in order to inspect it. Inside, Layard found a small brick compartment which was "mathematically correct in its bearings towards the cardinal points" in addition to some thin pieces of gold and a small bone.[72]

However, mathematical calculations did not provide the only avenue for Britons' engagements with artifacts. A desire to create records and collect specimens of architecture was also shared by British explorers. Observers of Anuradhapura tried hard to legitimate their findings; notebooks and sketches were the markers of antiquarian exploration. Chapman placed great stress on the importance of his notes, and was hesitant to positively identify a dagaba he had observed as the legendary Ruvanvalisaya because his notes had been "effaced."[73] A close reading of his paper also suggests how he relied on visual illustrations to provide supporting evidence for his claims. Of one of his plates he noted, in a manner reminiscent of a surveyor: "The sketch was taken from a spot which offered least difficulty, as the weather was oppressively hot and I was much hurried; it will however serve to convey some idea of its general character" (fig. 4.2).[74] Illustrations such as this may have circulated more widely than the texts that were produced alongside them. We know for instance that a lithograph was printed from a sketch of Anuradhapura made by Mr. C. H. Cameron of the Commissioners of General Inquiry, around the time Chapman's paper was published.[75] A plate entitled "Specimens of Sculpture at Anaradhepura," also from Chapman's account, illustrates the types of representations made of ancient artifacts, and how measurements created authenticity (fig. 4.3). In addition to this type of sketch, it was also common to make copies of inscriptions found at Anuradhapura.[76] The British were therefore as visual in their engagement with Anuradhapura as were the Kandyan kings.

The connection between these antiquities and others in India and Greece was a common subject of discussion among early British observers,

71. Chapman, "Some Remarks upon the Ancient city of Anarajapura," 473.

72. J. W. Bennett, *Ceylon and its Capabilities: An Account of its Natural Resources, Indigenous Productions and Commercial Facilities* (1843), 338.

73. Chapman, "Some Remarks," 479.

74. Ibid., 473.

75. Ibid., 466

76. See I. J. Chapman, "Some Additional Remarks upon the Ancient City of Anurajapura or Anuradhapura" (1850), 175.

Fig. 4.2. "Lanca-Rama." From I. J. Chapman, "Some Remarks upon the Ancient
City of Anarajapura," *Transactions of the Royal Asiatic Society* 3 (1832): plate 19.
Reproduced by kind permission of the Syndics of Cambridge University Library.

picking up the new global context in which indigenous traditions were
contextualized in this age. The elegance of the columns, the proportions of
the ruins, and the conical shape of the dagabas attracted the greatest comment. Cameron noted: "I saw here ornamented capitals and balustrades
and bas-reliefs of animals and foliage that have nothing of the rudeness
and grotesque in the modern Singalese sculpture. I cannot better express
my opinion of their elegance than by saying that had I seen them in a
museum I should without hesitation have pronounced them to be Grecian
or of Grecian descent."[77] The writer of an article in the *Colombo Journal*
observed that the dagabas were at least the height of the diameter of their
bases—the perfect Burkeian size for the sublime.[78] However, the contrary
view—that the proportions of these ruins were rude and stupefying—was
also expressed. This is well exemplified by a writer in the *Asiatic Journal*
for 1841: "The Dagobas have a ponderous and ignoble appearance; their
magnitude is, however, almost unparalleled, and elicits the admiration and

77. Private letter dated Colombo, 24 December 1833, CO/54/131, TNA.
78. *The Colombo Journal*, 21 November 1833, 537.

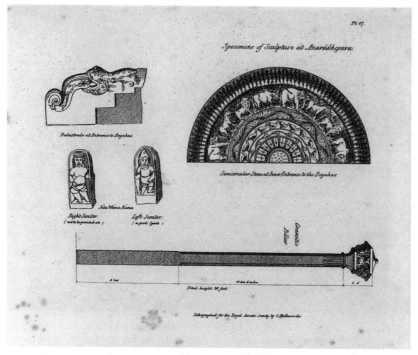

Fig. 4.3. "Specimens of Sculpture at Anaradhepura." From I. J. Chapman,
"Some Remarks upon the Ancient City of Anarajapura," *Transactions
of the Royal Asiatic Society* 3 (1832): plate 17. Reproduced by kind
permission of the Syndics of Cambridge University Library.

contempt of the European pilgrim, who may either applaud the perseverance or ridicule the injudicious taste of the ancient islanders."[79] As always
colonial rhetoric was not monolithic but multivalent and contradictory.[80]

This archaeological knowledge could not have proceeded without the
practices set in motion by Nayakkar kingship. British interest in these
sites encouraged Buddhist priests to present themselves as guides. Ievers
writes of this region in the 1830s: "Many of the caves and abandoned vi-
hares were reoccupied by priests—generally natives of the Seven Korales.
These priests frequently pretended to know the names of the places they

79. *Asiatic Journal* 35 (1841): 88–89.
80. This follows the line of a whole range of post-Saidian scholars of colonial
discourse; see, for instance, Nicholas Thomas, *Colonialism's Culture: Anthropology,
Travel and Government.*

occupied, and gave names which their reading of the chronicles sug-gested."[81] While the sources often present Europeans as lone discoverers, the brief mentions of guides, informants, and priests need to be studied closely. For instance, the carefully measured and annotated illustrations of sculptures in Chapman's paper include an image of the right janitor at a dagaba with the caption "not to be pointed to" and the left janitor with the label "a good spirit" (fig. 4.3). These incongruous titles arose out of a conversation between Chapman and a Buddhist priest who was showing him the ruins. Chapman relates that while he was making an outline of what he saw, "the priest informed me that I must not point the finger at one of them or I should get a fever."[82] It is unclear exactly why the priest said this, but the fact that Chapman had many measuring devices and that he was engaged in the act of representing ruins may have had much to do with it. A possible interpretation is that the priest objected to the aggres-siveness of British science, even though he served as Chapman's guide to the ruins.

In a later paper, Chapman recounted another visit to the ruined capi-tal: "The coolies who accompanied our party to Anuradhapura in 1829, when we arrived at a certain point in our journey, applied to be allowed a double ration of arrack, because they said on the morrow they should enter the 'Sacred District,' [in] which they could not taste spirits."[83] Chapman related this episode because it provided a means of calculating the limits of the sacred district. He noted that when his coolies had said this, they were still twenty miles from the sacred tree in the center of the settle-ment. Chapman's reliance on coolie knowledge provides us with an in-stance where a Briton relied on local and popular traditions to create co-lonial knowledge, while dismissing such discourses at least rhetorically. This notion of the "Sacred District" might be linked with an earlier his-torical memory of the settlement, illustrated for instance by the mural of the plowing of the limits of Anuradhapura commissioned by Kirti Sri.

By the late 1830s British authority in this region was becoming much firmer. As this occurred the colonizers moved beyond giving the impres-sion that they were following the precedent of Nayakkar rule. A case in point is Major Thomas Skinner's road to Anuradhapura. Governor R. W. Horton wrote to London in October 1832, explaining the need for the road.

81. Ievers, *Manual of the North-Central Province*, 213.

82. Chapman, "Some Remarks," 475.

83. Chapman, "Some Additional Remarks," 165.

He noted how a pretender to the throne, Doraisami, had secreted himself in this district until 1830 and that there had been a "partial insurrection" here. In conclusion he stated of Anuradhapura: "From its great distance from all our stations the difficulty of access, and the supposed insalubrity of the climate, which does not appear to be the case, it has been hitherto almost beyond the controul of the authorities in the interior." This letter may be read as an admission of British failure to overcome established patterns of engagement with Nuvarakalaviya. For instance, Horton admitted that the insalubrity of the district was exaggerated and merely "supposed." With the building of a new road Horton hoped that these older meanings would change. In a dispatch of the following year the governor added: "It is only since officers have been sent to trace the road that it has been fully ascertained to contain a much larger population than was suspected to exist, and which are in general, found to be in better circumstances than those of the neighbouring maritime provinces."[84]

Horton continued that despite the established view that this province was isolated, the inhabitants maintain a "very considerable commercial intercourse." Indeed, he had ascertained that some of Nuvarakalaviya's exports made their way to India. By making a road to the old capital, Horton hoped to further improve the commerce of the region. He envisaged that Nuvarakalaviya's grain could be used to alleviate the drought in the neighboring district of Mannar. By February 1833, Horton was convinced that an Assistant Agent should be stationed at Anuradhapura. The governor noted that the inhabitants have to travel fifty to one hundred miles to file a case when a crime is committed, and that they complain "with justice of the want of a resident magistrate." The decision to establish a firmer foothold of British authority in Nuvarakalaviya could not be postponed any further. Yet even in this period, it is important to pay attention to how newer forms of control arose out of the recycling of extant traditions. Major Thomas Skinner, who took charge of the building of the new road, noted how he came across "low over-grown jungle paths" used by Kandyans to access the ancient capital.[85] The newer form of British authority, exemplified by the road, arises here out of the pathways of Kandyan travel.

A similar case may be made in relation to the establishment of a District Court in Anuradhapura. On 28 June 1837, the District Judge of Anuradhapura wrote to the Colonial Secretary in Colombo: "The business of

84. Governor Horton to the Colonial Secretary, Frederick John Robinson, the first Viscount Goderich, dated Colombo, 14 January 1833, CO/54/127, TNA.

85. Thomas Skinner, *Fifty Years in Ceylon: An Autobiography* (1891), 162.

this court is much on the increase. . . . The district tho' not populous for its extent, occupies a large space of territory."[86] But despite the increasing work, it appeared that the older idea of Nuvarakalaviya as an unhealthy region filled with troublemakers persisted. This judge was keen to apply for a leave of absence on several occasions and at one such opportunity noted: "It is scarcely necessary for me to bring before His Excellency the particular disadvantages of this place which is known to him."[87] In the meanwhile, by 1842, the Medical Sub-Assistant could certify that the Assistant Government Agent at Nuvarakalaviya had been "taken ill with bilious fever accompanied with liver complaint."[88] On another occasion the District Judge noted that the province had a great many persons of "bad character" who had ended up in Nuvarakalaviya because they could not live elsewhere; he continued that many such individuals had no means of livelihood and appeared "altogether in extreme indigent circumstances."[89]

The record of cases brought before the court supports the claim of the recontextualization of the remnants of the precolonial. Just as ballads from an earlier period suggest how the reservoirs were the subject of lively legends, the proceedings of the court contain accounts of cases of accidental death, where a child was killed by falling into the water-channel of a reservoir, and where a man was killed by falling into a reservoir.[90] Another case of a woman who drowned with her child in the reservoir was reported to the court in September 1842. After being infected with measles and passing the disease on to her child aged two, "Menike Natchire" was said to have "eaten fish and foul" and become "deranged": "About four hours before daybreak, taking the child with her . . . without anybody's knowledge and having proceeded to Mahakamoomalle about the distance of a mile fell in the reservoir of that village and drowned in it."[91] This incident evidently created great consternation in the village, and in all eleven

86. District Judge to the Colonial Secretary, dated Anuradhapura, 28 June 1837, Lot 6/1445, SLNA.

87. District Judge to Colonial Secretary, dated Anuradhapura, 4 November 1839, Lot 6/1445, SLNA. See also letter dated 29 April 1840, where he notes that, after six years of residence in Anuradhapura, he is often liable to fever and other diseases.

88. J. Task, Medical Sub-Assistant, to Colonial Secretary, dated 14 June 1842, Lot 6/1692, SLNA.

89. District Judge to the Acting Colonial Secretary, Anuradhapura, dated 18 November 1839, Lot 6/1692, SLNA.

90. See papers in Lot 6/1692, SLNA.

91. All quotations from "Report submitted with due deference to the District Court of Nouwera Kalawia," dated 6 September 1842, Lot 6/1692, SLNA.

persons went to the reservoir and "saw prints of feet in the mire leading to the water." Though Manike's body was not retrieved, there was a dead alligator in the center of the reservoir, and the child's body was found. This gruesome tale is useful in hinting at how ballads such as those preserved in the Hugh Nevill collection arose. With the advent of the British and stricter surveillance of reservoirs and temples, the older idea of these ancient structures was not lost. Manike and her child drowned in the reservoir, and it is entirely plausible that this event was retold alongside the old ballads, for instance of Kadawara Deva.

THE SACRED MOUNTAIN

If the British followed and reinterpreted Kandyan traditions of engaging with Anuradhapura, such an entanglement is much easier to establish in the case of Sri Pada, which arguably attracted more attention on the part of both the last kings of Kandy and the early British explorers. One reason for its preeminence was its visibility from various locations, both in the interior and from the sea. This is the mountain variously called, Sri Pada, to denote the footprint at its top, said to be that of the Buddha or Siva, and also Samantakuta or Sumantakuta in honor of its guardian god Saman or Sumana. British travelers routinely called the mountain Adam's Peak, following the Islamic tradition of venerating the footprint as belonging to Adam. A *sannasa* dating from 1751 in the reign of Kirti Sri describes this mountain as

> a crown of blue sapphire gems, worn upon the head of the lady of the glorious Island of Lanka, beautified with various rivers and cataracts, filled with clear water of cool springs, adorned with groves of multitudes of noble trees, loaded with flowers, and enriched with much sweet fragrance of well blossomed filaments.[92]

Thus the Kandyans utilized a rich store of language to describe the mountain, and the British followed suit. From the confines of Colombo, Britons gazed at Adam's Peak and came to see it as an emblem of the interior. J. W. Bennett wrote of the "indescribable pleasure" of seeing "the lofty

---

92. William Skeen, *Adam's Peak: Legendary, Traditional and Historic Notices of the Samanala and Sri Pada with a Descriptive Route from Colombo to the Sacred Footprint* (1870), Appendix C.

Fig. 4.4. "View of Adam's Peak, and Point de Galle." From Viscount
Valentia, *Voyages and Travels to India, Ceylon, the Red Sea, Abyssinia
and Egypt in 1802–1803* (London, 1809), facing page 266. Reproduced by
kind permission of the Syndics of Cambridge University Library.

mountain of the *Sri Pada* or Adam's Peak" soon after daybreak.[93] Viscount
George Valentia in his much read *Voyages and Travels* provided a panoramic image from the harbor of Galle drawn by Henry Salt, the painter
who accompanied him, showing the mountain rising majestically in the
distance (fig. 4.4). Adam's Peak became an established feature of British illustrations and imaginings of Ceylon.

Sri Pada's importance as a Buddhist site is evident from the large number of pilgrims who in the eighteenth century sought to climb it and worship the imprint of the foot at its top. In traveling up the peak these pilgrims followed a long-established tradition, associated with the kings of
the island and recorded in Buddhist texts.[94] Nissamka Malla (r. 1187–96) is
the first monarch to be described as a pilgrim himself in the *Mahavamsa*:
"With the four-membered army the Ruler full of pious devotion, went

93. Bennett, *Ceylon and its Capabilities*, 158.
94. For a description of royal pilgrimages in Buddhist chronicles, see John M. Senaveratna, *Sri-Pada (Adam's Peak): Historic Pilgrimages to the Sacred Footprint.*

forth to the Samantakuta and performed there his devotions."[95] Kingly
pilgrimages to Sri Pada, like those to other outlying sacred sites, sought
to tie the island together as a spiritual and political whole. Extant rock
inscriptions suggest that these monarchical journeys were remembered by
those who followed. One such inscription from a cave called Bhagalavena
(or Bhagava), one hundred feet below the summit of Sri Pada, shows the
outline of a man, likely to be Nissamka Malla, with the inscription "This
is the manner in which King Nissamka Malla stood worshipping the foot
print."[96]

In the reign of Rajasimha I (r. 1581–93), a king who allegedly gave up
Buddhism for Hinduism, the mountain came under the control of Hindu
Saivites. According to the *Sulurajavaliya*, written around 1820, Rajasimha
transferred all the income from the peak to "heretic Andis," immigrants
from India and possibly from Andhra lands.[97] The *Mahavamsa* records
how Rajasimha I became a "dead tree truck in the cycle of rebirths" and
adopted "a false faith." "He placed miscreant ascetics of false faith on the
Sumanakuta to take for themselves all the profit accruing therefrom."[98] It
is perhaps from this period, and not from earlier, that the footprint became
associated with Siva rather than the Buddha.[99]

Later kings retreated to the peak several times, and the mountain receives repeated mention in the *Mahavamsa*'s description of the last decades of the Kandyan kingdom. The monarch whose pilgrimage receives
the most elaborate discussion is Vimaladharma Surya II (r. 1687–1707),
who held a festival that lasted seven days on the peak, and who made offerings of jewels and pearls, gold and precious stones.[100] The *Narendra Cari-*
tavalokana Pradipikava notes that this king covered the sacred footprint
with a "large silver parasol white in color."[101] Vimaladharma Surya's son
Narendrasimha, who went to Anuradhapura, went to the peak twice, and

95. Geiger, *Culavamsa*, 2:80, line 24. For an account of this inscription, see H.A.P.
Abeyawardana, *Heritage of Sabaragamuwa*, 125. The *Rajavaliya* notes how Parakramabahu I (1153–86) also became a pilgrim to the mountain; see Senaveratna, *Sri-Pada*, 16.

96. Senarat Paranavitana, *The God of Adam's Peak*, 15.

97. From a translation undertaken by Prof. Udaya Meddegama of the University of
Peradeniya, Sri Lanka from D.P.R. Samaranayake, *Sulurajavaliya*, under the description
of Narendrasimha.

98. Geiger, *Culavamsa*, 2:93, lines 11–13.

99. Paranavitana, *The God of Adam's Peak*, 21.

100. Geiger, *Culavamsa*, 2:97, lines 16–17.

101. From a translation undertaken by Prof. Udaya Meddegama of the University of
Peradeniya, from P. A. Hewavitarana, ed., *Narendra Caritavalokana Pradipikava*.

this is mentioned in the context of his journey to Anuradhapura.[102] It is interesting that there is no mention of how early Nayakkars came to terms with the Hindu Saivites on the mountain, given their own heritage. Yet with Kirti Sri's reign, the *Mahavamsa* provides the following narrative:

> The wicked king known by the name of Rajasinha, in the town of Sitavaka who had committed patricide and destroyed the Order of the Victor, as he could not distinguish what it was right to do, adopted a false faith, was devoted to the adherents of the false faith and ordered them to take for themselves the income accruing from the worship of the sacred footprint of the Enlightened One on the Sumanakuta. . . . When the highly famed Great king [Kirti Sri] heard of these things, he realised, reverently devoted to the Enlightened One, that this was unseemly. He commanded the adherents of the false faith from now onwards not to do so, and charged the sons of the Buddha to carry out in the right way the sacrificial ceremonies which should be performed there.[103]

A critical reading is necessary here, given that the *Mahavamsa* was written by Buddhist priests and that the Nayakkar line were Hindus by birth who took on Buddhist beliefs and the patronage of Buddhism. Evidently, the peak was tied up with the complicated relationship between the court and the priests, and with the tense cohabitation of Hinduism and Buddhism in the Kandyan kingdom.

Kirti Sri's intervention in the history of the mountain is further complicated by the fact that he placed it in the charge of Valivita Saranamkara, the monk who reorganized Buddhism in Kandy. The 1751 *sannasa* notes that Kirti Sri granted the village of Kuttapitiya as an offering to the peak. It makes special mention of how the village contains a "sowing extent of which is one hundred and sixty-five amunams of paddy," which presumably indicates how it could provide for pilgrims on the way to the mountain.[104] It ends by noting how "Lord Saranankara of Weliwita" was given in charge of the peak, and that his pupils would keep up offerings to the mountain. The fact that Saranamkara was appointed to the peak in 1751 suggests how the Silvat Samagama that he led was rising in courtly esteem; two years later it would be instituted as the Siyam

102. Geiger, *Culavamsa*, 2:97, lines 30–34.
103. Ibid., 2:100, 220–27.
104. Skeen, *Adam's Peak*, 299.

Nikaya with the arrival of Siamese priests. But in these crucial years of this fraternity's extraordinary evolution, it is plausible that the court was still dominated by the aristocratic monks who were Saranamkara's rivals. Because of this, the granting of an outlying sacred site such as Sri Pada to Saranamkara made perfect sense in terms of both religion and politics.

Under Saranamkara's instructions the peak was given to the management of three of his low-country pupils: Malimbada Dhammadhara, Vehalle Dhammadinna, and Kumburupitiye Gunaratna.[105] Eventually the peak came to the lot of Karatota Dhammarama, who we encountered in the last chapter as a key mediator and who was deprived of the charge of the mountain because of his friendship with Europeans. At this point, the peak reverted to the authority of the center and to the Malvatta priests in the city of Kandy.[106] The Kandyan history of Sri Pada shows how the mountain, by its position in the peripheries of the kingdom, could serve as a place for new religious traditions to be affirmed and for political differences between center and periphery, allies and foes, to be played out. Kirti Sri could use it to propagandize his glorification of Buddhism; Saranamkara could have it appointed to his charge in the time of his increasing authority, and Karatota could lose it because of his closeness to Christian colonists. Sri Pada was thus an icon that could be wedded to the assertion of Buddhist piety.

As in the case of Anuradhapura, a series of other rituals and symbols were important in publicizing the status of the mountain as a sacred site. These included artifacts, the perahara, temple paintings, and popular ballads. Two artifacts were particularly important in the later Kandyan kingdom and in the early years of British control: Kirti Sri's *sannasa* of 1751 and a pair of elephant tusks, said to have been given to Kirti Sri by the Saivite priests who had control of the mountain in an effort to make the king change his mind.[107] Kirti Sri's *sannasa* of 1751 was taken to Kandy, when Karatota was deprived of the peak, as a title deed of ownership and placed in the Malvatta Vihara. John Davy, in his account of his ascent of Adam's Peak, provided a transcription of a "curious Sanus," which may have been this one, which shows how a *sannasa* could become an object of

---

105. K. Malalgoda, *Buddhism in Sinhalese Society, 1750–1900: A Study of Religious Revival and Change*, 85.

106. Letter from George Turnour to the Board of Commissioners of Kandy, dated 27 February 1827, reproduced in Skeen, *Adam's Peak*, 359.

107. Skeen, *Adam's Peak*, 40.

antiquarian interest and could come to symbolize the mountain.[108] Meanwhile, Kirti Sri accepted the pair of elephant tusks as a gift and in turn gave them as an offering to the mountain. When the British took control of the kingdom the tusks were taken back to Kandy, where they could be viewed. George Turnour reported that "Sree Pada" was carved on them, presumably indicating an image of the peak or its name; he recommended that they be restored to the priest who was to be appointed in charge of the peak.[109] The movement of the *sannasa* and the tusks between the mountain and Kandy, or the periphery and the center, show how objects could signify the Buddhist site and come to be tokens—or even relics—of ownership.

Again as in the case of Anuradhapura, the mountain might have become linked with the region in which it lay; Sabaragamuwa constituted a distinct *disawani*, which was made up of various other *korale* and symbolized by a set of banners. The province of Sabaragamuwa took a place of pride in the annual perahara in Kandy. Of particular note here is the role played by the people of Hatara Korale. This *korale* was well known for its brave men, who provided security and protection to the royal palace at Kandy. In the perahara, the people of Hatara Korale took a prominent place as an acknowledgement of this bravery.[110]

Kandyan temple paintings point to how the mountain was adopted into a tradition of popular pilgrimage. Sri Pada represented two of the *solosmasthana*: the Buddha's footprint counted as one sacred site, and the cave in which it is said that he rested at the foot of the mountain, called the *Divaguha*, counted as a second. Given that Sri Pada was a mountain, and that most of the *solosmasthana* were stupas, its depiction presented an opportunity for the Kandyan artist to experiment with new techniques. At the Ridi Vihara, for instance, the footprint and the cave were combined into one sacred site and the unusual colors of grey and green were utilized to depict them. The guardian god Saman is displayed hovering over the mountain and worshipping the footprint. It is interesting that the Buddha is not depicted here: he is represented by the footprint, which appears almost as part of the summit, and by the curtain which is spread across the entrance of the cave where he rested. It is as if nature has molded its

108. John Davy, *An Account of the Interior of Ceylon and of its Inhabitants* (1821), 348–49.

109. See Skeen, *Adam's Peak*, 40; letter from George Turnour to the Board, reproduced in Skeen, *Adam's Peak*, 360.

110. These details are taken from Abeyawardana, *Heritage of Sabaragamuwa*, 3.

form to receive the sacred visitor. The depiction of the footprint and the cave at Madavala Vihara again shows how nature was compliant to the teachers' wishes: trees stand above the sleeping Buddha, and the footprint is safely deposited at the top of the mountain, as if forming part of it.[111] The image of Sri Pada at Lankatilaka Vihara, from the reign of Kirti Sri, is rich in both animals and plants ranging from fruit trees to deer, elk, wild boar, monkeys, squirrels, and parrots. The steps of the mountain are shown being climbed by an elephant, which symbolizes Saman.[112] These striking images and others like them in regional temples would no doubt have drawn the attention of those who visited the temples and would have placed the mountain quite firmly in a typology of sacred sites.

In addition to its connection to the visit of the Buddha, the mountain also supported a range of other legends. Foremost among these was probably its association with its guardian god Saman, who is also one of the four guardian gods of the whole island. This deity is not merely a regional or local god, like those discussed in the case of Anuradhapura. Saman continues to symbolize the mountain to pilgrims from all regions, who seek to worship him. A statue of Saman was installed at the peak in the thirteenth century, and a small shrine to Saman still exists on the mountain, at a lower level to the footprint. About nine miles from the mountain, there is a shrine devoted to Saman at Sabaragamu, called Saman Devale. At this devale, there is an annual perahara in honor of Saman. This perahara is only second to that undertaken in Kandy, and was noted by early British explorers; in its contemporary incarnation, it combines a procession in honor of the Tooth Relic and processions in honor of Pattini, Bisodeva, and Kumaradeva, in addition to one in honor of Saman. Saman is represented in the rituals and procession by the figure of an arrow.[113] Senarat Paranavitana has argued that the god Saman is equivalent to the Yama of the Indian ancient literature, the destroyer, and that it is likely that Saman was worshipped by the early settlers of the island prior to the advent of Buddhism. Further, Paranavitana contends that settlers from North India would have seen Sri Pada from the sea and would have thought the mountain to be the home of Yama, the guardian of the South.

---

111. This paragraph relies on images from SinhaRaja Tammita-Delgoda, *Ridi Vihare: The Flowering of Kandyan Art*, 91–94.

112. See, for instance, the image on the cover of Abeyawardana, *Heritage of Sabaragamuwa*.

113. Paranavitana, *The God of Adam's Peak*, 23, also 40. For a contemporary account of the perahara, see Abeyawardana, *Heritage of Sabaragamuwa*, 127–28.

For our purposes what is important is how Saman's worship on the mountain and the mountain's proximity to Saman Devale point to the hybrid layers of religious meaning attached to this site. These meanings were accessible on site but were also dispersed across the island.

A late-eighteenth-century *sandesa*, a poem called the *Katakirili Sandesa*, imagines a bird going from the city of Kandy to pray to Saman, to get permission to worship at the pilgrimage site of Kataragama and to pray to the god Skanda. The author describes the path taken by the bird from Kandy to the mountain. Lord Buddha is cast as a universal monarch in conversation with the god Saman. "When he placed the Foot Print that is like the full moon / Rays emitted from his [body] and illuminated the entire world."[114] The description of Saman points to how the god, despite his status as protector of the entire island, takes on board aspects of geography specific to Sri Pada. His dark blue complexion is said to resemble the ornamental arches of *indranila* stones in Sabaragamuwa. The popular sixteenth-century poet Alagiyavanna Mohottala notes, "The pair of long blue eyes of the divine king Sumana, which shines like priceless *indranila* gems and *indivara* flowers [blue water lilies], are verily two beautiful fishes which move about in the river of his compassion extended equally towards all beings at all times."[115] It is suggestive to consider whether the appearance of precious stones in descriptions of Saman here, and also in the description of the mountain in Kirti Sri's *sannasa* of 1751, which was cited at the head of this section, arises out of the fact that the region around the peak is well known for its precious stones. It is also pertinent, given the metaphor, to note that four rivers, including the island's longest, take their origin from Sri Pada.

The various meanings attached to the mountain, and the resonance of those meanings both at the court and in various regions of the island, are borne out in relation to the accounts written by early British explorers of the observances of the pilgrims they witnessed climbing the peak. Lieut. Malcolm of the first Ceylon Rifle Regiment is the first Briton to have climbed the mountain, a few weeks after the fall of the kingdom of Kandy on 26 April 1815. He saw about two hundred pilgrims on his passage up the peak, who were traveling to "the sound of Tam-a-tams and

114. Verse 16, *Katakirili Sandesa*, from a translation undertaken by Prof. Udaya Meddegama of the University of Peradeniya, Sri Lanka. This poem may be compared with a late-sixteenth-century *sandesa*, the *Savul Sandesa*, which portrays a cock traveling to the shrine of Saman at "Saparagamua."

115. Paranavitana, *The God of Adam's Peak*, 53.

other instruments of Singhalese music."[116] Two years later, in a letter to
his brother, John Davy described his ascent and documented the number
and diversity of the pilgrims who ascended the peak:

> The influence of religion on the minds of the natives is well exem-
> plified in the immense number of pilgrims that annually ascend this
> steep and rugged mountain. The number must amount to many thou-
> sands. We saw, at least, two or three hundred. They were of all ranks
> and descriptions of people, from the highest to the lowest casts, women
> as well as men: all ages, from the child that was carried on its father's
> back to the grey-headed tottering old man, that could not ascend with-
> out support. The object of their worship is a strong example too of the
> lowness of their faith, and their amazing credulity.[117]

Despite the constant scorn that Britons poured on the belief that the impression at the top of the mountain was indeed a footprint worthy of devotion, British ascents of the mountain were framed by the history of previous
pilgrimages. The account of an officer who climbed the mountain shortly
after Lieut. Malcolm bears this out very clearly. This officer, Captain Anderson, made ready use of the acts of kingly patronage that had sought to
ease the passage of travelers up the peak. He resided at spots where travelers regularly sheltered, made use of chains in his climb, which he noted to
be "donations to the Temple, and the name of the donor is engraved on one
of the links made solid for that purpose." He also observed the *ambalam*
or resting place on the way, and the ruins of a building erected by Ahalepola when he was the Disava of Saffragam, and who we encountered in the
previous chapter as the minister who fled Kandy. Anderson's awareness of
all those who had gone before him is perhaps best illustrated by his observation that there were "many holes in the face of the rock" formed "by the
feet of the numerous pilgrims who have ascended."[118]

Major Forbes, who we encountered as an explorer of Anuradhapura, ascended the peak after residing with George Turnour, the translator of the
Mahavamsa, who was Agent at Ratnapura, the closest town to the peak.
His account of ascent emphasized the number of traditions associated
with the peak, and in an aside he commented that every stone and bank
had a legend attached to it. He also included an account of the god Saman

116. Skeen, *Adam's Peak*, 338.
117. John Davy, "A Description of Adam's Peak" (1818), 27.
118. All quotations from Skeen, *Adam's Peak*, 341–44.

and Saman Devale, and noted the presence of "a cheerful party of respect-
ably dressed Mohammedan pilgrims of both sexes" and two "ill-fed, and
certainly ill-favoured, sinister-looking persons, in the dress of Hindus."[119]
Charles Pridham's narrative of the peak also pointed to its hybrid religious character and in particular made the point that when the pilgrim
or traveler was "on a throne of clouds" one could not resist being "led to a
contemplation of the source whence all this grandeur [had] originated . . .
whether Buddhist, Mahommedan, Hindoo or Christian."[120] This goes back
to a verse in the *Katakirili Sandesa*:

> Talking happily in different tongues such as Tamil, Sanskrit, and Pali,
> Many learned men endowed with knowledge like Brahaspati,
> Having seen the famous [foot] print, sprinkle scent over it.
> Crowds have gathered for Bodhi puja from the ten directions.[121]

Britons who climbed Adam's Peak could not see themselves as lone discoverers given the sheer mass of pilgrims who kept them company. If approached from this context, the rather bold statement that Lieut. Malcolm
made at the end of his ascent can take on a new meaning. In his journal,
Malcolm wrote: "We were not, I regret to say, provided with a 'Union Jack,'
but we fired three vollies, to the great astonishment of the Buddhists, as
a memorial to them that a British armed party had reached the summit,
spite of the predictions of the priest of Palabadoolla."[122] Such a statement
can quite easily be interpreted as a gesture of power and aggression, and it
was probably interpreted in that light by onlookers. But it may also be seen
as an expression of British insecurity—might Malcolm have been forced
to stamp his presence on the summit, and have been challenged to do so
by the number of climbers and the wealth of traditions attached to the
peak? In referring to a priest who had predicted that he would not ascend
the peak, Malcolm returned to an earlier event in the ascent, where he had
met a priest who told him that "no white man ever did and never could
ascend the mountain."[123] Rather ironically, immediately following the fir-

119. Forbes, *Eleven Years in Ceylon*, 1:169.

120. Charles Pridham, *An Historical and Statistical Account of Ceylon and its
Dependencies* (1849), 2:617.

121. Verse 26, from a translation of the *Katakirili Sandesa* undertaken by Prof.
Udaya Meddegama.

122. From Skeen, *Adam's Peak*, 340.

123. Ibid., 338.

ing of volleys, Malcolm decided to honor another priestly prophecy, which this time was about impending rain. He wrote: "We *had* some faith in that warning, and made the best of our way down the mountain."[124] British authority on the mountain was clearly no straightforward imposition.

The role of Buddhist priests in facilitating and serving as a context for British ascents of the mountain is also evident from the account of the surgeon Henry Marshall. Marshall wrote of how he climbed the peak with Simon Sawers in 1819, taking a route seldom used by pilgrims. Yet he notes that "two chiefs" with about one hundred followers climbed the peak two years earlier on this route; this is interesting because it suggests again the continuance of supposedly precolonial traditions in the very period when Britain took Kandy. Marshall's engagement with the Buddhist priests appears in this extract:

> Immediately upon our reaching the summit of the Peak, the senior priest waited upon us, and made many inquiries respecting our health, &c. Having learned that we intended to remain there all night, he most earnestly recommended and entreated us to alter our determination in that respect; he said we should certainly be visited by sickness if we remained on the hill all night. . . . But in a very short time he returned, bringing with him a handful of dried plants, a portion of which he gave to each of us. He took great pains to impress us with a belief in the virtues they possessed to prevent disease, when worn as an amulet.[125]

This encounter is similar to that recorded by Malcolm—there is a warning that harm may come to the British from the mountain. But when this warning is not heeded, there is no retreat on the part of the Buddhist priest, but rather the presentation of other information and assistance to the visitors. Marshall's ascent, as much as that of Malcolm, was characterized by both a critical meeting of traditions of travel and an entanglement of indigenous and colonial meanings.

As in the case of Anuradhapura, Britons on their way up Adam's Peak and at its summit displayed their usual obsessions: hunting, measurement, and science. A later climber, William Knighton, who served as Secretary of the Royal Asiatic Society of Ceylon, takes on an idiosyncratic and dramatic style in his account, which is full of derision for the footprint. He is

124. Ibid., 340.
125. Henry Marshall, *Ceylon: A General Description of the Island and its Inhabitants* (1846), 236.

also distracted by the possibility of hunting along the way.[126] The importance of measurement is borne out from the very start of British explorations. Malcolm measured the summit of the peak and the footprint; Davy kept up measurements of the barometer and thermometer while on the peak and sought to better the measurement of its height by estimating that it was 7,000 feet.[127] Lieutenant De Butts wrote of how the mountain, because of its conspicuous position, had been of "eminent service as a trigonometrical point in the survey of the island."[128]

Whereas Britons turned to archaeology in the case of Anuradhapura, on the mountain the favored science was natural history. Davy wrote of the sacred trees on the mountain and in particular of a "new species of rhododendron," and provided a description of its appearance.[129] Another climber who came to have her account circulated widely was Mrs. Walker, the wife of Col. Walker, who climbed the mountain for the second time in 1833, having done so in 1820. Her journal was reprinted by William Hooker, the eminent botanist, in the *Companion to the Botanical Magazine*.[130] The Walkers were inveterate plant hunters and went up the peak equipped with all the necessary instruments. Their description of the peak was theorized in relation to current ideas of biogeography: "In different regions, of equal height, we have observed plants of the same family, and even genus to abound, but rarely of the same species." Looking back down at the coast, they observed: "With the glass we could distinctly observe the fringing of the cocoa nut trees round the sea coast."[131] The interest in natural history took up the idea that the mountain was a sacred garden.[132]

The Walkers' own climbing party was composed of forty people, including thirty coolies, four servants, a soldier, a headman, and rather tellingly a Buddhist priest with a boy who was his attendant. There are several anecdotes from the journal which bear out how this journey was enrolled as an account of piety by those who witnessed it. When visiting a temple near Ruvanvalla, on the way up, Mrs. Walker noted how its incumbent priest may well have believed her to be a proselyte, because she was climbing the peak for a second time, and therefore presented her with a gift.

126. William Knighton, *Forest Life in Ceylon* (1854), vol. 1, chap. 7.

127. Davy, *An Account of the Interior of Ceylon and of its Inhabitants*, 347.

128. Augustus De Butts, *Rambles in Ceylon* (1841), 237–38.

129. Davy, "A Description of Adam's Peak," 26.

130. Mrs. Walker, "Journal of Mrs. Walker and Col. Walker to Adam's Peak in 1833."

131. Ibid., 12, 9.

132. See Pridham, *An Historical and Statistical Account*, 2:613.

The couple's mode of travel is also significant: Col. Walker sat in a chair, which was elevated on the shoulders of his bearers, while Mrs. Walker was carried in a palanquin. The Walkers thus traveled like the nobility of the Kandyan kingdom. They used the *ambalam* on the way, like other pilgrims, and even noted a "rude figure traced on the rock, said, by the natives to be the picture of pious Raje, who had the steps cut for the benefit of the pilgrims." This was probably Nissamka Malla. Their party saluted the peak with "salaams, and sometimes prostrations, whenever it came in sight." Yet the most interesting point in the narrative where pilgrimage and exploration are intertwined is the account of one of the Walkers' treasured scientific instruments, a thermometer, which they left behind by mistake near the footprint at the top of the peak. Their cook told them that he had noticed the thermometer by the sacred footprint, "but he did not touch it, as he thought [the Walkers] had left it there designedly (as an offering to Boodh)."[133] Even the relics of the new mode of modern travel are here inscribed with radically alternative meanings.

Why was the climb of Sri Pada so iconic? In writing of the reason why he wanted to climb the peak, Knighton noted: "I had determined, from the first moment that I saw Adam's Peak, when at Point de Galle, to ascend it, if the ascent were possible."[134] The ascent of Sri Pada became a staple in British travelogues, and it attained status among the sites that adventurous Britons sought to discover. It is curious to recall the thesis that it was its visibility that may have led to its eminence in the traditions of Saman, Buddhism, Hinduism, and Islam in turn.[135] In seeking to climb the mountain, and in being prompted to do so by seeing it from the coasts, Britons followed an established mode of reverential sight. The difficulty of access made this climb a sacrifice, or, in other words, an "unpleasant sedative on the ardor of the unbelieving but inquisitive Christian."[136] Their climbs may not have been motivated by religion, but perhaps unwittingly at times, their journeys became bound up with religious observances. From the point of view of their helpers, the British climbs of Sri Pada may well have fitted into a tradition of chiefly, noble, and kingly pilgrimage.

133. Walker, "Journal," 4, 6–7, 10, 11, 13.

134. Knighton, *Forest Life in Ceylon*, 1:218.

135. See Paranavitana, *The God of Adam's Peak*, Conclusion.

136. This is the view of De Butts, *Rambles in Ceylon*, 236. See also Pridham, *An Historical and Statistical Account*, 2:612, for why the route to the peak was not made easier, for fear of decreasing the "merits of pilgrimage."

CONCLUSION

If translation and language study were central to the lineage that linked Saranamkara to the British scholars, the practice of travel as a means of asserting piety and political control was another way in which Kandy and Britain were imbricated. Yet it is tempting to treat the ritualized motion of pilgrimage and the seemingly detached traditions of scientific travel as opposing norms separated by religiosity and secularism. Jaś Elsner and Joan-Pau Rubiés write: "The cultural history of travel is best seen as a dialectic of dominant paradigms between two poles, that we might define as the transcendentalist vision of pilgrimage and the open-ended process which typically characterizes modernity."[137] Recent work in the field has shown how pilgrimage has left a long legacy on modern cultures of travel in the search for spiritual fulfillment, the imposition of mythic readings and interpretations on physical terrain, and the act of collecting mementos. At the same time both pilgrimage and travel, by their kinetic qualities and by their aim of interpreting the new by way of the familiar, are related practices of knowledge creation among communities.[138] This movement was a central feature of colonial transition, both in literal terms as travelers sought to consolidate new political power, but also in metaphorical terms as they recycled the indigenous in the passage to colonialism.

Such a claim works well for this study of the links between Kandyan and British practice. The making of British narratives of objective discovery—in genres as distinct as archaeology and mountain climbing—fed off the culture of pilgrimage in South Asia. This is reminiscent of the importance of the accounts of pilgrimage, in particular of Chinese pilgrims, for British archaeology in India in the same period.[139] One aspect of pilgrimage that left a stamp on British colonial politics was its characteristic of linking geographically dispersed regions around an idea of collective unity in the island form. Early British travel in Lanka was also an attempt to bring unruly territories such as Anuradhapura to order. Yet alongside the operation of such a metropolitan discourse of territorialization was the

137. Jaś Elsner and Joan-Pau Rubiés, eds., *Voyages and Visions: Towards a Cultural History of Travel*, 5.

138. See, for instance, Simon Coleman and John Elsner, *Pilgrimage, Past and Present: Sacred Travel and Sacred Space in World Religions*, esp. "Epilogue," 196–220. Also, Simon Coleman and John Eade, eds., *Reframing Pilgrimage: Cultures in Motion*.

139. See Upinder Singh, *The Discovery of Ancient India: Early Archeologists and the Beginnings of Archaeology*.

fact that pilgrimage sites accommodate—and perhaps even encourage by the act of drawing diverse pilgrims—multiple traditions of religious and cultural meaning. Sri Pada's meanings, ranging from Buddhism to Hinduism and on to the cult of Saman and the idea of Adam's Peak, are an excellent instance of this. The local, the regional, and the metropolitan pulled at the monopoly of meaning over the island space.

When this chapter is placed alongside the last, it is possible to see that the kind of historical work that the British undertook in Ceylon encompassed a wide range of material forms—textual, visual, and artifactual. In Kandy, history was linked to acts of performance whereby the king associated his greatness in relation to a past line through processions, translations, and artistic and religious patronage. Following Kumkum Chatterjee's work on the transition to colonialism in India, history was a heterogeneous enterprise for both sides of the encounter, and there was not a radical break in the culture of history in the subcontinent with the advent of the British.[140] It is important to expand the history of history to take account both of eclectic materials and of the diverse traditions both colonizing and colonized, in various combinations, that were encompassed under understandings of the past. Scientific history certainly became dominant with the passage of the nineteenth century. Yet European historiography did not arrive to a blank slate, and history as a practice was not exclusively a product of the European Enlightenment or of nationalist invention in the later nineteenth century.

The study of Sri Pada provides a good introduction to the direction of the discussion that follows, which takes us into another order of knowledge politics, namely that connected with land and nature. Yet at this point of transition in the argument, it is important to keep in mind that religiosity was itself landed and spatialized, and the distinction between the religious and the environmental should not be overdone. In other words, "the site" should not be defined too narrowly or in disaggregated ways where the physical and the cultural are separated. The connection between religiosity and natural history has already been pointed to by the fact that Adam's Peak became a prime location for the botanists and natural historians of Ceylon, who are in central view next.

140. Kumkum Chatterjee, *The Cultures of History in Early Modern India: Persianization and Mughal Culture in Bengal.*

CHAPTER FIVE

Gardens

In the nineteenth century, islands were vital places for the emergence of new scientific ideas and of a consciousness of nature, partly because they held so many treasures which astounded naturalists.[1] Sri Lanka was no exception; indeed, the manner in which the island was naturalized by practitioners of science, around ideas of luxury, tropicality, and paradise, served as one of the means through which it was separated off from the mainland. Yet the making of this natural knowledge, like colonial understandings of what was called Buddhism, fell into existent ideas and practices within eighteenth-century Lanka, even as it recontextualized the terms of what counted as the "indigenous."

The story of the rise of modern science has too often been narrated through the lens of the European Enlightenment.[2] The story tells of the emergence of a number of discrete disciplines in the nineteenth century, such as botany or physics, out of the broader framework of early modern natural history and natural philosophy. This meta-narrative is still enveloped in colonialist rhetoric, which saw science as one of the West's benefactions to the rest of the world. One way in which the story can be retold is through what I have called elsewhere a process of "cross-contextualization," where the copious theories and archives of European scientific history are read alongside the more scarce material traces of non-European

---

1. See, for instance, Richard Grove, *Green Imperialism: Colonial Expansion, Tropical Island Edens and the Origins of Environmentalism, 1600–1860.*

2. For a recent widely reviewed contribution, see Kapil Raj, *Relocating Modern Science: Circulation and the Construction of Knowledge in South Asia and Europe, 1650–1900.*

peoples.[3] The palm-leaf texts of Lanka present a valuable opportunity to reconsider the arrival of British science in relation to existent ideas of the body, the land, and the stars. Such an enterprise should not count as an attempt to find the "indigenous" in these sources, but to come to the place of science in colonial transition from a diversity of points of view.

A running theme of the discussion thus far has been the idea of mediation—this was evident in the role of Karatota Dhammarama, or indeed of the other Buddhist priests who served as guides and helpers to the explorers of Anuradhapura or the climbers of Sri Pada. This chapter focuses on one intermediary—a Sinhalese botanical artist, Harmanis De Alwis Seneviratne. Indigenous knowledge—a term which has had some currency in the literature—is ultimately an unhelpful label in coming to terms with the information provided by such guides, for this term creates a monolithic category, which is set against Western traditions. There was a variety of trajectories of knowledge, both colonizing and colonized, which came into intensive contact and combat as the British took over Ceylon. As James Duncan, the historical geographer of Ceylon, writes: "Knowledge practices and techniques flowed into the colonies and were transformed there not as a new imposed order but as newly evolving local and extra-local forms."[4] In Ceylon, slippages between different orders of natural knowledge gave way to a consolidation of British state power. De Alwis Seneviratne himself rose up the state structures and eventually traveled outside Lanka.

In line with the general argument of how Lanka was islanded by the British, who took up and deepened extant notions of unity as they pertained to the land, this chapter studies the consolidation of a contradictory discourse to islanding, which sought to segregate the highlands from the lowlands. This discourse was used by the Kandyans, for despite their claims to universal sovereignty on the island, they sought to guard access to the higher altitudes. The island's inhabitants more generally used a distinction between *uda rata* and *pata rata*, and these words might be translated literally as "up country" and "low country." This rhetoric was stretched further by the British and aligned with a colonial distinction between the temperate and the tropical. Yet why is this an indicator of

3. See Sujit Sivasundaram, "Sciences and the Global: On Methods, Questions and Theory," as well as my introduction to the special section "Global Histories of Science," *Isis* 101 (2010): 95–158.

4. James Duncan, *In the Shadow of the Tropics: Climate, Race and Biopower in Nineteenth-century Ceylon*, 190.

islanding? Throughout the function of this discourse, the highlands and the lowlands were knit together as a united field with oppositional topographies, both on the part of the islanders and the British. For the British, the taking of Kandy was supported by the rhetoric of freeing its subjects from isolating tyranny, and once the entire island was in their power, the opposition of the highlands and lowlands was inverted so that the higher reaches became the terrain of civility and healthfulness. Political power sought to reach beyond the topography of the highlands or the lowlands, even as it utilized the distinction to frame the other side, and the evidence of supremacy lay in political mobility across the land.

To return more particularly to knowledge about nature: one Kandyan king, Kirti Sri Rajasimha, hoped to unite his kingdom under his own spiritual rule, in part by establishing gardens attached to Buddhist temples and utilizing natural historical symbols, alongside other acts of benefaction. At the same time, the British also set up botanical gardens on the island, and these followed their political ambitions. The first garden was in Colombo, the second was based south of the city in Kalutara, and then after the taking of Kandy, the main botanical garden moved to the interior. The political fallout between the British and the Kandyans might be cast in parallel to the entanglement of two different networks which sought to encompass the natural history of the island. One way of expressing or consolidating a hold on territory was by mobilizing natural historical knowledge from the political center to new areas of control.

Yet to pick up the problem of using "indigenous knowledge" as a category, the distinction between Kandyan and British natural history should not be overplayed. For these were not discrete bodies of knowledge; on the side of the islanders, engagement with nature varied from peasant to king. Similarly, British natural history was not exclusively British, including a fair number of individuals of mixed ancestry. The British network of natural history came also to be linked to existent sites of natural knowledge and the expertise of islanders. The hybrid character of the natural history that predated the British framed and fed into the arrival of their own natural history. This isn't to deny that the British brought new knowledge to the island, or that the island's natural historical knowledge was exported into a wider network of colonial knowledge centered in London. Rather, it is to counterbalance a traditional emphasis on the power of the center and the submission of the periphery by an attention to the importance of locality. British natural history did not arrive to a tabula rasa, especially in an island such as Sri Lanka, where Britons sought to discover themselves by attending to local culture.

The chapter begins with an account of the recycling of the indigenous in British natural history. It will then shift to an analysis of two of the icons of the low country and up country, namely the coconut tree and the hill station of Nuwara Eliya. It should be stressed that, though these icons have been chosen from among many possibilities, they capture the quintessential divide between the tropics and temperate zones that came to frame British understandings of the island. The discussion begins several decades before Kandy fell to the British and ends with establishment of Nuwara Eliya as an ideal retreat for the British in the 1830s, which saw the conquest of the deepest reaches of the land by the colonists. Therefore temporal shifts will be brought alongside the spatial politics of knowledge in this age of colonialism.

LANKAN NATURALISM

Just as Kirti Sri Rajasimha was a patron of Buddhist traditions and artistic practices, he also took up an interest in nature. The king's efforts at building a temple at Gangarama, just outside the capital, occupy a prominent place in the *Mahavamsa*. A statue of the Buddha, a temple, courts, and paintings were executed at great expense "on a fair spot" near the island's principal river. To aid the clergy's comfort, this temple was granted villages, fields, and gardens.[5] Kirti Sri's commitment to attaching gardens to temples emerges elsewhere too. For instance, southeast of the capital of Kandy he erected a temple and made certain to give it an attractive view of a garden. This garden was "adorned with bread-fruit trees, mango trees, cocopalms and other fruit trees."[6] In this way, the monarchical practice of natural improvement was closely tied to the site of the temple in the Kandyan kingdom.

In eighteenth-century Kandy, religious painters commonly depicted trees along with flowers and even strange mythical beasts (fig. 5.1).[7] Some of the topics which were staples of this religious art necessitated the repre-

---

5. See W. Geiger, trans., *Culavamsa, Being the Most Recent Part of the Maha-vamsa*, 2:100, line 201.

6. Ibid., 2:100, line 218. For the restoration of another temple with natural historical symbols, 2:100, lines 250ff.

7. For commentary on this, see Senake Bandaranayake and Gamini Jayasinghe, "Late Period Murals," in *The Rock and Wall paintings of Sri Lanka*; and SinhaRaja Tammita-Delgoda, *Ridi Vihare: The Flowering of Kandyan Art*, esp. 112.

Fig. 5.1. An eighteenth-century temple mural showing the Buddha gazing at the
Bo-tree (*Ficus religiosa*) in the weeks after his enlightenment, set upon a lotus
pedestal. From the Dambulla temple, Sri Lanka. Photograph by Roshan Perret.
By permission of *Studio Times*, Colombo, and SinhaRaja Tammitta-Delgoda.

sentation of trees; for instance, the Bo-tree (*Ficus religiosa*) was constantly
depicted, because it was the tree under which the Buddha was said to have
attained enlightenment; and Sri Pada was depicted in the manner described
in the previous chapter.[8] Elephants regularly appeared in depictions of the
jataka stories, which retold the lives of the Buddha; for instance, the *Ves-
santara jataka*, a popular subject of painting, necessitated the painting of

8. For examples of this, see Tammita-Delgoda, *Ridi Vihare*, 58, 92–93.

a white elephant which could create rain. Some of the recurrent decorative motifs of Kandyan art utilized natural elements; these included the entangled creeper and the lotus flower.[9] This interest in naturalia is also evident in other objects of craft from this period, such as doorways, boxes, panels, and pillars within the temples, which were made of wood and ivory.[10]

The style of natural depiction in Kandyan art might usefully be contextualized within the rituals of the temple. The prominence of natural emblems matched the fact that devotees brought flowers to the temple as offerings. This is borne out by the fact that one temple, Danakirigala, which was particularly renowned for a royal garden called Daswatta which supplied a thousand flowers for worship, contains a striking scene of a landscape with trees in bloom.[11] The free illustration of landscape also appears in a very beautiful image from the middle of the nineteenth century from the temple at Dambulla.[12] The traditional colors of Kandyan art were red, yellow, black, and white, and yet natural subjects were sometimes rendered in green and grey, and would therefore have attracted special attention.

Attitudes to nature on the island mirrored the multiple registers of religious sensibility. In addition to the Buddha's Bo-tree, there is a range of others such as *erabadu (Erythrina indica)* and *diwul (Feronia elephantum)*, which are believed to be the abodes of certain gods and devils. Tales of plants that grow in the home of the gods, such as *Parasatu* and *Kusa*, were well known. Indeed *Parasatu* flowers occur often in temple paintings. There were also myths surrounding particular trees. For instance, there is the *Kapruka*, which is thought to come into existence once in a *kalpa* or millennium, and which in effect is a wishing tree. There is also the *Kalunika*, a tree which when met in the depths of the jungle restores an old man at once to health and youth.[13]

Some of the best-known natural historical riddles and proverbs were

9. See J. B. Disanayaka, *Paintings of Kalani Vihara*, 12 for the elephant, and 22–23 for flowers.

10. Tammita-Delgoda, *Ridi Vihare*, 133ff.

11. Bandaranayake and Jayasinghe, *The Rock and Wall Paintings*, plate 68.

12. Ibid., plate 92.

13. This information has been extracted from W. A. De Silva, "A Contribution to Sinhalese Plant Lore." See also Andreas Nell, "Some Trees and Plants Mentioned in the Mahavamsa." Nell points, for instance, to how there was a well-established tradition of the island's kings establishing gardens, and also how both Kirti Sri and his successor donated robes made of cotton to monks.

brought together by the Royal Asiatic Society of Ceylon at the end of the nineteenth century. This example refers to the names for rice in Sinhala:

> The plant has one name while living and another when dead; the fruit is known by one name, but when eaten by another; the coat has a name and the kernel has another; the person who propounds the meanings of these lines will indeed be wise.[14]

That islanders engaged with nature in temples, through religious rituals, and by utilizing different types of narrations must not allow us to place the wider Lankan traditions of natural knowledge in opposition to that which emerged from Western Europe. Or in other words, this is not an attempt to separate the natural awareness on the coasts from the highlands of Kandy. There were some links between Kandyan and European engagements with nature. For instance, Dutch records from 1720 tell of how the reigning king at Kandy demanded presents of rare birds and animals for his royal garden. When a request arrived for a *hamsa*, a mythical water bird, whose symbol adorns the palaces of Kandy, the Dutch were dumbfounded. They asked the king what the bird looked like and concluded that he wanted an ostrich![15] The documents relating to British-Kandyan diplomacy also point to the trade in natural historical specimens. For instance, John D'Oyly exchanged specimens with Kandyan nobles. His diary includes the following entry for 18 March 1812, with a reference to Sri Pada or Adam's Peak:

> Mudalnayaka Mohottale of Kuruwita & Mudduwe Vidan attend with an Ola (No. 79) & a Basket of 12 Species of Plants from the Forests surrounding Sripade—They inform, that they are directed by Eyheylepola 1st Adikar to deliver these to me with his best Compts., & await Orders & bring back what they recd.—
> N.B. The species sent are, Kuda Heydaya, Maha Heydaya, Kuda Sudana, Maha Sudana, Wada Manchal, Sita Manchal, Wana raja, Ira raja, Yet hawari, Jata makuta, Kapuru, Nagawalli[16]

14. De Silva, "A Contribution to Sinhalese Plant Lore," 140.

15. James S. Duncan, *The City as Text: The Politics of Landscape Interpretation in the Kandyan Kingdom*, 69.

16. H. W. Codrington, ed., *Diary of Mr. John D'Oyly*, entry dated 18 March 1812. The sender of these plants was the first adigar, who carried on a secret corre-

There were thus vibrant traditions of natural knowledge in the island, which were aesthetic as well as religious, courtly as well as popular.

THE POLITICS OF BRITISH GARDENS

Before seizing the highland kingdom of Kandy, the British began their own program of natural history on the coasts, following in the footsteps of the Dutch.[17] Given that the kings of Kandy kept their own gardens, it was fitting that the first Crown Governor, Frederick North, kept his own too, outside Colombo, under the charge of a Frenchman, Joseph Jonville, whom he brought out to the island as "clerk for natural history and agriculture," and who went on to become Surveyor-General of the island. Jonville assembled a rich collection of natural historical illustrations, and he also sent "three chests" filled with objects of natural history to the East India Company. General Hay Macdowall, the leading military man on the island, also had his own garden, and received boxes of exotics from the botanist William Roxburgh in Calcutta.[18] He had attempted to cultivate English vegetables, without success. By 1812 the island had an official botanic garden, which was established in Slave Island in Colombo, with the blessing of Joseph Banks, the leading man of science in London, who was probably consulted by Alexander Johnston.[19] Banks made particular note

---

spondence with the British. The number cited (No. 79) is a reference to a palm-leaf manuscript.

17. I do not have space here to explore the undoubted legacy of the Dutch for British natural history. For more on Dutch natural history, see K. D. Paranavitana and C. G. Uragoda, "Medicinalia Ceylonica: Specifications of Indigenous Medicines of Ceylon Sent by the Dutch to Batavia in 1746," and Rohan Pethiyangoda, *Pearls, Spices and Green Gold: An Illustrated History of Biodiversity Exploration in Sri Lanka*, 40–56.

18. James L. A. Webb, *Tropical Pioneers: Human Agency and Ecological Change in the Highlands of Sri Lanka, 1800–1900*, 55; J. C. Willis, "The Royal Botanic Gardens of Ceylon and their History," 2–3; Pethiyangoda, *Pearls, Spices and Green Gold*, 59; R. Desmond, *The European Discovery of the Indian Flora*, 160. For detailed discussion of the possible sites of the Ceylon garden, see T. Petch, "The Early History of Botanic Gardens in Ceylon with Notes on the Topography of Ceylon" and "The Early History of Botanic Gardens in Ceylon." See also Anon., *Analyses of New Works of Voyages and Travels Lately Published in London*, 106–7, and James Cordiner, *A Description of Ceylon: Containing an Account of the Country, Inhabitants and Natural Productions* (1809), 1:386.

19. For Johnston's involvement, see Petch, "The Early History," 63; also *Saturday Magazine* (1834), 90.

of the importance of topography in writing of the principles on which a botanic garden ought to be founded in the island:

> In all hilly countries near the Line, there are a variety of climates in which the Plants of different countries will if properly attended to, succeed to perfection. The European Strawberry is abundant in Jamaica, towards the summit of the lowest ridge of Hills; at a somewhat higher elevation Apples, Pears and the Fruits of cold climates attain a considerable degree of perfection; in their arrangement of the Intertropical Plants we find that Coffee and Pimento thrive best in elevated stations, while sugar requires low land and the Cocoa Nut which bears abundantly near the sea, becomes sterile when removed to the first slope of the hills. . . . These observations point out the necessity of small subsidiary Gardens in various parts of the Island under the direction of Native Foremen, but under the superintendence of the Royal Gardener.[20]

The Royal Gardener was picked by Banks; he was William Kerr, a collector for Kew Gardens, who had served well in Macau and Canton. But Kerr did not last long in Ceylon: he was dead within two years. By 1815, the Slave Island garden was criticized for being ill conceived. Banks was informed that its "situation was very flat and not sufficiently elevated above the surface of water for the purposes of a garden." For a short time plants were said to "thrive," but then when their roots extended deeper they found a soil that was too damp.[21] The search for an ideal space for the British botanical garden continued.

Before his death, Kerr oversaw the inauguration of a new garden, south of the city in Kalutara, on a site of 560 acres of land, which the government bought from Messrs. Layart and Mooyart and which included a plantation of coconut trees.[22] When Kerr's successor, Alexander Moon, a working gardener from Kew recommended by Banks, took up his post on the island, he developed the Kalutara gardens.[23] The British annexation of the Kandyan

---

20. Webb, *Tropical Pioneers*, appendix 1.

21. Desmond, *European Discovery of the Indian Flora*, 161–62; Pethiyangoda, *Pearls, Spices and Green Gold*, 60–62.

22. William Kerr to the Governor, dated 11 May 1813, Lot 6/281, SLNA. This evidence goes against claims that this was the site of a sugar estate; see Petch, "Early History," 68.

23. *Asiatic Journal* (1826), 91.

kingdom provided Moon with a new terrain for botanical exploration. In 1818, a year after his arrival, Moon proposed a journey "through the Kandyan provinces for the purposes of obtaining more correct knowledge of the vegetable productions of that interesting country." The following year he wrote of the botanical promise of the Kandyan kingdom: "I entertain great hopes that if the Kandyan country (where European plants of every description thrive beyond every expectation) is supplied with garden seeds for a few years it may at no distant period be rendered independent of extraneous aid."[24]

By 1821, Alexander Moon's interest in the interior came to fruition in the creation of another botanical garden, this one located outside the Kandyan capital in the highlands in Peradeniya (fig. 5.2). In making the case for this garden, Moon differentiated the climate of the Kandyan kingdom from that of the "maritime provinces": "for never or seldom any sort of European vegetables can be brought to such perfection on the sea coast as to produce seeds, and in my humble opinion a garden in Kandy under the liberal views of Government offers the only source."[25] This program of experimentation bears out emerging British senses of the scientific difference between the highlands and the lowlands, in accord with the increasing popularity of biogeography, the science of mapping species distribution according to altitude and landform to posit "nations" of living things. However, even as the distinction between the highlands and lowlands as temperate versus tropic is evident here, by 1835 the search for the ideal climate for botany carried on. One of Moon's successors, J. G. Watson, wrote comparing the climate of "Poosalave" with Peradeniya and Nuwera Eliya, and urged that the former was superior.[26] He reasoned in an imprecise way that it was in "an intermediate station and climate."[27] In this way the highlands were themselves fragmented into zones.

The Peredeniya botanic garden provides an important point of reference for the claim that the British recycled extant traditions, for the establishment of this garden was an act of literal recycling (fig. 5.3). Though both contemporary and recent histories of the Peradeniya botanic garden often align its inauguration with Moon and the date of 1822, in fact the

24. Alexander Moon to Deputy Secretary of Government, dated 7 January 1819, Lot 6/282, SLNA.

25. Alexander Moon to John Rodney, dated 26 September 1821, Lot 6/283, SLNA.

26. No garden was established in Pussellawa, though one was later established in Hakgala in 1860.

27. J. G. Watson to Colonial Secretary, dated 19 November 1835, Lot 6/1322, SLNA.

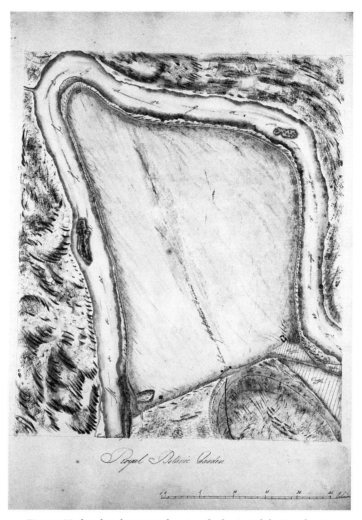

Fig. 5.2. Undated early image showing the layout of the Peradeniya
Botanical Gardens, with the Kandy road, coffee plantations, and entrance
to the Gardens. From a volume of drawings entitled *Sinhala Drawings*.
Courtesy Lindley Library, Royal Horticultural Society, London.

new botanical garden was established on a historic site, part of which
belonged to the Dalada Maligawa, the Temple of the Tooth Relic in the
capital of Kandy, and a part of which belonged to another shrine (fig. 5.4).
Moon wrote: "I am of the opinion that the site of the late Kandyan King's
Garden at Peradenia is better adapted than any other place for the proposed Botanic Establishment." In the original proposal for the garden's

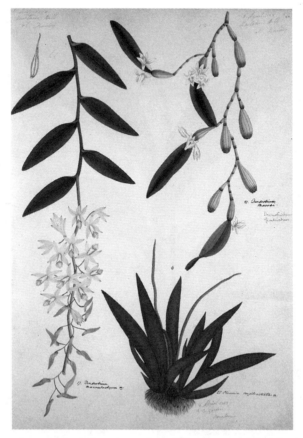

Fig. 5.3. A comparison of specimens from different parts of the island, both
from the lowlands and highlands. Specimens are marked "Nuwara Eliya"
and "Kalutara." From a volume of drawings entitled *Sinhala Drawings.*
Courtesy Lindley Library, Royal Horticultural Society, London.

establishment it was envisaged that some stones from the ruins in the
area could be reused to build Moon's quarters.[28] The establishment of a colonial botanical garden on the site of an existent Kandyan garden supports
the claim that there are symmetries in how both Britons and Kandyans
used natural history, for both sides sought to exert their control over terri-

28. See Petch, "Early History," 68–69. The Secretary for the Kandyan Provinces
noted that the plot of land utilized for the Peradeniya gardens comprised four gardens, one of which belonged to the temple, and another of which paid tribute to the
temple.

Fig. 5.4. The Peradeniya Botanic Gardens. Author's photograph.

tory by establishing a network of gardens. The emergence of the network of British botanical gardens needs to be associated less closely with the European Enlightenment and the model of Kew Gardens. A recent visitor's guide to the Peradeniya botanic garden tells this story in a way that no European history has ever done. The guide dates the origin of the gardens to 1371; notes Kirti Sri's making of a "Royal Garden" on this site; states that Rajadhi Rajasimha resided at the gardens in a temporary residence; asserts that the British destroyed a temple on the site; and claims that a priest lived at the site until Moon's arrival.[29] In addition, supporting this narrative is one likely derivation of the name Peradeniya from valley or glen of pears or guavas.[30]

Alexander Johnston was again a key person through whose advice the Peradeniya gardens were established. Johnston suggested that a royal garden be established on the island after his encounter with a Muslim physi-

29. *Visitors' Guide to the Royal Botanical Gardens, Peradeniya,* from the author's visits in 2005 and 2008, in author's collection. Please note that some of the details here cannot be verified.

30. For the question of derivation, see P.R.A. Dharmawansa, "Peradeniya."

cian, Mira Lebbe Mestriar, who we will consider in greater detail in chapter 7. Mestriar compiled for Johnston a list of plants Muslim physicians used for medical purposes. Muslim physicians had been utilized in the Kandyan court, and so appears a web of contacts that links the court to the British botanists.[31]

The British sources that follow the establishment of the garden need to be teased very heavily for any reference to the prior use of the site of the Peradeniya botanic garden. In the copious correspondence in the Sri Lankan national archives, any reference to this amounts to a few lines. One such reference appears in a report of 1828 written to investigate whether land adjacent to the botanic garden should be given over to the cultivation of coffee. Here reference is made to the site of the "old palace," which is said to be opposite the botanic garden on the road from Colombo to Kandy.[32] A report written two years later by one of Moon's successors, James Macrae, begins:

> On my arrival in the Island in March 1827 I found the Botanical Garden in its infant state. It having only been established in the early part of 1822 and was formerly a Royal Garden of the Kings of Kandy, in a delightful situation, on the banks of one of the largest Rivers in the Island within Three Miles of the City of Kandy. . . .[33]

The difficulty of using British sources to get to the prior history of the site of the Peradeniya gardens suggests how sure the colonizers were that they were bringing a new genre of knowledge to the island, when their dependence on existent traditions lay before their eyes. Indeed as late as 1838 there were residents at the garden who could recall when the garden was used by the kings of Kandy for "their royal gardens."[34]

Another reason why Alexander Moon is cast as the founder of the Peradeniya gardens lies in the fact that he published a botanical work that drew some attention. The complete title of this book hints at the multiple

---

31. For Mira Lebbe Mestriar, see Lorna Dewaraja, *The Muslims of Sri Lanka: One Thousand Years of Ethnic Harmony, 900–1915*, 128–29. I thank Lorna Dewaraja for pointing this out to me.

32. "Extract from the Proceedings of the Board of Commissioners held at Kandy, 25 November 1828," in CO 54/84, TNA.

33. "A Report on the Present State of the Botanic Garden at Ceylon," 26 April 1838, CO/54/107, TNA.

34. "A Report on the Present State of the Royal Botanic Gardens of Ceylon by J. G. Lear, dated September 1838," Lot 6/1405, SLNA.

purposes that lay behind its writing and publication: *A Catalogue of the Indigenous and Exotic Plants Growing in Ceylon, Distinguishing the Several Esculent Vegetables, Fruits, Roots and Grains together with a Sketch of the Divisions of Genera and Species in Use amongst the Sinhalese, also an Outline of the Linnaean System of Botany in the English and Singhalese Languages for the use of the Singhalese*. The book was published by the Wesleyan Mission Press of Colombo in 1824. The *Catalogue* is a curiously hybrid work at a number of levels. It contains both English and Sinhala script, as well as plants organized according to Linnaean and Sinhala modes of classification. It ends with two indexes, one in Latin and one in Sinhala, to all the specimens catalogued in the book, both according to the Linnaean and the Sinhalese classificatory systems. The Sinhala index appears both in Sinhala script and as a transliteration in Roman script. A total of 1,127 species were brought together in the *Catalogue*.

In the preface to this work, Moon displayed contradictory attitudes toward Sinhala knowledge of botany. Taking on board the typical orientalist thesis of degeneration, he postulated that in the glorious past the Sinhalese would have had "a complete and general system of Botanical arrangement" but that this was lost in time. He noted the "dependence of the natives upon the *Vegetable Kingdom* for almost every necessary of life" and theorized that this interest in natural history led to a "desire to extend that knowledge" that "very generally prevails amongst them." It is suggestive to consider this as a reference to the popular traditions of natural historical narration outlined above. At the same time, Moon derided the Sinhala system of classification as a "*natural* arrangement by *families*" rather than one that adhered to a "more *artificial* method." The grasses, the esculent greens, and edible roots, for instance, had their own divisions. Furthermore Sinhala names gestured towards nothing more than the qualities of the specimen. Moon noted how *rat-mul* translated merely as red-root; *kiri-wael* corresponded to milky-creeper. Moon's presentation of the Sinhala catalogue also pointed to the use of qualities to categorize a plant. For instance, on the third page we find a list of different types of yam, under the Sinhala *ala*, taking the qualities of black, rat-snake, lime, snake, great, wild, white, small and knotty, small and indigestible, small and "elk," small and deep, and small and from Java etc.[35] The dismissal of contemporary Sinhalese botanical knowledge may have arisen from the fact that Moon presented himself as a botanist. The clergyman and writer James Cordiner took a

---

35. For a critique of Moon's presentation of the Sinhala system of botany, as merely a collection of names, see Petch, "Sinhalese Plant Names," 169–79.

different view in praising "the most illiterate peasant" who could "tell the names" as well as "the qualities of the minutest plant that is to be found within the precincts of the district which he inhabits."[36]

Governor Edward Barnes wrote approvingly about Moon's published work and took some share of the praise in noting that his government had subscribed a sum of five hundred rix dollars in making its appearance possible. One of his dispatches to London included six copies of the work.[37] Indeed, it is relatively easy to trace how Moon's work was incorporated into European botanical knowledge. By 1817, Moon had prepared "hundreds of species of the choicest seeds" for the Emperor of Austria and for the Kew Gardens; the next year he transmitted the "journal of [his] proceedings" to Banks, together with specimens kept safe in coconut shells, enclosed in baskets made from "Ceylon Rattan for the convenience of slinging them on board ship, as well as for the protection of the plants from injury."[38] By the next year, he had received botanical books from England and a collection of English garden seeds, including cabbage, cauliflower, carrot, turnip, and parsley, and wrote of his hope of "great success in bringing the cultivation of European fruits and vegetables to a more superior degree of perfection."[39] By 1820, he had to apologize for spending too much on almirahs to hold "dried plants and seeds" to be transported to Kew and the Calcutta gardens.[40] In this manner, with each successive year, the gardens became more firmly situated in the colonial network of botany, and this included connections to other colonial gardens as well as Kew.[41]

However, the ease of locating the flow of this information out of the island into a wider sphere should not mean that its location within the island itself is misplaced from the narrative. In the preface to his book, Moon acknowledged the help of an important early orientalist, Rev. B. Clough,

36. Cordiner, *Description of Ceylon*, 1:106.

37. Dispatch dated 26 October 1824, from Governor Edward Barnes, CO 54/86, TNA.

38. Alexander Moon to Sir Joseph Banks, dated 16 February 1818, Lot 6/282, SLNA.

39. Alexander Moon to Deputy Secretary of Government, dated 7 January 1819, Lot 6/282, SLNA.

40. Alexander Moon to Deputy Secretary of Government, dated 23 August 1820, Lot 6/283, SLNA.

41. One decade later, one of Moon's successors, James Macrae, informed Governor Edward Barnes that he was in "constant correspondence" with the Superintendent of the botanic gardens in Mauritius, another Indian Ocean Crown colony, in addition to the Directors of the botanic gardens at Java, Poona, and Pondicherry; he had sent more than fifteen hundred specimens of different kinds to the Horticultural Society in London. Dispatch dated 14 January 1830, from Governor Edward Barnes, CO 54/107, TNA.

who we have encountered for his work in compiling a Sinhala dictionary.[42] However there was no mention of Moon's closest aide Harmanis De Alwis Seneviratne, who had served as his "writer" from 1818, first in Slave Island, then in Kalutara, and finally in Peradeniya.[43] There is no information about De Alwis's background, but he was an Anglican at the time of his death.[44] De Alwis worked closely with Moon in putting the material together for the *Catalogue*. Moon may have omitted De Alwis's name from the preface because he paid personally for De Alwis's expenses, including his lessons in art; Moon may have felt that he owned De Alwis's labors.[45] By 1823, the colonial state promoted De Alwis from the position of Moon's "writer" to that of "draftsman," which post he held for thirty-eight years, so that he became a good example of a "native" who was institutionalized in state structures.

In 1831, Governor Barnes took note of De Alwis's accomplished drawings in awarding him the title Mohandiram. It has been suggested that this image of De Alwis dates from this point (fig. 5.5). In 1836, De Alwis met the naturalist Robert Wight, Company surgeon in Madras. Wight was so taken with the illustrator that he arranged for him to visit Madras, where he received instruction in botanical dissection and in the making of drawings with the aid of a microscope. It was during this period that De Alwis drew some of the plates for Wight's *Icones Plantarum Indae Orientalis*. The new botanical superintendent in Ceylon, George Gardener, who arrived in 1843, took De Alwis with him on every botanical excursion as did his successor, G.H.K. Thwaites. In the preface to *Enumeratio Plantarum Zeylaniae*, Thwaites acknowledges De Alwis's "intelligent and hearty cooperation." It was only in 1861 that De Alwis retired, and he lived on for a further thirty-three years. It was De Alwis's long career that ensures that some traces of his life have been preserved, and that the colonists' talent of overlooking their assistants didn't expunge him from the record.

It is difficult to pin De Alwis down in the early correspondence concerning botany, as Moon had several assistants of Lankan origin working under him. Only his helpers of European or Malay origin were ever named.

42. Alexander Moon, *A Catalogue of the Indigenous and Exotic Plants Growing in Ceylon*, preface, 3.

43. All uncited biographical information from Henry Trimen's obituary of Seneviratne, *Journal of Botany* 32 (1894): 255–56.

44. Rohan Pethiyagoda, "The Family de Alwis Seneviratne of Sri Lanka: Pioneers in Biological Illustration," 100.

45. Ibid., 100.

Fig. 5.5. Portrait of Harmanis De Alwis Seneviratne, the frontispiece
from a volume of drawings entitled *Sinhala Drawings*. Courtesy
Lindley Library, Royal Horticultural Society, London.

In 1820, his indigenous assistants at the garden of Kalutara included a
"Mohandiram" who was paid eighteen rix dollars per month; an "Over-
seer" who was paid the same; and a "Native Gardener" and "Clerk" who
were paid fifteen rix dollars. The "Clerk" may correspond to De Alwis.[46]
De Alwis is easier to identify once he took on the title of "draftsman."
In 1828 Macrae asked for "cartridge paper" for his use;[47] by 1830, G. Bird,

46. Alexander Moon to Chief Secretary of Government, dated 5 August 1820,
Lot 6/283, SLNA.
47. James Macrae to Deputy Secretary of Government, dated 9 December 1828,
Lot 6/1027, SLNA.

his successor, requested "150 sheets of drawing paper," cakes of "carmine red," "indigo blue," "lake red," "king's yellow," a "white chalk," "6 drawing pencils" and "1 balle of Indian rubber," and the next year he requested a "colour box." The draftsman was expected to attend work "precisely from 10 till 5 o'clock," and his pay was to be reduced if he was five minutes late, or absent for half a day or all day.[48] In 1840, Superintendent H. F. Normansell wrote to the Colonial Secretary of the island that the draftsman completed seventy drawings in three months on average (fig. 5.6).[49] By 1845, Superintendent George Gardener noted that the draftsman was paid only £3 a month: "His equal as a native botanical artist, I am certain does not exist in India."[50]

In 1859, John Lindley named "a curious and minute leafless orchid," *Taniophyllum alwisii*, after the illustrator.[51] De Alwis's two sons, William and George, and three of his grandsons became botanical illustrators, taking up this family tradition until after the First World War, which suggests that it became like a caste occupation for this family.[52] De Alwis was also held up as something of a model by Macrae to the Commission of Inquiry, which visited the island in the 1830s. The Commissioners inquired whether more local people might be interested in botanical science, and how the Superintendent envisaged creating such an enthusiasm for botany. Macrae endorsed their idea that young apprentices, of eight or ten years, should be attached to the gardens, who would in time become overseers. It was expected that they would also be taught English: "Such persons would also learn and collect the various indigenous plants for the garden and disseminate throughout the country such plants as it would be desirable to promote the cultivation of, they should preferably be of good caste, as they would command more influence and respect as overseers."[53]

By the mid-nineteenth century, with the emergence of an English-

---

48. G. Bird to Deputy Secretary of Government, dated 28 July 1830, Lot 6/1028, SLNA, and J. G. Watson to Chief Secretary of Government, dated 6 April 1832, Lot 6/1028, SLNA.

49. H. F. Normansell to Acting Colonial Secretary, 5 October 1840, Lot 6/1487, SLNA.

50. Cited in Pethiyagoda, "The Family de Alwis Seneviratne of Sri Lanka," 102.

51. Trimen's obituary of de Alwis, 256.

52. Pethiyagoda, "The Family," 101. The idea of caste is my own. See *The Ceylon Almanac* (1854), 44, for a mention of two native draftsmen who are at work at this time at the gardens.

53. "Evidence of Mr. James MacRae, Superintendent of the Royal Botanical Gardens in Ceylon," Lot 19/1, SLNA.

Fig. 5.6. *"Rhododendron aboreum."* A drawing in all probability completed by
Harmanis de Alwis Seneviratne. From a volume of drawings entitled *Sinhala
Drawings.* Courtesy Lindley Library, Royal Horticultural Society, London.

speaking indigenous intellectual elite, local people took on a new role in
the practice of botany as collectors and describers of specimens in their
own right. One "indigenous" botanist held up by the periodical *Young
Ceylon* was "our countryman Dr. Kelaart," and another was Simon Casie
Chetty, who published on botany in the *Journal of the Royal Asiatic Society.*[54] For the former, the ambiguity in the definition of indigenous is manifest; Kelaart was likely to have been of at least part European descent. The
rise of these figures was also tied to the emergence of a range of voluntary
associations such as the short-lived Literary and Agricultural Society, the

54. Simon Casie Chetty, "The Pursuit of Natural History in Ceylon," 105ff. For
Kelaart, see chap. 7 below.

Horticultural Society of Colombo, and the Ceylon Agricultural Society, established at the Kandy Library in 1842.[55]

The importance of complicating the narrative of the expansion of British botany on the island, by pointing to how it drew from multiple directions and redefined the nature of the indigenous, is also clear with respect to the physical layout of the botanical gardens at Peradeniya. It is important to note that the gardens were not arranged strictly according to the Linnaean system, or for the study of botany.[56] In 1830, James Macrae noted how the garden had divisions for ornamental gardening, an orchard for the improvement of fruit trees and for the introduction of species, an area for medicinal plants peculiar to the island, "of which a great many are used by the Natives," a nursery for raising coffee, cocoa, and other valuable plants, a kitchen garden "for the culture of European and Native Vegetables," and a considerable portion of land devoted to raising teak "and other valuable timber."[57] In 1838, J. G. Lear wrote of how it was the "original intention" for the gardens to have "a systematical arrangement usual in the general formation of such establishments," but that the death of Moon had meant that this intention was not met. Lear also noted the lack of experience of those keepers who came after Moon, and how the gardens had "frequently" been left to the charge of "native people" and how it now "exposes such a confusion," that would "place a stranger in a situation to ask for what it was intended."[58] The pines, for instance, were said to be planted "after the native system." It is intriguing to reflect on whether there is a relationship between the layout of these gardens during the time of the kings of Kandy and during the early British period.[59] Unfortunately there is no definite

55. See for instance J. W. Bennett, *Ceylon and its Capabilities: An Account of its Natural Resources, Indigenous Productions and Commercial Facilities* (1843), 160, and *Proceedings of the Ceylon Agricultural Society for the Half Year Ending 1st July 1842* (Kandy: Baptist Mission Press, 1842).

56. The need to organize botanical gardens for study rather than entertainment, pleasure, or utility was a point of debate in colonial stations in this period. For the need for a Linnaean layout to the Calcutta gardens and plans to improve it, see Richard Axelby, "Calcutta Botanic Garden and the Re-ordering of the Indian Environment."

57. "A Report on the Present State of the Botanic Garden at Ceylon," dated 26 April 1838, CO 54/107, TNA.

58. "Report on the Present State of the Royal Botanic Garden, J. G. Lear, September 1838," Lot 6/1405, SLNA.

59. For instance, Moon had written in originally establishing the garden: "There are already a number of fruit and Forest Trees, common to the island dispersed all over the grounds, which will afford immediate shade and shelter to the more tender Exotic

evidence for this, but it is clear that the British did not successfully impose the model of botanical arrangement with which they were familiar.[60]
In 1853, in a rare description, the character of the "Singhalese Garden"
was set out as follows:

> Everything indicates rude simplicity. It appears to have been the prac-
> tice of the ancient kings to have a separate garden for trees of each
> kind. Thus there was a garden of Jack; of Cocoanut; of Palmyrah; of
> Arecanut; of Mango; of Betel; of Flowers &c. and a lakh (or one hun-
> dred thousand) of each kind was generally planted in each garden; a
> cistern for the priests to bathe in, and a stone cell for the purpose that
> a summer-house is intended to serve in an English garden, were also
> added. A native garden differs according as it is in the Maritime dis-
> tricts, or in the interior.[61]

New colonial knowledge did not cause a natural historical revolution,
where existent botanical knowledge and expertise was replaced by a new
order of information. The sites, publications, and collaborations of early
British science make it necessary to attend to pre-British traditions. However, this is not to argue that the British did not shift the nature of botany
in the early nineteenth century; it is to suggest that colonial science had
a complex parentage and that its novelty should not be assumed. If there
was a shift, an important element of it lay in the recycling of inherited
traditions.

THE TROPICAL COLONNADE

The importance of locality to the project of British colonial science has
thus far been discussed primarily in relation to how colonial knowledge
fitted into and remolded existent ideas, institutions and practices. Yet this
story has also hinted at the role of another idea of space. The botanical
garden intervened in a debate about the distinction between the highlands
and the lowlands, which arose in part out of indigenous traditions which

---

and Indigenous Plants on their introduction." See Petch, "Early History of Botanic
Gardens in Ceylon," 68.
    60. This might have resulted from the general state of disarray at the botanical
garden after Moon's death. For a complaint about the disappointing status of botany in
Ceylon after Moon, see W. Champion, "Remarks on the State of Botany in Ceylon."
    61. *The Ceylon Almanac* (1853), 6.

differentiated *uda rata* from *pata rata*. The colonists followed an existing classification of space, a classification that is shown especially well
in how the temperate and the tropical became symbolized by particular
icons. The discussion will turn first to the re-creation of the coastal lowlands and the use of the icon of the coconut tree.

The coconut tree was an important subject of British natural historical research, as is evident from a bibliography of articles on the island's
agriculture and botany published in 1915, which listed a total of 185 essays and articles on this tree. [62] An important early essay was written by
Henry Marshall, military surgeon in Ceylon, and entitled *Contribution to
a Natural and Economical History of the Coco-nut Tree*. It was based primarily on experience on the island, though it brought in knowledge about
the tree from other colonial territories. Indeed, Marshall commented that
the tree was "nowhere so extensively cultivated as in Ceylon."[63] A case
can be made that from the early period of British expansion the colonizers were obsessed with this tree and that it was one of the prime icons of
the coastal belt. In chapter 2 the centrality of the tree to the aesthetics of
the islanding discourse was touched upon. For instance, for J. W. Bennett,
a "verdant island" was signified by coasts populated with "myriads" of
coconut trees "to the very verge of the sea."[64]

Frequently, popular writers on Ceylon made the link between the sight
of coconut trees and ideas of the romantic and picturesque. The clergyman James Cordiner noted that the southeast coast was particularly "picturesque and romantic" because of the border of coconut trees, which
then framed a range of hills and mountains.[65] There were two rival arrangements of coconut trees that were said to be the most aesthetically
pleasing: the well-arranged grove or *tope*, and the isolated tree that rose
in graceful majesty.[66] In the words of the traveler Maria Graham: "Sometimes the straight tall trunks of the palm-trees . . . seemed to represent a
magnificent colonnade; sometimes . . . the creeping plants had entwined

62. T. Petch, *Bibliography of Books and Papers relating to Agriculture and Botany
to the End of the Year 1915*, 23–30.

63. Henry Marshall, *Contribution to a Natural and Economical History of the
Coconut-Tree* (1836), 113.

64. Bennett, *Ceylon and its Capabilities*, 14.

65. Cordiner, *Description of Ceylon*, 1:7. For a similar idea of coconut trees serving
as the foreground of a tropical scene, see Robert Percival, *An Account of the Island of
Ceylon* (1803), 32.

66. Bennett, *Ceylon and its Capabilities*, 82–84, and Percival, *An Account of the
Island of Ceylon*, 32.

themselves round them and hung from tree to tree."[67] Visual descriptions of coconut trees often stressed their elegance. In keeping with the idea of the picturesque, Henry Marshall compared a grove of coconut trees to "the long aisles and Gothic arches of a cathedral."[68]

The aesthetic pleasure that observers enjoyed in looking at coconut trees sat alongside a fascination with their utility. Popular accounts on Ceylon are full of details about how each element of the coconut tree can be put to use.[69] However, visions of the tree's usefulness could be inverted in line with colonial rhetoric. Rev. Reginald Heber, the Bishop of Calcutta, on visiting the island noted the statement of a British resident: "Give a man a coconut tree, and he will do nothing for his livelihood; he sleeps under its shade, or perhaps builds a hut of its branches, eats its nuts as they fall, drinks its juices, and smokes his life away."[70] Tropical nature was thus chastised for promoting indolence by its luxury.

Coconut trees were not common in the interior; Marshall stated that trees which grew in the immediate neighborhood of the sea were more luxuriant. The natural history of the tree fitted then into the classification of tropicality: "The Coco tree seems to require for its perfection a mean temperature of not less than 72° FAHRENHEIT, the proper climate for it will therefore be from the Equator to the 25th parallel latitude, and in the Equinoctical Zone to an altitude of about 2900 feet."[71] The absence of coconut trees from the interior, where Europeans had only recently ventured, meant that the tree was often taken as a sign of civilization.

For instance, James Emerson Tennent, who later became a Belfast poli-

67. Maria Graham, *Journal of a Residence in India* (1812), 93.

68. Marshall, *Contribution to a Natural and Economical History of the Coconut-Tree*, 114.

69. In a typical list of the uses of the coconut tree, Robert Percival wrote of how the rind of the nut could be converted to rope, cables, and cordage; how inside the nut there is a very cooling milky liquid; how the oil extracted from the nut is highly esteemed; how the top of the tree is useful for producing a liquor called toddy; how this toddy can be converted into sugar; how a web grows along the branches that can be converted into gunny cloth; how the branches are used by local peoples for roofing; how vessels can be built from the wood. See Percival, *An Account*, 318–20. Marshall devoted most of the space in his essay to the uses of the tree; he was particularly keen to provide detail on indigenous techniques, for instance in relation to the making of the alcohol, arrack, from the tree's flowers.

70. Reginald Heber and Amelia Heber, *Narrative of a journey through the Upper Provinces of India from Calcutta to Bombay, 1824–1825*, 2:229. Amelia wrote the section on Ceylon.

71. Marshall, *Contribution*, 112.

tician, in perhaps the most important nineteenth-century account of Ceylon, published in 1859, wrote of how a coconut tree is always indicative of "the vicinity of man." In the deepest jungle, when a traveler was lost, Tennent wrote that he should look for a coconut tree, as it invariably pointed to the approach of a village.[72]

The meaning that the British gave the coconut tree, as an icon of the coastal belt, may be linked to preexistent traditions of this tree, which should not be classed as strictly Buddhist. British orientalist travelers were fascinated by a statue that was connected to the coconut tree, which still stands outside Weligama. Major Jonathan Forbes, in his account of Ceylon, wrote:

> Before entering Belligamma (which is seventeen miles from Galle,) I stopped to examine a figure called the Kustia Raja: it stands on the road-side, is twelve feet in height, and forms part of a great mass of rock in which it is sculptured. One tradition affirms that the statue represents the Prince from the continent of India, who introduced the cocoanut-tree, and taught the Cingalese its many uses: another and more probable account states, that a King afflicted with leprosy established himself at this place for the convenience of worshipping at the neighbourhood wihare of Agra Bodi.[73]

This second account of the statue was also connected to the coconut tree; palm-leaf sources point to how Kusta Raja, the prince variously said to have arrived on the coast from India or from the interior of the island, drank the water of the coconut and found that he was cured of leprosy.

In another account, J. W. Bennett wrote of how he came to know the tale of Kusta Raja from the head priest of the Karangoda Vihare in Saffragam (fig. 5.7). Tellingly, Bennett had been taken ill with what he called

72. James Emerson Tennent, *Ceylon: An Account of the Island, Physical, Historical, and Topographical, with Notices of its Natural History, Antiquities, and Productions* (1860), 1:119.

73. Jonathan Forbes, *Eleven Years in Ceylon: Comprising Sketches of the Field Sports and Natural History and an Account of its History and Antiquities* (1840), 2:170. Possibly the first reference to this statue in British sources appears in Colin McKenzie, "Remarks on Some Antiquities on the West and South Coasts of Ceylon, Written in the Year 1796." McKenzie notes on 434: "Whatever degree of credit we may give to this story, the name of the COUTTA raja seemed to be familiar to all ranks, and is no doubt connected with some historical event."

Image of the Kousta Rajah or Leprous King.

Fig. 5.7. "Image of the Kousta Raja or Leprous King." From J. W. Bennett,
*Ceylon and its Capabilities: An Account of its Natural Resources, Indigenous
Productions and Commercial Facilities* (London, 1843). Reproduced by
kind permission of the Syndics of Cambridge University Library.

"a severe and intermittent attack of fever" and had relied on the priest
and his attendants to nurse him through. Bennett devoted four and a half
pages to the high priest's tale of Kusta Raja. The priest said that the coconut was a cure for leprosy because it was uncommon—Kusta Raja was a
Sinhalese king from the interior, where coconut trees do not grow. While
in his leprous condition, Kusta Raja fell into a trance which lasted for
several days, and while in it he saw the ocean, "a large expanse of water,
which he tasted, and found both salt and nauseous, and although it was of
a fine green color near, and blue in the distance; its margin [was] covered
with groves of trees of a rare kind, such as he had never seen; for instead
of branches in various directions, as trees had in his country, their tops
appeared crowned with a tuft of feathery leaves." In the end the father
of Buddha appeared to Kusta Raja and gave him directions to go and seek
after the fruit of the coconut tree: "The inside of transparent liquid, and
of innocent pulp, must be thy sole diet, till thrice the Great Moon (Maha

Handah] shall have given and refused her light."[74] In language reminiscent of British travel writing, we are told of how the king tasted the coconut and found it "sweet and delicious, and pure as crystal itself . . . whilst the fleshy part of it was a cooling and grateful food."[75]

This account is constrained by the problem of translation. It is clear that the colonial voice dominates that of the high priest, and indeed the visions of the coconut and the idealization of its fruit go back to colonial modes of representation. Nonetheless, Bennett's account of Kusta Raja is a hybrid which emerged out of a process of discussion with the high priest, and it shows how a local legend pictured the tree as a fabulous feature of the coastal belt.[76]

The palm-leaf manuscripts collected by Hugh Nevill contain further evidence of the iconicity of the tree among Lankans. The *Pol Upata* is said to date to the eighteenth century and contains twenty verses, and its title translates as "The Birth of Coconuts."[77] It retells how the nine rishis went to the land beyond the seven seas and fetched the coconut tree; in that land beyond the seven seas the coconut grew up in seven days with a crown of fronds given by the god Gana Deva [Ganesha]. The verses tell how the first kind of coconuts were golden colored ones, *tambili*, the next were *gon tambili* or pale yellow coconuts, then *nawasi* or nuts with edible husk, etc. Another song or *kavi* preserved in the palm-leaf collection is titled *Pol Wismaya*, which translates as "The Coconut Wonder" and dates to the nineteenth century.[78] It tells the curious tale of how in the garden Dematagolla at the village of Koralegama, one coconut sprouted into ten plants:

> In the Four Korale or Seven Korale [regions], have you heard of a wonder
> like this?
> Was it God King Sakra, who prescribed the auspicious moment by his
> divine eye?

74. Bennett, *Ceylon and its Capabilities*, 334. For another rendition of the tale of Kusta Raja in all probability written by J. W. Bennett, see "The Cocoa-Nut Tree."

75. Bennett, *Ceylon and its Capabilities*, 333.

76. For an interpretation of the sculptural tradition that this statue arises out of, see J. E. Van Lohuizen-de Leeuw, "The Kustaraja Image: An Identification." See also W. A. De Silva, "A Probable Origin of the Name Kushtharajagala," 86, who suggests that the name originates from the deity of the coconut in Travancore, south India; and also H. E. Ameresekere, "The Statue of Kusta Raja at Weligama."

77. Hugh Nevill, *Sinhala Verse (Kavi)*, vol. 2, no. 396.

78. Ibid., no. 467

Was it the astrologer of the Durakannadige that found this amazing
  moment?
Are there any other countries where ten plants have sprouted from one
  coconut?[79]

Both these kavi, like the account of Kusta Raja, point to the alterity of
the coconut. The *Pol Upata* was meant to be recited to cure sickness. The
medical efficacy of the tree was predicated on a narrative of its arrival in
the island. Similarly, the *Pol Wismaya* looks back to a time when the tree
was a rare, and magical, feature of the coastal belt. In this way these verses
also see the coconut tree as exotic rather than indigenous, and they suggest an awareness of the coconut tree's arrival from elsewhere, across the
beach.

The coconut tree and plantation might be cast as peculiarly romantic,
scientific, and utilitarian interests on the part of British colonists, linked
to the emergence of the modern empire and its exportation of natural historical regimes out from the center.[80] Indeed, the mobility of the new science of natural history is clearly discernible in this narrative. Yet this account of colonial science does not tell the whole story, for the iconicity
of the coconut tree was also shared by those who inhabited the island,
from the coastal belt to the highlands. In seeing the tree as a curiosity of
the tropical coasts, and casting it as a valuable commodity, the British followed existent patterns even as they extended them.

CREATING NATURE ANEW IN THE HIGHLANDS?

The British classified topography according to temperature, altitude, and
distance from the ocean. The coconut tree thus became an emblem of a

79. This is from a translation undertaken for me by Prof. Udaya Meddegama of the
University of Peradeniya, of a copy taken by me from *Pol Wismaya*, Or 6611 (200), British Library.

80. I do not have space here to explore in detail the colonial economy of the
coconut. For a beginning, see Colvin R. De Silva, *Ceylon under British Occupation*,
2:471–73. Anthony Bertolacci in his account of 1817 cited here records that the British
had 100,000 acres under coconut cultivation. A good tree produced fifty to sixty nuts
annually, and the quantity of nuts exported increased from an average of 2,750,000 nuts
in the period 1808–13 to about 4,000,000 nuts in the period 1825–33. Marshall wrote of
the coconut, in *Contributions*, 139: "Indeed they are therefore now said to be as common in the shops and streets of London as the orange."

distinct zone, and in symmetric fashion icons were also found for the highlands. Since the interior of the island was integrated into British control
later than the maritime provinces and since by this time British political
control had become more assured, the imaginative geography that Britons
attached to this region could be more inventive. The symbolism attached
to the highlands became linked to British landscapes rather than to colonial ones. Yet even in the higher altitudes the meanings that the colonists
used did not operate on a land that was empty of earlier historical memo-
ries, or which was as "unindigenous" as they pretended.

This line of argument is borne out very clearly in the events leading up
to the creation of the hill station at Nuwara Eliya. The Briton who is usually given credit for having "discovered" the plain of Nuwara Eliya is John
Davy, the climber of Adam's Peak and the younger brother of Humphry
Davy, the chemist.[81] While on a tour through the highlands in 1819, Davy
came across what he recorded as "Neuraellyia-pattan," a tract of land
which was said to be fifteen or twenty miles in circumference lying about
5,300 feet above the level of the sea.[82] It took another decade for Britons
to pay due attention to Nuwara Eliya: in August 1828 Governor Barnes
wrote to London stating that he was building a new military post there.
He noted the post's strategic significance; at a time of urgency, it could
supply troops "on either side of the great range of mountains that separate
the island."[83] Barnes also praised the utility of this spot as a place of con-
valescence: "I look to it as being the means of saving many lives as well as
the inconvenience and expense to individuals of a Voyage to Europe for the
recovery of Health."[84]

Two years later, Barnes's vision of Nuwara Eliya had already come to
some realization. A report in the papers of the Colonial Office provides evidence of how quickly invalid soldiers started being sent to Nuwara Eliya
to recover from illness. The report illustrated the positive effects of the
climate of Nuwara Eliya by contrasting the soldiers who recently arrived
from the "low country" with those who had resided there for a while: "The
former appear sallow and debilitated having that characteristic unhealthy

81. Robert Knox, the sailor in the captivity of the king of Kandy, noted Nuwara
Eliya in the map that accompanied his work; see Robert Knox, *An Historical Relation
of the Island of Ceylon* (1681).
82. John Davy, *An Account of the Interior of Ceylon and of its Inhabitants* (1821), 457.
83. Dispatch dated 11 August 1828, from Governor Edward Barnes, CO 54/101, TNA.
84. Ibid.

countenance of Europeans living in tropical climates, whilst the latter, seem to possess that robustness of frame which we commonly meet with among the natives of an English agricultural district."[85]

Nuwara Eliya also became an ideal retreat because European vegetables could be grown there. According to one military captain, writing in 1831, "the finest vegetables grown in England" were said to "thrive" in Nuwara Eliya.[86] Governor Barnes established his own residence, "a large and substantial country-house," in Nuwara Eliya and cultivated a variety of vegetables, flowers, and fruits.[87] In a dispatch of 1833 Barnes sent to London an estimate for a carriage road to Nuwara Eliya, and by 1836 this road had been completed, making access to the hill station much easier.[88] The station quickly became a small town; one commentator noted how Nuwara Eliya consisted of "military and public buildings, many whitewashed cottages and smoking chimneys."[89] By 1891, the population of Nuwara Eliya reached 2,726.[90]

The early successes inspired an eccentric attempt at "settlement" in Nuwara Eliya in 1845, led by Samuel Baker, who later became well known as an African explorer. Baker bought land from the government and transported machines, stock, seeds, and hounds from London on board the *Earl of Hardwicke*; he persuaded his brothers to join him. In his own idiosyncratic account of the setting up of his settlement, he records the tale of "a beautiful beast," the "well bred Durham cow" which came on board the ship. Baker described his attempts at transporting the cow to Nuwara Eliya thus: "A van was arranged for her, which the maker assured me would carry an elephant. But no sooner had the cow entered it than the whole thing came down with a crash, and the cow made her exit through the bottom."[91] When it was forced to make its way by foot to Nuwara Eliya, the poor animal eventually met its death; the journey from London to Ceylon proved less difficult than that from the lowlands to the highlands. While

85. Report titled "Some Remarks on the Soil, Climate and Temperature of Nuwera Ellia," dated 29 November 1831, from M. McDermott, CO 54/114, TNA.

86. Letter from F. C. Barlow, dated 9 December 1831, CO 54/114, TNA.

87. Forbes, *Eleven Years*, 133; Webb, *Tropical Pioneers*, 184 n. 58.

88. For estimate, see dispatch dated 19 December 1833 from Governor Edward Barnes, CO/54/130, TNA; for completion of the road, see Forbes, *Eleven Years*, 133.

89. Forbes, *Eleven Years*, 133.

90. C.J.R. Le Mesurier, *Manual of the Nuwara Eliya District of the Central Province of Ceylon* (1893), 75. For material relating to the governance of Nuwera Eliya, see also Lot 47, SLNA.

91. Samuel Baker, *Eight Years in Ceylon* (1855), 16–19.

Baker's idealistic plans initially ran into difficulties, in the end his family managed to build a relatively successful estate in Nuwara Eliya, which remained through the course of the century.

The British saw the town of Nuwara Eliya as a new and welcome addition to the island. Lieutenant Augustus De Butts wrote: "Nuwera Ellia is, in truth, a new creation, and still in a state of transition from the majesty of 'nature unadorned' to the less sublime, yet equally pleasing, charms that belong to cultivation."[92] It was thus a site which in discursive terms was devoid of the "indigenous." Their conception of the settlement of the highlands was linked then to the vocabulary of how nature could be improved through human intervention. Samuel Baker wrote:

> How changed are some features of the landscape within the few past years, and how wonderful the alteration made by man on the face of Nature! Comparatively but a few years ago, Newera Ellia was undiscovered—a secluded plain among the mountaintops, tenanted by the elk and boar. . . . Here where wild forest stood, are gardens teeming with English flowers; rosy-faced children and ruddy countrymen are about the cottage doors; equestrians of both sexes are galloping round the plain, and the cry of the hounds is ringing on the mountain-side.[93]

Civilizing the highland wilderness was a romantic project which drew its peculiar charm from the possibility of making in Ceylon a place which was a little Britain, at times very English and at others very Scottish, a home away from home. Nuwara Eliya was expected to provide a refuge from the tropics. A letter published in the *Asiatic Journal* for 1834 on the "health-station" of Nuwara Eliya is striking for its tone. The author hoped to convey his sense of revelation in finding Nuwara Eliya: "I am now clothed from head to foot in broad-cloth, with flannel next to my skin, the room closely shut up, and I declare to you that I can scarcely hold my pen my fingers are so cold."[94] He carried on, detailing the temperature, the need for fires and blankets, and the availability of familiar vegetables. A running idea was the Englishness of Nuwara Eliya: English grass was said to thrive in perfection and the hams and bacons were much finer than in England.

This representation of Nuwara Eliya was consciously set against its

92. Augustus De Butts, *Rambles in Ceylon* (1841), 21.

93. Baker, *Eight Years in Ceylon*, 38. The "elk" would be Sri Lankan sambar deer, *Rusa unicolor*.

94. *Asiatic Journal* (1834), 172–73.

opposite, the coastal belt of the island which embodied tropicality. It is noteworthy how often popular writers on Ceylon used the word "tropics" in opposition to what Nuwara Eliya represented.[95] One example comes from Charles Pridham, who wrote how the station was "so entirely dissimilar from any view or sensation within the tropics, that the novelty is at first delightful and exhilarating."[96] This makes the point that the station's identity derived from its place in an all-island space. However, there were some features of Nuwara Eliya which questioned casting it as a new Britain. Foremost among these was the presence of elephants. John Davy in the first report on the plains grappled with the contradiction thus: "Reasoning *a priori*, would have led to a different conclusion; and, at first, it appears not a little singular, that the most elevated and coldest tract of Ceylon . . . should be the favourite haunt of an animal that it supposed to be particularly fond of warmth."[97] Another point at which the rhetoric of Britishness was halted was in relation to the constant failures which met attempts at large-scale cultivation. Samuel Baker moaned how the soil of Nuwara Eliya was not as luxuriant as first thought. "Every acre of land must be manured. . . . With manure everything will thrive to perfection except wheat. There is neither lime nor magnesia in the soil."[98] Yet the idea of Nuwara Eliya as a temperate "paradise" was strong enough to surmount these niggles.[99]

It is important not to cast Nuwara Eliya simply as an icon of British power and of the increasing reach of British ideas of nature. It is worth keeping in mind that, even though the area was not inhabited, local people were familiar with it and had a name for it. John Davy wrote: "Our guides call it Neuraellyia-pattan."[100] He also commented that while the plain was

---

95. See for instance, Forbes, *Eleven Years*, 132: "Perhaps there is in no country a climate more congenial to the natives of Great Britain, both as regards salubrity and comfort, than Nuwara-ellia; the temperature of the air never approaching to what is called tropical heat."

96. Charles Pridham, *An Historical and Statistical Account of Ceylon and its Dependencies*, 678. For another example of how Nuwara Eliya was placed in opposition to the tropics, see Anon., "Remarks upon the Comparative Healthfulness and Other Local Advantages of Nuwera Eliya in the Island of Ceylon and the Neilgherry Hills in Hindoostan."

97. Davy, *An Account*, 458.

98. Baker, *Eight Years*, 31.

99. For the use of "Paradise," see De Butts, *Rambles in Ceylon*, 212.

100. Davy, *An Account*, 457. "Neuraellyia-pattan" could mean "plains outside the city."

not inhabited, it was visited by blacksmiths from Kotmale looking for iron and by those looking for precious stones. This same observation appears in the manual to the district published at the end of the century.[101]

More information about Nuwara Eliya's pre-British history appears in Baker's narrative; he was adamant that it was known to the Sinhalese. Nuwara Eliya was said to denote "Royal Plains," because the kings of Kandy had themselves sought refuge here in a bygone age in times of political turmoil. Baker wrote: "There are native paths from village to village, across the mountains, which, although in appearance no more than deer-runs, have existed for many centuries, and are used by the natives even to this day."[102] Baker confessed that he had learned the lay of the country by using these paths and by observing "notches" on trees which had been made by travelers in the past who had traversed the plain. He went on to present a theory of why Nuwara Eliya had been important to the Kandyan kings: he stressed water and gems. In ancient times, according to Baker, the waters of Nuwara Eliya were conveyed by complicated waterways down the hills. From examining the remains of such waterways, Baker postulated that "more than fifty times the volume of water was then required than in use at present, and in the same ratio must have been the amount of population."[103] The view that the plains were a refuge in times of trouble was also taken up by James Holman in his *Voyage Round the World from 1827 to 1832*. Holman noted that the kings retreated there whenever they feared invasion, and temporary huts were built to accommodate the retreating parties.[104]

Baker's theory is no doubt exaggerated and in keeping with the ways in which Britons historicized lands in the subcontinent. Nevertheless, it is ironic that the very man who reveled in the thrill of settlement and the creation of a new town at the same time documented what had existed before. An interesting account of the pre-British history of Nuwara Eliya was also presented by James Rutnam in the mid-twentieth century. He argued that the area around present-day Nuwara Eliya is linked with the epic historical poem the *Ramayana*, which tells of Rama rescuing Sita from the clutches of the island-king, Ravana. Sita is said to have been kept captive in Nuwara Eliya. In addition to this, Kotmale, within reach of Nuwara Eliya, may be the spot where one of the island's hero-kings,

101. Le Mersurier, *Manual of the Nuwara Eliya District*, 64.
102. Baker, *Eight Years*, 40.
103. Baker, *Eight Years*, 48.
104. James Holman, *A Voyage Round the World from 1827 to 1832* (1834), 264.

Duttagamani, resided for a while, and the discovery of a lithic inscription points to the possibility that there was a Buddhist temple in the plains of Nuwara Eliya.[105] Despite the difficulty of verifying myths, there is, from information gathered from sources ranging from the guides who accompanied Davy to Baker's exploration of vestiges from the past, enough evidence to support the claim that the development of colonial Nuwara Eliya relied in part upon an earlier historical engagement with the highlands. At the same time Nuwara Eliya provides rich evidence for how the British utilized a confident set of tropes in remaking the land. Even where they sought to make nature anew, and to find a place without the "indigenous," these crept back into their narrations.

CONCLUSION

In recovering the politics of colonial knowledge, it is important to keep in view a range of variables, among which temporality and spatiality are important.

It is too often the case that the character of colonial knowledge at the *end* of the nineteenth century is stretched back and made to describe the early expansion of the British empire. This is sometimes the case in the historiography of hill stations in India, where the late-nineteenth-century moment, when hill stations became places of inaccessible government, where the British retreated in season, and which were linked to a strict regime of race, climate, and class, is taken to stand for the character of these urban spaces more generally. Judith Kenny writes: "The hill stations thus served a particular role within the imaginative geographies of imperial discourse, a role that enabled the imperialist mind to intensify its own sense of itself by dramatizing distance and difference."[106] This chapter is an attempt not to deny this narrative but to show that the hill station was not built on a blank slate. In one sense this argument extends Dane Kennedy's interesting claim about how the later history of the hill station encapsulates a story of failure, in that they were ultimately unable to sustain the envisaged program of racial segregation.[107] It is important to attend to the lim-

---

105. James T. Rutnam, "Ancient Nuwara Eliya." For more on Dutugemunu's link to this region and for Kotmale's place in the Kandyan kingdom, see also Anuradha Seneviratna, *Sunset in a Valley, Kotmale: Record of a Lost Culture and Civilisation.*

106. Judith K. Kenny, "Climate, Race and Imperial Authority: The Symbolic Landscape of the British Hill Stations in India," 696.

107. Dane Kennedy, *Magic Mountains: Hill Stations and the British Raj.*

its and contexts of colonial knowledge: in the early phase of expansion in Ceylon, British knowledge was hemmed in by existent senses of the land, and this applies to the account of the making of Nuwara Eliya. The discourse of separateness must not be taken too seriously.

Yet this has not been a chapter which simply presents the history of the hill station of Nuwara Eliya. Rather, it has been an attempt to come to terms with the ways in which British natural programs of islanding were entangled with Lankan ones, even as they redefined the indigenous. The history of the hill station therefore makes sense only—and this brings in the second variable of space—in relation to what was happening in the lowlands. The distinction between the lowlands and the highlands was an established dichotomy of spatial imagination connected to the island polity. The British inherited this set of associations, and these fed into British views of tropicality and its separation from the temperate. Both eighteenth-century inhabitants of Lanka and the British were aware of the potential of natural icons, such as the coconut tree, and utilized a rich store of language and narrations in coming to terms with the variation of nature across space and altitude. While botany, as a science of natural historical diversity, was institutionalized by the British, here too there is a remarkable synergy of what islanders and colonists knew. The main botanical garden was built on the site of an established garden, and it relied on local knowledge and labor for its operation. The indigenous was recontextualized within the structures of the Crown colony.

The application of science in imperialism has been a central theme of this discussion. By coming to the history of botany from Ceylon, it is possible to view the history of Kew Gardens in a different way. Richard Drayton's important work showed how Kew Gardens was a center for the collection of natural historical information from across the world, an entrepôt through which plants moved south to north and east to west.[108] Botany was an engine of agrarian patriotism and natural improvement, driven as it was by a natural theological language of Christian paternalism. Yet when the subject is turned inside out, it looks very different with Peradeniya in central focus rather than Kew. Here, the fragility of Kew's reach is more apparent. The nodes of this network of information were not connected seamlessly, for Peradeniya's origin and early history show the persistence of established forms and modes of classification, alongside the insistent attempt to assert Linnaean botany. It was from these weak foundations that the later power of the plantation complex arose.

108. Richard Drayton, *Nature's Government*.

The next turn in the book expands this attention to the land by bringing in other types of knowledge, connected to mapping, road building, and
irrigation. There is development of one line of thought, which has been
evident already. In making the claim that this early period of British expansion saw slippages of knowledge—relating to texts, travel, sites, history, and nature—it has not been my intention to present these connections simply as unconscious ones. Earlier, I suggested that the discourse of
kingship could have been inherited unwittingly, and also that the lineage
connecting earlier scholarship to British orientalism was not orchestrated
by the British. From this it is clear that there was an element of serendip-
ity to how the British followed established routes; yet this is a telling symmetry, for it shows how knowledge works by default through established
channels. Yet it would be incorrect to present the entanglements between
British and Lankan practice to be entirely fortuitous. The strength of using diverse traditions of knowledge was something that the British were
aware of, and may indeed have silenced with intent, as in the case of the
establishment of the Peradeniya gardens. At other times, in borrowing and
yet surpassing established knowledge, the British trumpeted their power
in public fashion. Some of these occasions—which surface in the next
chapter—are telling instances of how slippage could lead to contestation,
power struggles, and claims of supremacy.

CHAPTER SIX

Land

Knowledge was critical to the sinews of power. Without proper information British armies struggled to take Kandy. At the same time, the regulation of workers, soldiers, and laborers in the expanding colony depended on information about allegedly ancient laws and customs. Some of the radical examples of the workings of the colonial government in Ceylon in the first half of the nineteenth century, such as the integration of the Kandyan provinces with the coastal areas so as to make an all-island colony, and the development of the plantation complex, can be seen as episodes in the state's creation and utilization of knowledge. Yet if information was at the heart of governance, it is necessary to attend to how islanders were aware of this link and responded to the rolling out of colonialism. While the last two chapters focused on how topographies and localities within the island shaped the islanding project, this chapter and the next turn to the resistance of Lankans to the colonial state.

The examples that were cited above of the Ceylonese government at work have one thing in common—they are connected to the land and the attempted knitting together of the land into a united infrastructure. At the heart of the discussion that follows are microhistories of this broader process—firstly, the building of a road from Colombo to Kandy, and secondly the surveying of the ancient irrigation reservoirs of the north-central provinces. In both cases the British recycled precedents by charting existing paths and structures. The ambition of these programs was radical: the British exercised aggressive intent, in linking provinces together and in seeking after new means of making arid land cultivable in a part of the island that receives most of its rain in a couple of months of the year. However, both these projects were dogged with disappointment, delay, and even failure. The road to Kandy took longer to build than anticipated and was on occasion washed away by the monsoon rains; the reservoirs of the north-central provinces were not restored as planned in this period. The confidence of British interactions with the very land of the island were thus constrained by the reality of applying knowledge in a new terrain. While it is easy to see the arrival of the British as marking a moment of great ecological change, such a thesis needs to be more nuanced.[1]

Despite these symptoms of failure, these programs of "public works" were performances of governing power.[2] Discovering the vibrancy of extant geographical ideas amongst local peoples, the British sought to display the superiority of their information order by intervening in the land in a way that was visible to the emerging public. Such public contests about knowledge mean that in addition to the unwitting slippages and ready borrowings that have been explored thus far, the making of British colonialism in Ceylon also involved contests and clashes between systems of information. This is where violent resistance arose: alternative knowledges could be used for insurgency and combat in war, and Lankans could respond to public works by plans of sabotage or by attacking those who labored on them.[3] In this context it is less surprising that much of the practice of technical information in Ceylon was directed and dominated by military men on the British side. Exploration, surveying, and medicine were all part and parcel of questions of military security. But more broadly,

1. For the traditional story of how colonialism creates ecological change, see Alfred Crosby, *Ecological Imperialism: The Biological Expansion of Europe, 900–1900.* For recent ecological histories of India which present a more nuanced story, see Mahesh Rangarajan, *Fencing the Forest: Conservation and Ecological Change in India's Central Provinces, 1860–1914,* and Ajay Skaria, *Hybrid Histories: Forests, Frontiers and Wildness in Western India.* For the ecological history of Sri Lanka, see James L. A. Webb, *Tropical Pioneers: Human Agency and Ecological Change in the Highlands of Sri Lanka, 1800–1900.*

2. For a wonderfully evocative account of the role of technology in colonialism and nationalism, see Rudolf Mrazek, *Engineers of Happy Land: Technology and Nationalism in a Colony.* For other accounts of technology and colonialism, see Daniel Headrick, *When Information Came of Age: Technologies of Information in an Age of Reason and Revolution, 1700–1850.*

3. For Kandyan warfare, see Channa Wickremesekera, *Kandy at War: Indigenous Military Resistance to European Expansion in Sri Lanka, 1594–1818.* For more on resistance around the land in Ceylon, see James S. Duncan, *In the Shadow of the Tropics: Climate, Race and Biopower in Nineteenth-Century Ceylon.*

in the expansion of the British empire it was military knowledge that gave rise to the emergence of disciplines in the colonies, such as geography or medicine.[4]

This chapter is first a history of cartography and then a history of roads and bridges. As historians of maps tell us, to govern a territory it is essential to conceptualize land in the mind's eye, and mapmaking was one way of doing this, in addition to other representational techniques. Accordingly, the island became a cohesive whole through cartographic definitions and renditions of space. Yet following Matthew Edney's important work on the survey of India, it is suggestive to see the map as "an inherently flawed cognitive panopticon."[5] By this Edney suggests that though the British sought to create gridlike maps for India, which constituted India as a unit, in reality at the ground level these maps did not create such unified spaces. It is a shame that Edney does not develop this line of analysis in relation to extant ideas of topography and mapping in South Asia, and neither does Ian Barrow's recent work on Sri Lanka.[6] Colonial cartography in Ceylon is placed in such a context here, in addition to being placed in the narrative of the consolidation of the colonial state. The novelty of this argument arises from the collection of palm-leaf sources on topography in Lanka, which are brought into the history of cartography for the first time.

The discussion begins with an account of the wars of 1803 and 1815, and the importance of topographical information to their outcome, before outlining the sources of topographical information among rural communities in Lanka. It then traces the emergence of British cartography in the island and also highlights how this cartography was not purely an objective study, but one that had aesthetic qualities too. This allows us to come to a more balanced comparison of the modes of operation of different knowledges, without assuming reliance on the colonialist benchmarks of rationality. The second half of the chapter takes the analysis forward to the application of map-making to roads, bridges, canals, and reservoirs.

4. For the military and knowledge in South Asia, see R. Cooter, M. Harrison, and S. Sturdy, eds., *Medicine and Modern Warfare*, and Seema Alavi, *Sepoys and the Company: Tradition and Transition in Northern India, 1770–1830*.

5. Matthew Edney, *Mapping an Empire: The Geographical Construction of British India*, 331.

6. Ian Barrow, *Surveying and Mapping in Colonial Sri Lanka*.

THE LAND AS DEFENSE

In 1803, when the British waged war with the Kandyans, with disastrous consequences, the lack of accurate maps proved perilous. According to Major Arthur Johnston of the Third Ceylon Regiment, the guides that could be procured were often "in the pay of the enemy," and could "tangle the troops in the forest" and then leave them to their fate.[7] But even if the guide was a Muslim, and so friendly to the British, he might be perfectly well acquainted with the high roads that could be used for travel in daylight and yet be of no use in helping an army at night. "In these thick forests," the major wrote, in following the trope of playing up the isolation of Kandy, "even in the brightest moon-light it is extremely difficult, and often impossible, for one not perfectly acquainted with the track to discern the foot-path."[8] The usual technology of the compass could also provide no assistance, "it being impossible to march in a direct line through a thick forest, intersected in many places by rivers and swamps."[9] When Kandy did eventually fall in 1815, Captain L. De Bussche, who held the post of Deputy Adjutant General, wrote how geographical information had played a crucial role in the victory, as "the different divisions of the army were supplied with excellent charts, and the most distinct information respecting the strengths of the passes leading into the interior."[10]

Even as the British conquest of Kandy depended on the acquisition of geographical knowledge and guides, Kandyan resistance to encroachment involved the protection of information about the interior. According to Major Johnston, on the paths to the interior "gates [were] fixed and guards stationed, to prevent the entrance of strangers, and to examine all passengers," and this was so as to "watch the ingress and egress of their territory."[11] It was therefore the intentional strategy of the Kandyan kings to keep the roads nearly impassable so that their territory was inaccessible.[12]

---

7. Arthur Johnston, *Narrative of the Operations of a Detachment in an Expedition to Candy in the Island of Ceylon in the Year 1804* (1810), 129–30.

8. Ibid., 127.

9. Ibid., 128.

10. L. De Bussche, *Letters on Ceylon, Particularly Relative to the Kingdom of Kandy* (1826), 20.

11. Johnston, *Narrative*, 3.

12. See Jonathan Forbes, *Eleven Years in Ceylon: Comprising Sketches of the Field Sports and Natural History of that Colony and an Account of its History and Antiquities* (1840), 2:66.

The strategies of war the Kandyans deployed also point to geographically inspired resistance.

The colonial sources for this period have copious descriptions of how Kandyans went to battle, the most detailed being Johnston's account of the failed war of 1803. The major wrote that the Kandyans harassed the British army while they were on march by cutting off supplies, interrupting communication between the divisions, and "occupying the heights which command the passes, from whence they fire in perfect security behind rocks and trees."[13] While using the natural features of the land to hide themselves, Kandyans made a habit of trapping Britons inside the mountains so that they could not escape. They were "certain that the diseases incident to Europeans in that climate, and the want of provisions" soon obliged the British to fall back. So even as Britons were on the march, the Kandyans blocked up the routes of retreat, and, when the British attempted to return to their seaside strongholds "encumbered by a long train of sick and wounded," the Kandyans would finally attack the Britons and have the upper hand.[14] The Kandyans' familiarity with their environment was such that it was reported that they cut paths to right and left through the jungle so that they could outflank the invading British army.[15] Until 1815, a mere jungle track, where men had to march in single file, led from the British outpost at Avissawella into the Kandyan kingdom.[16]

In addition to using geography for military purposes, the Kandyans relied on nature for ammunition and supplies. Johnston notes how each village chief on receiving orders from his superior "summons every third, fourth, or fifth man, according to the nature of his instructions, and proceeds with his feudatory levies to the place of rendezvous." Each soldier in addition to a musket takes with him "a leaf of the talipot tree" which could serve as an umbrella and a tent, when the broad ends of the leaf were tied together. Their chief food was a grain that grew on the hills with little cultivation, therefore negating the necessity for supplies to be carried across distance. Johnston added that "two or three cocoa-nuts, a few cakes, made of the grain I have just described, and a small quantity of rice, compose the whole of the soldier's stock for the campaign."[17] In this narrative,

---

13. Johnston, *Narrative*, 5.

14. Ibid., 6.

15. Wickremesekera, *Kandy at War*, 141.

16. W. Ivor Jennings, *The Kandy Road*, 12.

17. Johnston, *Narrative*, 9. The grain is likely to be *kurakkan*, finger or caracan millet *(Eleusine coracana)*.

the Kandyans themselves appear as creatures of nature. A local soldier "crawls through the paths in the woods, . . . or climbs the mountains and places himself behind a rock, or a tree, patiently to await the enemy's approach."[18] Once battle was won, it was common practice among the Kandyan troops to heap the enemy's heads as war trophies; Sinhala war poems liken these collections of heads to piles of coconuts.[19]

Topographical information and territorial expansion operated side by side and resulted in a contest between the colonizers and the Kandyans about what they knew about the land and how they engaged with nature. Faced with this contest, the British portrayed the inhabitants of the interior as isolated and unhappy, awaiting British liberation when open commerce and exchange could begin.

MAKING BOUNDARIES

The Kandyans' deployment of geographical knowledge in war illustrates the fragility of European knowledge. However, it is still important to recover as much as possible about what Lankans knew about the land before British conquest. By using a series of sources concerned with boundary divisions in early modern Sri Lanka, which are just coming to the attention of Western academics, I hope to suggest the vibrancy and popularity of a set of sacred geographical knowledges.[20]

The sources which will be used in the discussion that follows are of two types. First, the *kadaim* or boundary books are state documents which describe the three main divisions of the island of Sri Lanka, and which were in the custody of the *lekam-gey-attan* or functionaries to the secretariat at court. [21] Though the tradition of fixing boundaries has an ancient history in Sri Lanka, it is striking that the making of kadaim books

18. Ibid., 10.

19. Wickremesekera, *Kandy at War*, 146.

20. I am drawing on the work of H.A.P. Abeyawardana, *Boundary Divisions of Mediaeval Sri Lanka* (1999), which is a translation of the Sinhala version first published in 1978. A selection of these texts is also currently being published in Sinhala under the editorship of Prof. Gananath Obeyesekere; see, e.g.: G. Obeyesekere, ed., *Bandaravaliya saha Kadaimpot; Vanni Upata, Vanni Vittiya saha Vanni Kadaimpot;* and *Ravana Rajavaliya saha Upat Katha.* Sources that have been translated into English may also be followed up in H.C.P. Bell, *Archaeological Survey of Ceylon, Kegalle District* (1892); A. C. Lawrie, *A Gazetteer of the Central Provinces of Ceylon* (1896); and W. A. De Silva, "Sinhalese Vittipot (Books of Incidents) and Kadaimpot (Books of Division Boundaries)."

21. Abeyawardana, *Boundary Divisions*, 9.

appears to be an early modern preoccupation. No kadaim book has been discovered which predates the period when Gampola was a capital in the fourteenth century.[22] Information about territorial boundaries, historical anecdotes, and comments on social organization are seamlessly intertwined in these books. I will also use a set of sources called the *kadaimkavi* or boundary verses, which were recited at village ceremonies or to entertain local chiefs, often as a means of publicizing erudition. Again these were mostly composed during the period prior to the advent of the British. The anthropologist Gananath Obeysekere has suggested that the preponderance of boundary texts in the early modern period is a reaction to the redrawing or rearranging of boundaries by the Portuguese.[23] It is plausible that the arrival of European geographical knowledge saw a resurgence and acceleration of interest in boundaries among people ruled by the king of Kandy.

The oldest kadaim book, *Sri Lamkadvipaye Kadaim*, divides the entire island into 114 *ratas* or countries. The three principalities of Tri Simhala are divided as follows: Maya has 28 ratas, Pihiti has 43 ratas, and Ruhunu has 43 ratas. Boundary pillars mark off the limits of particular ratas. Take the description of Bogambara-rata in the *Sri Lamkadvipaye Kadaim*: "For the four boundaries, stone pillars are placed on which are carved the figure of bō-leaves."[24] Elsewhere this text makes note of other types of boundary pillars: Mayadunna-rata is bounded by sixteen stone pillars bearing the figures of trees; Navayotna-rata has sixteen stone pillars bearing the inscription of cobras; Devamadda-rata has sixteen stone pillars on which were carved figures of parrots. In Amada-rata "to the east and west, three boundaries are marked by inscribed stone pillars placed on the tank bunds," and in Elasara-rata, "a giant canal has been built and in the four corners are placed boundary stone pillars upon which are indited the figures of an arrow."[25]

H.A.P. Abeyawardana, in his commentary on these texts, has argued that the meaning of these symbols was easily comprehended by observers: for instance, the sight of the cobra could denote protection, and the figures of the sun and moon could denote that the gods had decreed that these boundaries would last as long as the sun and moon.[26] But there are

22. Ibid., 7.

23. Gananath Obeyesekere, ed., *Rare Historical Manuscript Series 1*, xvii.

24. Abeyawardana, *Boundary Divisions*, 193.

25. Ibid., 202.

26. Ibid., 148.

exceptions to this pattern: some pillars were hidden and constituted secret knowledge about boundaries. In addition to this, certain boundary pillars denoted the name of the region they marked out. For instance, the figure of the horse on a stone pillar denoted a region taking its name from the horse, the figure of a squirrel a region taking its name from the squirrel, and the figure of the moon a place taking its name from the moon. The *Sri Lamkadvipaye Kadaim* notes an intention on the part of the rulers to have these boundary pillars last unaltered for eternity.[27]

In addition to boundary pillars, natural features of the land were also used as boundary marks, and this indicates that rocks, mountains, and rivers were seen as stable features of the landscape. Marks in trees and rocks may have been renewed from time to time in order to make them endure.[28] Take the example of this kadaim-kavi, which describes the boundaries of Muvatapattu:

> From the archers' rock to the Valimaluva (sand-strewn) rock,
> From that to the next upon which are engraved oxen hooves,
> Then to the nagara-letter carved rock,
> Past the three-pronged gorge to Kalukohovila pool,
>
> Beyond Dunukeyiya rock, proceed to a point,
> Below the rock fortress that flies the war flag,
> Past the three-swords-engraved rock, reach,
> The wooden pond frequented by elephants.

These verses illustrate that Lankans' view of the landscape was far from simplistic. Rather, local peoples engaged with the land symbolically and read the land in order to come to an idea of its natural boundaries. The rocks and waterfalls that appear in the kadaim-kavi are therefore laden with historical and religious meaning: there are references to a rock from which King Rajasimha watched a battle, a cave which sheltered a sacred relic, and a mountain which was full of incense.[29] At the same time peculiarities in nature were used as points of attention in marking a boundary or path. One poem refers to a creeper-laden tree and another to a stony cotton field frequented by a leopard.[30] The kadaim books weave together

27. Ibid., 151.
28. Ibid., 52.
29. Ibid., 219–22.
30. Ibid., 221.

a narrative of kingship, with very detailed local information: the *Laggala Kadaimpota* glorifies King Kirti Sri Rajasimha and presents a response to the king's question about who lives in Laggala by listing the people and bringing this together with folk tales of the area.[31]

The geographical knowledge demonstrated by the kadaim books and kadaim-kavi is closely linked with piety. The *Tri Simhale Kadaim* makes note of the sacred places in each of the regions. The misdeeds of King Kalanitissa are said to have changed the geography of Maya-rata: "the sea engulfed five leagues of land and kept only one league."[32] The "sacred places of Ruhunu-rata" include several *vihara* or temples and the "Sripadasthana on the Samanala mountain [Sri Pada] where Buddha placed his foot-mark which has 216 auspicious signs and the place is visited for worship by pilgrims."[33] The implication is that these boundaries are sacred divisions of the land. In fact the whole of Sri Lanka appears in these texts as a sacred land. *Lakvidiya* begins: "Adoration to the Buddha. Having subdued the yakkhas [demons] here/ In the past the Buddha made this land / The home of the Triple Gem / Hence the lineage of the righteous rulers / Firm in the Dhamma / Was established in the Simhala."[34]

This geographical knowledge was meant to be easily accessed. The versifier of one kadaim-kavi noted that he had derived his information from "following an ancient book," which was copied so that everyone could know the path.[35] It is possible that the kadaim-kavi fitted into a culture of entertainment. The kadaim-kavi which describes the boundary of Paranakuru Korale takes the form of a riddle:

> Reptiles and snakes gathered in a cloth bag,
> Bloody but sumptuous food kept atop a rock,
> Sweet and tasty betel leaves placed below the rock,
> And old elephants have added their rut to the heap.

This riddle is solved by the realization that each line contains a pun. For instance in line two, food *(bata)* with blood *(le)* upon the rock *(gala)* indicates the rock named Batalegala. And so the reply is given as follows:

31. Obeyesekere, ed., *Bandaravaliya*.
32. Abeyawardana, *Boundary Divisions*, 205.
33. Ibid., 207.
34. Ibid., 208.
35. Ibid., 224.

Friend! I reckon that what you are saying is that
From Nayiyankada is a difficult climb to Batalegala,
But, there is much fun to sport on the Rahala rock,
Is not the Parana kuruva amidst these three?

It is important to take on board the oral nature of this knowledge. It has
been argued that these types of texts were "written in simple prose or
verse by local intelligentsia (scribes, headmen, nobles and even ordinary
villagers)."[36] If this is confirmed to be the case, then we have evidence for
an extremely popular genre of geographical knowledge. The language of
these poems supports this view; the Sinhala is colloquial and related to
the diction of rural people.[37]

At the same time as pointing to popular culture, these texts also gesture to the way in which knowledge was codified. This is especially true
of some kadaim books which in a sense can be called censuses or route
surveys. The *Merata Kadaim* for instance, in describing the region of Ma-
hagalrata, notes its resources, religious statues, towns, population, natural
features, and boundary pillars. It is not too dissimilar from the colonial
travel journal.[38]

The interrelationship between topography, history, and religion suggests that geographical knowledge was deeply interwoven into Sinhala culture. It wasn't an elite form of information. These traditions allowed the
land to become a repository of meaning, and geographical features were
approached with veneration by local peoples. Though these traditions cannot be identified with Kandy, they provide a context for appreciating how
Kandyan troops did not merely know their environment better: they observed the land's contours, history, and sacredness.

AMATEURISM TO COMMERCE

As early as four years after the British took control of the maritime provinces from the Dutch, they had established a Survey Department to bring
their new acquisition into the bounds of known knowledge.[39] The *Government Gazette* of 5 April 1802 included the following: "Notice is hereby

36. Obeyesekere, ed., *Rare Historical Manuscript Series 1*, i.
37. Abeyawardana, *Boundary Divisions*, 67.
38. Obeysekere, ed., *Ravana Rajavaliya saha Upat Katha*, "Meratakadaimpot."
39. For the role of surveying under the Dutch, see Barrow, *Surveying and Mapping in Colonial Sri Lanka*, 20–30.

given, that in the measurement of land in the British territories of Ceylon, the standard measures of Great Britain, alone will . . . be made use of." The new units of measure were then specified: a chain was equal to four poles or perches, and could be divided into one hundred links.[40] After paying attention to the relatively poorly documented traditions of geography of Lanka, the confidence of this statement is rather startling.

Joseph Jonville, the first Surveyor-General, who took up his post in 1802, was entrusted with the task of surveying the land so as to assist the government in debates concerned with the ownership of property. Yet Jonville, being of the enterprising sort, was quickly distracted by the need to make "a general map of the Island of Ceylon with that exactness which alone can render it useful to Government and at least interesting to individuals fond of the Arts and Sciences."[41] The size and scale of this map was dictated by its intended use in administration. It was made to be as large as 90 to 100 feet in length, so that various provinces could be considered separately and individual villages could be inspected, and cultivated land marked as distinct from uncultivated. Of course, such a survey depended on trained observers and proper instruments, neither of which Jonville had in his possession. So he wrote: "Permit me to represent to you that it is impossible to procure instruments here, which are absolutely wanted." Undeterred, Jonville made his own instruments; by cutting a glass tube, he made several levels, and he manufactured small scales out of copper. This lack of organization came to characterize British surveyors from Jonville right up to the 1830s. For instance, the surveyor Lieut. Myliers reportedly spent much of his time having riotous, all-night drinking parties.[42]

Soon after the acquisition of the Kandyan kingdom in 1815, a military survey of the interior was undertaken by Captain John Fraser. Lieutenant Thomas Skinner, who was employed under Fraser, was directed to examine the great mountain chain surrounding the Upper Kandyan Provinces, "sketching its principal features and ramifications, with as much minuteness . . . and tracing and laying down its passes and defiles, and the paths and roads that lead through them."[43] The need to represent Kandy on paper was militarily strategic, especially given the impact that a lack of geographical information had on the British in the first decade of the century.

40. The *Ceylon Government Gazette*, 5 April 1802.
41. Cited in R. L. Brohier, *Lands, Maps, Surveys: A Review of the Evidence of the Land Surveys as Practised in Ceylon*, 2:15.
42. Barrow, *Surveying and Mapping in Colonial Sri Lanka*, 4.
43. Thomas Skinner, *Fifty Years in Ceylon* (1891), 170.

Skinner was directed to use the method of triangulation; he was given the best theodolite but no further training in its use. In his autobiography he writes:

> I was unfortunately too proud to acknowledge my incompetence, but gained immense strength from the faith my old chief, Sir Edward Barnes [the Governor] placed in me when he appointed me to the Quartermaster-General's department. In reply to my protest that I knew nothing of scientific duties, he said, "Do you think I do not know as well as you do. Will you try to qualify yourself?" "Yes certainly I will Sir." . . . With the recollection of his piercing eye which went through me . . . I shut myself up in a room with a theodolite and *Adams on Instruments*, took it entirely to pieces, put it together again, and learnt the use of all its parts. . . . I next took my new friend, the theodolite, to the western extremity of my base line, and took many complete series of observations round the circle, when I found that I gained by repeating them a certain coincidence in my angles.[44]

Behind the presentation of a theodolite to the untrained Skinner was a belief in the ability of a rational Briton to bring the unruly territory of Kandy under the cool gaze of science. Yet it was requisite that Skinner emphasize his hard work to gain his place as a hero. He wrote therefore that he "never knew the shelter of a roof between four or five o'clock in the morning till seven in the evening." He had often gone without provisions when surveying. Once over a two month period he had "only five miserable chickens, three of which had died from the rain and cold on their way up to the Peak." His heroism arose not only from his ability to map the environment but also from his aptitude at controlling his body. He observed: "All the liquid I took during the day did not exceed one imperial pint; this *regime* brought me into such splendid working condition that I could outrun everyone."[45] At night he lived in a tent composed out of "five sheets of tallipot leaf, stitched together" which contained a "little camp bed, a small camp table and chair."[46] So Skinner concluded: "It was precious hard work, delightful to think of in the retrospect."[47]

44. Ibid., 173.
45. Ibid., 178–79.
46. Ibid., 176.
47. Ibid., 174.

While this hagiographic rhetoric would have us believe that new maps and surveys were the result of lone individuals, this was never the case. In an aside, Skinner noted how the inhabitants of the interior referred to him as *Cannade Mahotmia*, or "Gentleman of Instruments." In opposition to his own persona as the objective man of science, Skinner presented his audience to be bemused by the apparatus of knowledge. Major Forbes also drew attention to the bewilderment of the local people with regard to the technological prowess of the British; the theodolite was said to have attained "an extraordinary character," and "the ticking of the watch, and the turning of the compass in the theodolite, seemed to the Kandians to be direct vitality."[48] Despite this stereotypical trope of distancing, it is important to stress how surveyors relied very practically on colonized labor for the purposes of conveyance. In 1812, George Hayster, Civil Engineer, wrote:

> I beg leave to represent to you that I have been obliged to advance between 4 & 500 Rix Dollars to my Palankeen boys, coolies &c &c about to attend me in surveying the Giant's Tank and as Government have been pleased to decide that my travelling expenses should be paid, in consequence of the low salary I receive (viz.) that of Assistant Civil Engineer, although I am now called upon to perform a more arduous and responsible service than has yet occurred in this island in the Civil Engineer's Department.[49]

The surveyor therefore did not travel alone; he was attended by a retinue of local laborers, who were eventually erased from the account. Yet, as this letter hints, the financial constraints of Crown rule meant that the Survey Department could not be expanded with the recruitment of skilled Europeans. In the context of drives for economy, surveyors had to train local people in the art of using instruments to map the land. In 1833, for instance, Captain Gaulterus Schneider, a Dutchman who decided to remain in Ceylon, reported how he had acted as a "teacher of mathematics and surveying" and that those who had been instructed by him had since been employed as surveyors.[50] When Skinner retired, 1,598 chiefs and headmen signed a memorial letter, which commended him for the encouragement

---

48. Forbes, *Eleven Years*, 2:87.

49. George Hayster to Chief Secretary of Government, dated 5 September 1812, Lot 6/259, SLNA.

50. Copy of letter from G. Schneider, dated 10 February 1833, CO/54/127, TNA.

he held out to indigenous talent "by the admission of young men into the Government Factory as apprentices, with a view to qualify them as practical Engineers."[51] Local peoples were thus enlisted to the task of extending British knowledge. Surveying the land did not see a straightforward influx of British observers; it relied on European and "indigenous" people of different backgrounds and training.

Despite the presentation of surveying as specialist science, it seems to have functioned as an aggrandized mode of colonial information gathering. While collecting information about the land, surveyors were instructed to observe people and their political organization also. Jonville required his surveyors to "keep a register containing the angles, distances and all possible details on the population, soil, and natural productions."[52] One copy of this register was to be deposited in the survey office and another in that of the Secretary of Government. The manner in which the science intervened in colonial systems of governance is best illustrated by some instructions given to surveyors by P. A. Dyke, the Government Agent of Jaffna. Dyke urged that headmen be included in surveys and noted that "more difficulties far more serious than will be readily supposed" might arise if they were excluded. He stressed that even if they proved troublesome or intrusive, "their formal participation or co-operation or at least cognizance" of the manner and results of the survey were crucial for the purposes of governance.[53]

While surveyors may have preferred to create grand maps that spoke of their expertise and heroism, the commercial advantages of the science were closer to the heart of the colonial regime. Indeed, from the 1830s onwards one of the main aims of the surveyor was to survey blocks of land for coffee plantations. This serves as another instance of how cartographic science was molded by the needs and structures of the British government. In January 1842, the Surveyor General advised the government that Crown lands in the neighborhoods of Colombo, Kandy, Galle, and Jaffna, and close to the high roads and navigable rivers, should be surveyed in small allotments and prepared for sale.[54] By 1844, Dyke could report that the "people of the country" were asking to purchase lands, but cautioned that inhabitants were generally occupied in "removing their fences and other

51. Skinner, *Fifty Years*, 227.
52. Brohier, *Lands, Maps, Surveys*, 17.
53. P. A. Dyke to the Surveyor General, dated 9 January 1845, Lot 10/115, SLNA.
54. Surveyor General to the Colonial Secretary, dated 5 January 1842, Lot 10/115, SLNA.

landmarks and taking in more or less of the waste lands adjoining."[55] An improved survey was said to be what could "check these parties and ward off the ultimate consequences from them of serious detriment to the public interest."[56]

British surveys were seen as testaments to the authority of the colonial state over its territories. Yet ultimately, cartography arose out of amateurism, and rather than becoming a demonstration of rationality in keeping with British discourse, it was fundamentally a way of adding to the coffers of the state.

CARTOGRAPHY'S AESTHETIC

It is easy to place the growing confidence of British surveying—as it moved from early amateurism to midcentury commercialism—in opposition to Lankan engagements with the land. Yet it is vital that we do not take the rhetoric of objectivity that surrounded colonial cartography too seriously. It is necessary to place any discussion of British surveying in the context of a wider account of colonial engagements with the land in Ceylon, for instance leisured travel.

Popular books on Ceylon from the first half of the century had the aim of advising travelers. J. W. Bennett advised that the best way for a visitor to travel across the island was by palanquin, usually a covered platform borne on the shoulders of local peoples, such as that used by Mrs. Walker, the climber of Adam's Peak. This would allow the "traveller to be at . . . ease, to stop when [they] please, to view the country, or to collect specimens in natural history." Since government regulations did not allow for a coolie to carry more than forty pounds for a greater distance than two miles from any town, Bennett advised that the provisions for a journey should be light. It was recommended that a tin box for breakfast should contain "supplies of tea, sugar, coffee, powder, shot, caps, a lamp fitted to a low candlestick, with a couple of glass shades, wax candles, &c. &c."[57] Travelers were told that they should "never strike a native, how much soever [their] temper may be put to the test," and that "for every days' halting, Sundays included, [their] coolies are entitled to *Batta*," an extra allowance. Bennett made certain to include the advice

55. P. A. Dyke to Colonial Secretary, dated 24 December 1844, Lot 10/115, SLNA.
56. Ibid.
57. J. W. Bennett, *Ceylon and its Capabilities: An Account of its Natural Resources, Indigenous Productions and Commercial Facilities* (1843), 174–75.

that an umbrella should be carried whenever practicable. While keeping the traveler in the cool, an umbrella could also be converted into a mosquito net for the night, and prove "an excellent defence against that inveterate enemy."

Travelers were encouraged to envisage the island as one vast garden awaiting exploration, which was in keeping with the island discourse. James Cordiner, the island's first British chaplain, noted that a journey in Ceylon may be "compared to an excursion in a large park or garden where there are no artificial walks."[58] Behind this was also a sense of ownership of the land. In a classic expression of how the introduction of British manners and customs could improve the land, Pridham noted that there was no scenery "more picturesque than the river near Baddagamma." The writer's peculiar interest in this site originated from a comparison with England: "The stream with its grassy banks, the green meads, and the woody hills around forcibly recall to mind the scene presented by the Thames in the vicinity of the Richmond."[59] This clue is important because the analogy with a park supposes an affinity between the green pastures of Britain and Ceylon.

Yet essential to this image was a belief in the importance of exploration. British travelers were in fact as attached to the mountains of the interior as the Kandyan inhabitants who lived there, for both sides sought to claim sovereignty over the entire island. The mountains marked the limits of their travels, and this attachment to mountains was a cartographic form of imagination. Note John Davy's description of the view from the summit of Adam's Peak: "It looked like a map, laid out on a magnificent scale, with a glow of coloring, warmth of light, and charm of landscape, that we rarely see combined, except in the paintings of the first masters."[60] British attachment to the view at the top of mountains is also evinced in their critique of Dutch houses built at the foot of hills. Leisure houses built by the previous colonizers were said to lie in "low, sheltered, often swampy positions, where the only object to be seen is a stagnant pool," while British governors and residents built their abodes "on the summits of cleared eminences, where refreshing gales allay the fervour of the torrid zone, and the eyes are delighted with the rich prospects and perpetual

58. James Cordiner, *A Description of Ceylon: Containing an Account of the Country, Inhabitants and Natural Productions* (1807), 16.

59. Charles Pridham, *An Historical and Statistical Account*, 2:606.

60. John Davy, *An Account of the Interior of Ceylon and of its Inhabitants* (1821), 361.

verdure."[61] Robert Percival noted that the towns of Ceylon built by local
rulers "look more like a number of distinct houses scattered up and down
in the midst of a thick wood or forest."[62]

The longing to make the mountains accessible led to Governor Barnes's
ambitious program of road construction. Striking here is the extent of
commentary on what could be seen from the road; the new highways
seemed to open up the park of Ceylon. Travelers were asked to contrast the
thick forests on either side with the powers of colonialism symbolized by
the road. "Far as the eye can reach, the road winding through the wilderness of jungle, is distinguishable; and this indication of the presence of
civilized man affords a striking contrast to the wildness of the mountain
scenery around. The more you gaze on the wild landscape, the more you
feel inclined to appreciate at its true value the enterprise and labor necessary for the formation of a carriage-road."[63] Roads without a view were
criticized for being dreary. Charles Pridham noted that even local peoples
referred to the region traversed by the road from Anuradhapura to Arippu
as desolate.[64]

The manner in which roads served as a platform from which to observe
the land, while providing the opportunity for a panoramic vision, is illustrated by this late nineteenth-century poem by V. M. Hamilton and S. M.
Fasson:

> They stretch like Roman roads of yore
> O'er swamp and plain, from shore to shore
> Up in the clouds they may be seen,
> Deep cut in side of steep ravine,
> Twisting and turning, in and out,
> The mountains rocky slope about;
> Now hidden from your wondering sight,
> Now terraced round some awful height,
> Whose massive overhanging brow

61. Cordiner, *A Description of Ceylon*, 184.

62. Robert Percival, *An Account of the Island of Ceylon Containing its History,
Geography, Natural History, with the Manners and Customs of its Various Inhabitants*
(1803), 174.

63. Augustus De Butts, *Rambles in Ceylon* (1841), 113.

64. Charles Pridham, *An Historical and Statistical Account of Ceylon and its
Dependencies* (1849), 2:527.

Fig. 6.1. "Amanapoora on the Road from Colombo to Kandy." From William
Lyttleton, *A Set of Views in the Island of Ceylon* (London, 1819). The
image shows a view seventy-six miles along the road from Colombo to
Kandy, at the military post of Amanapura. Author's photograph.

Threatens to crush the vale below
And on these roads the passer-by
May see much to delight the eye
If nature is his special craze
'Tis here displayed in every phase—
Jungle dark and palm-trees high,
Bold mountain peaks against the sky;
Glimpses of valleys far below,
All bathed in sunset's purple glow.[65]

These verses, which were written after much of the road network had been
completed, suggest that Britons sought to tame the landscape through
their engineering works (fig. 6.1). The evidence in this type of source of
a colonial aesthetic is very useful in destabilizing a dichotomy between
the modes of operation of what might be termed British and "indigenous"
geography, linked to the assumption that colonial knowledge was more
scientific and rational.

65. "On the Way up Country," in V. M. Hamilton and S. M. Fasson, *Scenes in Ceylon: Twenty Cartoons with Descriptions in Verse* (1881).

THE GRAND MILITARY ROAD

If there is one Briton who epitomized the program of roads in Ceylon, it was Governor Edward Barnes, who took up his position at the head of the colony in 1820, five years after Kandy had fallen. Barnes was trained as a soldier, and having taken a tour of the island, his immediate view was that a road to Kandy was the most crucial improvement necessary to secure the stability of Ceylon. While his predecessor, Brownrigg, had begun a program of road-building, this had been interrupted by the rebellion of 1817–18, and so to Barnes the task at hand appeared even more urgent.[66]

The need for this road arose in part out of a contest between British and local demonstrations of power over the land. One of the ancient accounts of the island held that it would never be subjugated "until the invaders bore a hole through a mountain that encircled the Kandian capital."[67] In building the direct Kandy road, the British made certain to construct a tunnel at Kurunegala, as was noted in the Introduction. Three months after taking his position, in setting the scene for the road, Barnes brought the Ceylon Pioneer Lascars, Engineer Pioneers, Quarter Master General's Pioneers, and Commissariat Coolies together into one unit. He wrote that this would enable local people to work under European supervision, thus circumventing the difficulties that Europeans encountered while being employed in construction in a tropical climate.[68] Having had the path of the proposed road traced, Barnes notified Lord Bathurst that the intended road "would consolidate the new with the old provinces; improve the commercial intercourse of the two, remove the principal difficulties that were experienced in the late military operations . . . and diminish the vast expense in the conveyance of the commissariat supplies." There were therefore several ambitions that motivated the Kandy road: the display of power in fulfilling a local myth, the efficient use of labor, the betterment of British finances in Ceylon, and the securing of Britain's military position in the interior.[69]

---

66. See Indira Munasinghe, *The Colonial Economy on Track: Roads and Railways in Sri Lanka, 1800–1905*, 13. Munasinghe has compiled a list of roads constructed or repaired by Barnes: a road from Colombo to Kandy via Kelaniya, Mahara, Henarathgoda, Veyangoda, Kegalle, and Peradeniya; a road from Colombo to Kandy via Kurunegala, the direct route; a road from Puttalam to Kurunegala; and the repairing and bridging of the old coastal road from Galle to Matara and from Colombo to Chilaw.

67. De Butts, *Rambles in Ceylon*, 169.

68. Dispatch from Edward Barnes, dated 19 May 1820, CO 54/77, TNA.

69. See also Munasinghe, *The Colonial Economy on Track*, chap. 2, for a description of the motivations of colonial road-building.

Despite this optimism, the road to Kandy took much longer to complete than anticipated. In 1822, while Barnes was away in India serving as Commander-in-Chief for ten months, Governor Edward Paget wrote to Bathurst: "The road now open for Carriages thro' the Seven Korles to Kandy requires much labor still to render it permanent throughout or capable of resisting the torrents of rain which fall twice a year."[70] Even in 1833, Governor Horton could still submit a bill for repairs to the Kandy road; he explained that it had been washed away by the rains.[71] The road was metalled only after 1841.[72] Britons therefore found it difficult to apply their science in a tropical climate. This is perhaps best illustrated by the plight of European officers, who continued to suffer in the heat despite the reorganization of the Pioneer Corps. Paget wrote for instance that they sacrificed their health and lives by "a continual exposure to the ardent heat of the sun in its highest elevation and never having any relaxation from fatigue, but when forced from the scene of their labors to confinement in sickness."[73]

The tale of Captain William Francis Dawson, one of the surveyors who had traced the line of the Kandy road and then expired from ill health in 1829, caught the imagination of the Anglo-Ceylonese community. Major Jonathan Forbes wrote that when he took the road to Kandy in 1828, Captain Dawson was still "in rude health and buoyant spirits," but that now "his lofty monumental column gleamed on the summit of the Kadeganawe pass."[74] This monument on the side of the road to Kandy became a regular stopping point for travelers, who could climb it and then behold a view quite like that which would greet their eyes atop a mountain (fig. 6.2). It commemorated Dawson's "science and skill," which had "planned and executed this Road and other works of Public Utility."[75] The remembrance of a road-builder, who had died under the strains of the task, with a column that shot high above the land that was to be tamed, symbolizes the anxiety displayed by Britons to acquire the requisite knowledge to govern Ceylon. Ironically, even British delight at their ability to use a local myth to present the powers of their science was short lived.

70. Dispatch from Edward Paget, dated 29 May 1822, CO 54/82, TNA.
71. Dispatch from R. W. Horton, dated 25 July 1833, CO 54/129, TNA.
72. Jennings, *The Kandy Road*, 12.
73. Dispatch from Edward Paget, dated 1 June 1822, CO 54/82, TNA.
74. Forbes, *Eleven Years*, 2:193.
75. P. M. Bingham, *History of the Public Works Department*, 1:13.

Fig. 6.2. Plaque on Dawson's Column on the road to Kandy. Author's photograph.

By the time Charles Pridham wrote his work in 1849, the tunnel had col-
lapsed and the Kandy road wound around a hill, adding two miles to the
journey.[76]

Britons therefore failed in the ideological task they set themselves: the
challenges of a tropical climate continued to be insurmountable. Despite
this failure, they remained keen on using the road to improve their financial standing in the colony. By 1822, the Deputy Secretary of Government
could write how the improvement in communication brought about by
roads had increased the consumption of British goods such as cottons,
iron, steel, cutlery, earthenware, glassware, haberdashery, umbrellas, and
bottles.[77] When there was a severe shortage of grain in Kandy in 1824, as a
result of a drought, Barnes noted that he would have been unable to make
supplies available without the road. Barnes's economic logic was to use

76. Pridham, *An Historical and Statistical Account*, 2:668.
77. Beaufort Shexham to Chief Secretary, dated 22 November 1822, CO 54/82, TNA.

tolls on public works such as roads and bridges so as to recoup the money
expended in their construction. By 1828, the law had been extended so
that Barnes had the right to put in place tolls on roads, canals, and ferries
wherever he thought appropriate.

Because the British presence in Ceylon was cast as a military form of
government, the civil and military affairs of the colony were intimately
linked, so that the making of roads, their repair, and the transfer of the
land over which they passed were all encompassed under direct orders
served up by the Governor. When Bathurst asked Barnes to separate civil
and military concerns, Barnes protested: "Here in most instances artifi-
cers, laborers and materials can only be procured through the medium of
the civil authority, and in many instances the Collectors of Districts must
actually cause the work to be performed." Defending the utility of military orders in the construction of roads and other public works, Barnes
wrote that local laborers "cheerfully obey such orders, they assure themselves of the protection of Government, not only for themselves whilst
employed in its service but for their families in the event of their dying
whilst so employed."[78]

By the end of the 1820s, there was agitation for reform of this militarized control. Peter Gordon, who described himself as a renter of farms
in India, and of a liberal temperament, published a letter on the political
condition of Ceylon in a volume of documents critiquing British presence
on the subcontinent. He wrote that "the MILITARY establishment with
which Ceylon is charged is permanent, expensive, and composed of foreign mercenaries. . . . Their task is to destroy her political existence. She is
therefore a conquered slave, and all her institutions are dependant on the
will of the conqueror."[79] Gordon noted how public service was legalized,
and that the law was merely the whim of the governor; property was said
to be insecure against the Crown. Writing of the system of compulsory
labor imposed on tenants of land, Gordon wrote that it was arbitrary and
undefined: "In the Candian provinces, from ten years of age the men may
be three months of the year in requisition, working under the white task-
master's lash. . . . The chief object to which this tax is applied, is, the mak-

78. Dispatch from Edward Barnes, dated 25 July 1825, CO 54/101, TNA.

79. Letter to the editor of the *Madras Courier*, dated 27 November 1827, reprinted
in Peter Gordon, *India; or, Notes on the Administration of the Establishments in India*
(1828), 59; Edward Barnes's comments on this letter appear in Private letter from Barnes
to Secretary of State, dated 22 January 1831, CO 54/112, TNA.

ing a grand military *road* from Colombo to Candy, a distance of 72 miles, which has employed 800 pressed men eleven years, besides three hundred miserables dressed as pioneers."[80]

In Ceylon, too, there were criticisms of Barnes's program of road-building and its use of *rajakariya*, the system of forced labor. George Turnour, for instance, noted how road service only fell on about a third of the landholders in the district of the Four Korales, and how this included workers who were infirm old men or young boys. His own recommendation was that road service be applied equally to all, but that it should only apply for one year from the date at which the road is commenced and that the poor have the option of paying a tax in rice or revenue. Yet Barnes took great umbrage at Turnour's boldness and in particular at Turnour's description of the rajakariya system as "oppressive."[81]

In 1830, this debate over the coerciveness of the British presence in Ceylon also entered into discussion in the British parliament. Mr. Stewart, M.P. for Beverley, proposed a motion for the appointment of a select committee to inquire into the financial state of Ceylon. Ceylon's debt at the close of 1826 was said to stand at £491,542, which was an increase of £28,000 over the previous two years, and the average excess of expenditure over revenue was above £97,000 in the preceding years. Stewart noted how the Kandyan people were compelled to keep the "military roads" to the interior in repair. This system of enforced labor was said to deter investment, given that it operated on a monopoly. "Until a more independent administration of justice is adopted, and a more enlightened commercial policy pursued, Europeans will not be inclined to settle in this fine colony."[82] Stewart also objected to how "military men" were appointed Governors of Ceylon: "Their education and previous habits unfit them to discharge the duties of [governor], with advantage to the public service."[83]

While the construction of roads was undertaken to prove the utility and power of British surveying, it speaks therefore as much of failure as success, both in financial terms and in the ideological expression of power (fig. 6.3).

80. Ibid., 63.

81. All citations from "Mr. Turnour's Report upon the settlement of the grain tax effected in certain parts of the Kandyan provinces, and upon the services required from the inhabitants in the construction and repair of roads," Lot 19/111, SLNA.

82. *Mirror of Parliament* (1830), 2:2016.

83. Ibid., 2:2014.

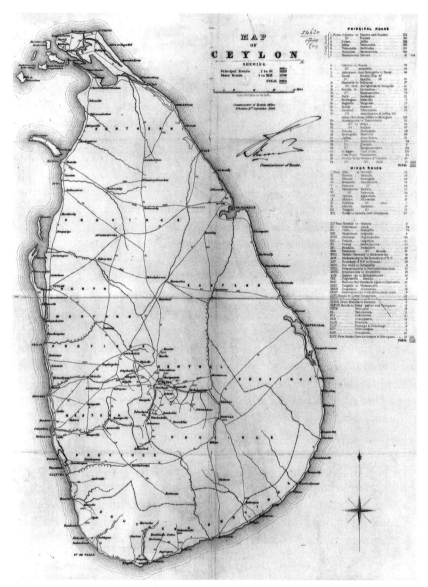

Fig. 6.3. *Map of Ceylon Showing Principal and Minor Roads*
(1850?), signed Thomas Skinner, Commissioner of Roads.
Photograph: © The British Library Board. Maps.54630.(11).

LABORERS AND RESISTANCE

In order to recover how the Lankans responded to the Kandy road, it is useful to pay attention to how the system of compulsory labor worked and how it incorporated allegedly ancient laws.

Barnes had two sources of labor available to him in building roads. First, the reorganized Pioneer Corps consisted of six divisions, each in the charge of a subaltern with a commanding officer; the corps consisted in part of laborers brought to Ceylon from Africa and India.[84] Second, tenants of Crown land were expected to work gratuitously, as a service attached to tenancy. Interestingly, the King of Kandy's forces, who had so successfully used geography as a means of defense, were composed of these same tenants of the land.[85] Major Skinner describes how he built roads with the help of "raw untaught Kandians" who were relieved of their labor after two weeks in accordance with the law, thus necessitating the regular training of new men in the requirements of road-making.[86] While this scheme of *rajakariya*, literally "service to the king," had been utilized on the coast by the Dutch for the purposes of the building of canals and roads, it had never been applied like this in the Kandyan kingdom.[87] Rajakariya was abolished as illiberal in 1831, under the recommendations of the Commissioners of Inquiry sent to Ceylon by the Colonial Office. Yet by 1848 a new Road Ordinance had been passed by Governor Torrington (George Byng, the seventh Viscount Torrington), requiring every male, with some exceptions, to perform labor on the roads or to commute such labor by paying a tax. This was perceived to be a return to the traditions of rajakariya, and was probably one of the immediate reasons for a rebellion in that same year.[88]

The conditions under which local people labored on the roads may easily be described as miserable, and disease struck more readily because laborers were far from home and did not have a ready means of subsistence or accommodation.[89] The way rajakariya was implemented, with no account taken of the size of the plot of land, meant that the burden fell most heavily on poorer families, who had to provide the same labor as those

---

84. Bingham, *History of the Public Works Department*, 1:193.
85. Skinner, *Fifty Years in Ceylon*, 215.
86. Ibid., 29–30.
87. See Munasinghe, *The Colonial Economy on Track*, 31.
88. See ibid., 43ff; for more on this see chap. 8 below.
89. See ibid., 34.

with larger plots.[90] Turnour provided this description to the Commission of Inquiry:

> The road service is allotted to the people in each District upon the old Kandyan tenure of land. The proprietor of each estate called Pangua, has the same extent of service to perform without reference either to the extent or to the number of persons participating in the property. In a district where the full resources are required to be called out for road service, the holders of these estates are formed into two divisions relieving each other every week or fortnight. As the estates are of various extent the burthen falls very unequally on the inhabitants. . . . During the performance of it they receive no provisions from Government. They reside in contiguous villages and provide their own subsistence.[91]

Mrs. Heber wrote of how they passed a whole village depopulated by fever. The British engineer who accompanied them explained how he had "built it for the accommodation of a gang of workmen who were employed in erecting a bridge; and on his return, after a very short absence, found it a desert."[92] J. W. Bennett asked why the government meted no compensation to the families of those who died while constructing roads.[93] Lieutenant De Butts wrote of how Kandyans were "dragged from their homes to toil in a service for which they received no sort of remuneration, the wretched Cingalese in many instances failed, from actual inanition and died at the feet of their Christian task-masters."[94] In response to criticism, Barnes noted that Britain itself had possessed a similar law and that "in the early stages of civilization of all countries there must be calls upon the people for personal service."[95]

In fact, Britons consolidated ancient laws by using them, in keeping with the argument about the recontextualization of the "indigenous." Mr. M. G. Colebrooke, one of the commissioners who recommended the

90. See ibid., 36.

91. "Evidence of George Turnour," in Lot 19/106, SLNA.

92. Reginald Heber and Amelia Heber, *Narrative of a Journey through the Upper Provinces of India from Calcutta to Bombay, 1824–1825, with Notes upon Ceylon Written by Mrs Heber* (1828), 255.

93. Bennett, *Ceylon and its Capabilities*, 173.

94. De Butts, *Rambles in Ceylon*, 262.

95. Barnes comments with respect to Commissioners' Report, dated 10 September 1830, CO/54/112, TNA.

revocation of this law, noted that the employment of headmen to oversee the system of labor had "tended to foster the prejudices of caste, and to favour abuses of authority."[96] To counteract what were seen as abuses of power, Colebrooke urged that headmen should be encouraged to improve their lot, by the government holding out to them the prospect of advancement in the public service, or the possibility of becoming independently established in agriculture or commerce.[97] The manner in which Barnes shored up the role of the headmen in the course of constructing public works is clear from the example of the Kirime canal. In this instance, he gave honorary titles and gold medals to "deserving headmen" to publicly approve their conduct and services.[98]

The decades following the abolition of compulsory labor in 1831 saw the influx of a greater number of European settlers, who brought with them private capital to be deployed in cultivation. There was also an attempt to reorganize the Survey Department, which was not fully realized.[99] As was the case with surveying, commercial interests rather than military control started to dictate the construction of roads. The emergence of the plantation economy and the extension of the road network went hand in hand. By this time, the Ceylonese also began to assert their ownership of the land more forcefully. Debates ensued therefore over the question of who had the right to define the entitlement to land. In 1841, Messrs. A. and R. Crowe applied to the government for assistance in making a road from their estate at Allagalla to the Kandy road. The capitalists declared that they were willing to bear the entire cost of the road and asked the government to investigate the ownership of the land and instruct the Agent of Kandy to order the local headmen to allow the road to be carried through.[100] But one of the local landlords, in the Amarapura korale, instituted a case in court to prevent the road from being built on his land. The government had to acquiesce in the end to his demands.[101]

The Government and the cultivators had to come to terms with the rights of this individual, if the road were to go ahead. The Agent of Kandy attempted to contain the situation by asking the *rattemahatmaya* of

96. Report of Mr. M. G. Colebrooke, dated 12 January 1831, CO 54/112, TNA.

97. Ibid.

98. Pridham, *An Historical and Statistical Account*, 2:594.

99. See Barrow, *Surveying and Mapping*, 15.

100. A. & R. Crowe to the Colonial Secretary, dated 31 August 1841, Lot 10/67, SLNA.

101. See J. Baybrooke, Surveyor, to A. & R. Crowe, dated 31 January 1842, and Colonial Secretary to Crowe & Co. dated 23 November 1842, Lot 10/67, SLNA.

Yattinuwara, a local headman, to read a government declaration in the province:

> No Kandyan who is disposed to exercise common sense can be igno-
> rant of the benefit which all the inhabitants derive from the number
> of English gentlemen becoming settlers in the country and spending
> large sums of money among the population. . . . In the construction of
> roads it is not always practicable to combine both public and private
> convenience. In order to gain an advantage a corresponding sacrifice
> has to be made and for this reason the smaller interest must give way
> for that which affects the people generally. . . . Compensation fair and
> reasonable will be made to all the owners of grounds through which
> any public road passes, but in consideration of such an equivalent the
> Government is determined to enforce its unquestionable right.

Despite the government's use of the headmen and its return to the older language of feudal control evident in this passage, the landowner in Amarapura korale did not give up his right to the land. And even though the government threatened to exercise the powers once used by the Kandyan monarchs in the declaration above, the Colonial Secretary wrote to A. & R. Crowe that the Governor was unwilling to exercise "a power so arbitrary in character."[102] This case suggests that the Ceylonese were able to exercise more resistance to public works in the later period, once compulsory labor had been abolished, when the land was no longer encompassed under a military style of governance and when commercial interests came to the fore. Yet some aspects of militarism continued to characterize the construction of roads, tanks, and public utilities. For instance, the Pioneer Forces retained a system of ranks and divisions even after 1833, when they ceased to function under the Military Department.[103] During the disturbances of 1848, discussed in chapter 8, the Pioneers were armed and drilled in order to contain resistance.[104]

In this context it is not surprising to find that Pioneers were attacked. In 1823, during the period of tense control following the British taking of Kandy, Bathurst was told that six Pioneers had been attacked by a party of

102. Letter dated 23 November 1842 from Colonial Secretary's Office to Crowe & Co., Lot 10/67, SLNA.

103. Bingham, *History of the Public Works Department*, 1:182; see also 2:203, for the military drills which characterized the Pioneers.

104. Skinner, *Fifty Years in Ceylon*, 262.

Kandyans five miles from Kandy; one of these Kandyans was said to have been armed with a musket, which he fired at the Pioneers, "the ball punctuating the Serjeants coat and grazing his side."[105] A rumor that this attack resulted from the Pioneers "committing depredations on the inhabitants" was also noted. John D'Oyly was asked to investigate the circumstances surrounding this attack, and he reported on the basis of the testimony of a prisoner called Kadarate, who attacked the Pioneers, and from interviews with the Pioneers. D'Oyly proposed that the attack arose out of one of the Pioneers being intimate with Kadarate's wife; yet, the Pioneers who were attacked did not include the individual who was said to have carried on with Kadarate's wife.[106]

The response to the construction of the Peradeniya Bridge also illustrates the manner in which local peoples were threatened by new public works. Like the Kandy Road, the Peradeniya Bridge was built in part to challenge local knowledge (fig. 6.4). The Kandyans allegedly held from "their ancient tales and legends" that the bridging of the *Mahavali ganga*, the great river that runs from Sri Pada to the eastern coast, was impossible. When work commenced in 1826, "with this persuasion they were in the habit of daily assembling to gaze on the gradual progress of the work, and laugh to scorn the vain and impotent labors of the *pale faces*."[107] Indeed, the construction of this bridge was a rather large infrastructural endeavor taking 1,039 days, and involving twenty five Europeans as carpenters and one hundred and fifty convicts each day, rising by the end of the project to one thousand.[108] That the bridge was perceived as a threat is evident from how a plot was formed to destroy it by fire in the course of the disturbances in 1834.[109]

Though compulsory labor was abolished in 1831, a military form of control over land and labor continued, while commercial interests came to the fore. Even then the building of public works continued to be viewed with suspicion, while necessitating the reorganization of structures of governance and the economy. The Ceylonese expressed their rights to the

---

105. Dispatch from James Campbell, dated 5 March 1823, CO 54/84, TNA.

106. Letter dated 17 March 1823 from John D'Oyly to G. Lusignan, Lot 21/56, SLNA.

107. De Butts, *Rambles in Ceylon*, 116.

108. For an account of the work that was taken, see "Correspondence between the Commissioners of the Inquiry and the Government of Ceylon relative to the expences incurred . . . ," in Lot 19/58, SLNA.

109. Pridham, *An Historical and Statistical Account*, 1:210.

Fig. 6.4. "Peradenia Bridge, Ceylon." Colored lithograph by W. Purser after a sketch
by Lieut. J. Braybrooke, 1834. Photograph: © The British Library Board. P16722.

land and resisted more overtly by obstructing the course of projects and by
attacking those who played a part in constructing them.

ANCIENT WISDOM

The program of public works relied heavily on the Survey Department.
Surveyors traced the projected course of roads and mapped the terrain of
channels, canals, and tanks. The appropriation of geographical knowledge
by these surveyors was not restricted to the early wars with Kandy; it extended to the period which witnessed the consolidation of British rule
and the making of the bridges and roads that constituted the new colonial
infrastructure.

Since the Sinhalese were seen to be a civilization in decline, the ancient roads of Ceylon were treated as the relics of a civilization that should
be restored under colonial rule.[110] Britons made use of knowledge about extant paths in planning the course of their own modes of conveyance. Governor Brownrigg reported in 1815, prior to Barnes's road-making, that it

110. For the ancient path to Kandy, see Bingham, *History of the Public Works
Department*, 1:1–4.

had been the policy of government to navigate to Kandy by using the Kalu-tara river and traveling on its left bank: "but it has been found on a survey made by Captain Schneider, the Island Engineer, that there is on the right bank of the River an old road which tho' now overgrown with jungle may be easily cleared and made passable for troops and cattle." Brownrigg also observed that Schneider was putting all the old roads of Kandy on paper, with the help of "some country born assistants."[111]

But the prime example of how Britons relied on local expertise in rela-tion to public works concerns colonial engagements with the ruined tanks of the north-central provinces. In this region, Britons discovered a series of reservoirs for the collection and diffusion of water, built by ancient kings to make arid land cultivable. Pridham noted that there were two types of tank in Ceylon. One kind was formed by vast mounds and the water was supplied by a channel or channels cut from some adjacent stream, while the other type used two sides of a valley for the purpose of embankment. "When industry was checked in Ceylon by intestine commotion, the tanks were neglected, morasses formed, the jungle rapidly encroached on the cultivated land, the climate became permanently deteriorated, the popula-tion diminished, and beasts of prey simultaneously multiplied."[112] Britons believed that if they renovated these tanks, the agricultural productivity of the region could be multiplied several fold. For instance, M. G. Cole-brooke wrote in 1831 that "there can be no question" of the "utility and importance" of the tanks as "a means of securing the inhabitants from the effects of droughts and floods, and of improving largely the agricultural resources of the Island."[113] By 1830, with the increasing commercializa-tion of the British presence in Ceylon, it was hoped that capitalists could be encouraged to invest in the repair of these tanks and in the cultivation of the desolate provinces in which they were found.[114]

Captain G. Schneider's early surveys of the tanks of Ceylon in 1806–7 form an important moment in British knowledge about the ancient reser-voirs. Upon the instruction of Governor Thomas Maitland, Schneider was dispatched to survey the tanks of the districts of Vanni, Mahagampattu,

111. Dispatch from the Governor to the Secretary of State to the Colonies, dated 26 September 1815, Lot 5/8/102, SLNA. For more on the old road to Kandy, see R. Raven-Hart, "The Great Road."

112. Pridham, *An Historical and Statistical Account*, 2:549.

113. Report of Mr. M. G. Colebrooke, dated 12 January 1831, CO/54/112, TNA.

114. See for instance, private letter from the Governor to the Secretary of State to the Colonies, dated 24 December 1833, CO 54/131, TNA.

Matara, and Galle, and also to report on the tank that attracted greatest attention in this period, namely Cattukara or Giant's Tank, now called Yoda Wewa, in Mannar. In each of these cases, Schneider commented on why the various tanks had deteriorated. In the Vanni, he noted how the channels for the distribution of water had been clogged by rotten tree trunks. He added that oxen and bullocks were allowed to go over these channels and destroy them; in addition unjust headmen were blamed for exercising arbitrary power while neglecting the tanks' upkeep.[115] The alleged slothfulness of the Ceylonese was in prime view in these reports. Britons hoped, however, to encourage the colonized to be industrious and to attend to the upkeep of their ancient tanks. They believed that if this were to happen the island would become self-sufficient in rice once again.

Schneider's reports contain a repeated formula of observations, which link topographical information to finance. Comments on the types of crops that may be grown in particular provinces were followed by the prospective size of such a crop and the extent of the reservoirs. Here is a typical entry from his report on the tanks of Matara:

> To this province belong Belligam Tottemanne having 14 villages and provided with about 444 Ammonams of paddy fields consisting of Ande, Maelpaloe, Devil and Otto fields, a part of the fields are served with the aid of water of Polatto River, another part with water that descends from the Hills, another part has the assistance of the Tanks, and the remaining that of the River. In this province are many gardens planted with cocoanut trees, jack trees and coffy trees, and pepper branches and other useful trees. Many empty grounds are here which can be used for making plantations. . . . When all the paddy fields situated in this province be properly sown, the share of government may be calculated at about 3500 Parrahs of paddy per annum.[116]

The prospective program of improvement encompassed the transportation of peoples, the restructuring of the local economy, and the reestablishment of an allegedly ancient state of progress. Yet, the conclusion that Britons sought to radically transform already existent structures is not wholly accurate. In fact, British plans for the renovation of ancient reservoirs fitted into an already established pattern of improvement that predated their

115. For Schneider's reports, see CO 54/126, TNA. For this comment on Vanni, see p. 20.

116. Ibid., p. 119.

arrival. For instance in the Vanni, Schneider found a tank that had been renovated about twenty years before his arrival. Families had then been brought from Kandy and remained in the region for eighteen years as cultivators. But the tank had deteriorated once again, leaving the land desolate. Schneider wrote: "The Modlier [Mudliyar] of this Province declared to me for a truth that he was informed that whenever the people who had so gone away should hear of this tank being repaired they would again come and settle there."[117] This bears out that, as with the restoration of monuments, the precolonial and colonial occupied the same temporal space. The fact that the British wished to resurrect ancient reservoirs, without building their own dams, attests also to their recycling of local knowledge.

The most striking instance of this reuse appears in a remarkable exchange recorded between Schneider and "the chiefs and principal inhabitants of the provinces of Mantotte and Nanathan" respecting the Cattukara Tank. Driving this conversation was the assumption that local peoples possessed information worthy of the colonizers' attention. Yet the inhabitants of this region are presented as ignorant of the arts of surveying and measurement, while eagerly awaiting the liberal improvements of British science. Here is an extract from the exchange between Schneider and the people:

Whether they do not know from any writings or otherwise when and by whom the Tank Cattoecarre and the stone bear or dam in the Moesele river have been made.

We have no writings but we heard from our parents that the tank has been made by giants.

Whether they think that it will be any benefit to bring the Tank Cattoecarre in a compleat state of repair.

It will be a very good work as well for the country as for the inhabitants.

Whether they are of opinion that when the Dam of Cattoecarre be brought to a sufficient height, the same will contain a sufficient quantity of water to irrigate the lands without or round about the Tank Cattoecarre.

117. Ibid., p. 14.

This we cannot state with certainty as we do not understand to make an exact calculation thereof but we do think, that the extent of Cattoe-carre is sufficient to contain the necessary quantity of water to irrigate the canals on the outside of the same.

Whether they do not know from experience, that the water will dry up sooner in the small tanks than in the larger.

We know well from experience that the water will be sooner dried in the small tanks than in the larger tanks.

Whither there are any springs, rivers or watercourses in the great wood beyond Cattoecarre that run into sea from the side of Wirtellivo and others in the Moesely river.

Several springs which can very easy be led on the side of Cattoecarre.

How many marcals of ground can be ploughed and sowed by one man only.

From 20 to 25 Marcals.

As it is the intention of His Excellency the Governor to promote the agriculture as far as the same be practicable, and to present the inhabi-tants under the British Government from suffering wants in unprofit-able years, His Excellency has been pleased to assist the inhabitants of these provinces with paddy seed and at the same time offers to advance of the expences for the repairing of the said tank Cattoecarre or the other small tank, whether they, in either case are inclined to pay to Government 1/5th instead of 1/10th part of the crop, until the expences be reimbursed.

With much pleasure and willingness we will and at the same time we thank for His Excellency's good intention towards us and our agriculture.

British attempts to revive the arteries of Ceylonese agriculture in fact met with initially limited success. In 1833, Governor Horton could still la-ment the dilapidated condition of the tanks; he suggested that a joint stock company be formed to aid their reconstruction.[118] He noted also that in the early phase of British presence in Ceylon an attempt was indeed made to resurrect a set of tanks in the Vanni, by George Turnour, a civil servant in Ceylon and possibly the father of the George Turnour who translated the *Mahavamsa*. It was said to have "been attended with complete success, though the death of that Gentleman and the impossibility of directing the efficient attention of Government to such an object prior to the capture of the Kandyan Provinces led to the tanks being again permitted to fall to ruin."[119] As in the case of road-making, the repair of the tanks of Ceylon was planned ambitiously, though follow-through was less confident.

CONCLUSION

How is it possible to balance the two arguments of this chapter that Britons relied on existing knowledge about the land provided by guides, informants, and inhabitants of the north-central provinces where the reservoirs lay, for example, while at the same time they trumpeted the superiority of British knowledge in seeking to make public improvements on the land? It is important to qualify generalizations about the relationship between colonialism and knowledge. Knowledges certainly passed from one regime to the other, but this does not preclude the fact that there was competition between local peoples and colonizers to guard knowledge and to use it strategically. Information and power interacted in multiple ways in this phase of colonial expansion, and therefore collaboration and resistance were equally available modes of operation. The undoubted contradiction here in using competition and slippage, appropriation and inheritance, mirrors a nineteenth-century debate about the relative viability of reformist and orientalist models of British colonialism. Historians who theorize the relationship between knowledge and empire may keep in view how contemporaries themselves articulated the difference between colonial and colonized. At the same time it is critical not to reify the colonial as

118. Private letter from the Governor to the Secretary of State to the Colonies, dated 24 December 1833, CO 54/131, TNA. For later comments on the possible benefits of restoring the Gina's Tank, see "Replies of Mr. Huskisson, Collector of Mannar, relative to the repair of certain tanks in that district," Lot 19/7, SLNA.

119. Dispatch from R. W. Horton dated 23 February 1833 to the Secretary of State to the Colonies, CO/54/127, TNA.

a strand of knowledge which may be distinguished from the colonized or even precolonial.

It is also useful to hold to the complex results that emerged out of the application of colonial knowledge. There was success of course, as is clear from the British taking of Kandy; yet, colonial projects were often characterized by failure. At the same time, colonial interventions could generate the kind of religious praise in evidence in the palm-leaf ballad about what pilgrims could gain from the road to Kandy, cited in the Introduction. The question that needs answering is not whether there was dialogue or imposition, or collaboration or resistance, or indeed success or failure; but how colonial practice could move between these rival nodes. The suggestion of this chapter is that the land could serve as an especially fraught site of celebration, competition, and resistance because it was a public space in a way that the chambers in which translators and scholars worked together or the Peradeniya botanic gardens were not. Roads and bridges became part of the everyday infrastructure used by both colonizers and colonized, and they relied on a new regime of labor. Both of these factors explain the presence of a rich spectrum of modes of engagement, running from resistance to collaboration. In bringing in the notion of competition alongside the slippages of the previous chapters, this has been an attempt to come to a more robust view of indigenous practice and a far less optimistic view of colonial science and technology's achievements.

While Buddhist priests, skilled guides, and assistants have been in view thus far as mediators of colonial knowledge, this chapter brings in the large gangs of laborers who were essential for the making of the British colony of Ceylon. Even as labor on the roads proceeded, it became more exploitative and cruel, and the British stretched and consolidated established traditions in utilizing men to work the roads. The creation of the state depended on large numbers of indigenous peoples, especially when the projects in hand had such public compass. The difference between the physical skills that laborers brought to the enterprise of road-building and those skills of translation and conveyance provided elsewhere should not be distinguished too quickly. Though laborers on the roads experienced more force in being deployed to work, the transparency of this force should not disguise the fact that all British collaborations with indigenous assistants in the end gave rise to a regime of entrenched colonialism and to a new context for the "indigenous."

The next chapter takes further this attention to the aggression of British intervention, and the changing character of collaboration, by considering the theme of British medicine. As in the case of surveying, for medicine as well the indigenous assistants shifted with time from guides who descended from the structures of the Kandyan court to people who were trained by the British to serve as assistants. Yet the practice of colonial medicine was not shaped by the land alone; it extended to a new site of operation, in the very bodies of the colony.

CHAPTER SEVEN

Medicine

B y setting itself the task of creating a healthy population, the colonial state sanctioned its existence and sought to create consent to its rule. But it was on the subject of medicine that the colonists issued some of their most provocative dictates, requiring the forced confinement of victims of smallpox and actively promoting vaccination. These encounters between the state and the body met with resistance for the reason that they impinged on the domain of tradition and individuality, even as they generated calls instead for what counted as public benefit.

In Lanka, state-making and medical institutionalization were linked in a constantly changing and contingent process of engagement, so that the originality of the colonial state's measures should not be seen as radical. First, there wasn't a simple rupture between a precolonial monarchical system of patronage over medicine and the modern colonial state which determined the public's interests in matters of health. While these two structures were different, there were a series of convolutions which connected kingdoms to colonial states. Second, the colonial state of the early nineteenth century was not a fully formed object. In perfect symmetry to medicine, the state in Ceylon in the 1830s was spawning a new field of power, namely the civil sphere, and this was a painful process, involving dissent and dispute between the governors and the governed, and also between the colony and the distant authorities in London.

The relationship between medicine and the military is a well-trodden path in South Asian historiography, though attention to Sri Lanka in this field is marginal.[1] The model that has been sketched is that of the East In-

---

1. See, however, the work of Margaret Jones, *The Hospital System and Health Care: Sri Lanka, 1815–1960*, and *Health Policy in Britain's Model Colony: Ceylon,*

dia Company's army in India, where military surgeons were at the forefront of research on tropical conditions, and the application of their theories to mass contingents of troops secured the hold of statistical and empiricist traditions in medicine, with interest in spatial distribution, and saw an objectification of both patients and diseases, prefiguring the spread of the clinical method in Europe.[2] The hierarchical structure of the Company's medical establishment is said to have bred conformity in therapy; for instance, in setting the context for "heroic treatment," such as purgation with large doses of mercury. It is also said to have led to an "enclavist" character, with focus on areas such as barracks and jails.[3] Though medicine used to be seen—following Daniel Headrick's work—as a tool of empire, the emphasis has now shifted, in a manner consistent with the literature on cartography, to a focus on how its discourse of instrumentality was limited in practice.[4]

One area which has arguably not attracted as much attention is the placement of Indian traditions of medicine in the military therapeutic practice of the East India Company. David Arnold writes of the dangers of imagining "a free and open exchange between equals" and adds that whatever passed from India to Europe was "largely a case of Europe taking from India whatever appeared useful for its own understanding and practice."[5] For him the story is broadly consistent with the evolution of orientalism into Anglicism: by the 1830s, any interest in Indian medicine, even if predicated on an interest in surpassing it, had given way to the need to dispense European medicine. Yet more recently, this story has been critiqued: there is in Arnold's work a tendency to unify the diversity of medical practice in the subcontinent under the umbrella of a religious and cultural system which opposed the onset of the secular, colonial other.[6] This

---

1900–1948, and also C. G. Uragoda, *A History of Medicine in Sri Lanka from the Earliest Times to 1948.* Since this chapter was written the following has also appeared and provides a further empirical survey: S. A. Meegama, *Famine, Fevers and Fear: The State and Disease in British Colonial Sri Lanka.*

2. See, for instance, Mark Harrison, "Disease and Medicine in the Armies of British India, 1750–1830: The Treatment of Fevers and the Emergence of Tropical Therapeutics," and *Public Health in British India: Anglo-Indian Preventive Medicine, 1859–1914.*

3. David Arnold, *Colonizing the Body: State Medicine and Epidemic Disease in Nineteenth-century India.*

4. Daniel Headrick, *Tools of Empire: Technology and European Imperialism in the Nineteenth Century.*

5. Arnold, *Colonizing the Body*, 46–47.

6. Sanjoy Bhattacharya, Mark Harrison, and Michael Worboys, *Fractured States: Smallpox, Public Health and Vaccination Policy in British India, 1800–1947*, in

chapter is a contribution to this debate; it makes the case that we need a better vocabulary in dealing with precolonial practices, in their diversity, and also that these practices should not be set against Europe in a simple fashion. The suggestion of this chapter is that historians might consider how the meaning of the "native" and the "indigenous" in medical terms was reconceptualized by the British state and its state-making practices. The colonial state incorporated within itself prior forms of indigenous practice, even as it evolved at a quick speed.

As in the case of other fields of intellectual history, it is vital to urge the specificity of the island as well as its connection to developments in India. Because of the newness of the British state in Lanka and its centralization, without obligations to vassal rulers, it could take an even greater role in determining medical priorities, in line not only with the needs of the military but with economic ones more broadly. Additionally, the cosmopolitanism of the island, with regard to medical traditions, marked it as distinct. Unlike Indian histories of army medicine, the discussion begins with the heterogeneous context of medical culture of eighteenth-century Lanka, including Buddhist, Muslim, South Indian, and European traditions. This diversity did not emerge in a political vacuum, for medical cultures had come to take prominence in accord with political opportunism and need. When the British took the island, and when the colonial state expanded, it had to encompass this diversity within itself rather than erasing it. This meant that Dutch and Portuguese descendants, Buddhist monks, Muslim physicians, and Malay helpers all had to be co-opted, taking account of their prior status as gatekeepers of medicine. The hierarchical character of the military medical outfit sought to fit its assistants into a well-ordered ladder of responsibilities, one that evolved as the state became more defined and colonialism more confident. This meant the continuation of hybridity in a new statish mold, which extracted medical work for poor wages from those at the bottom.

THE MEDICAL MARKETPLACE

In 1805, King Sri Vikrama Rajasimha, the reigning monarch in Kandy, was taken with smallpox.[7] This was an event of unprecedented symbolism,

---

particular the introduction and chapter 1, and also the review of this book by Niel Brimnes in *Modern Asian Studies* 44 (2010): 666–70.

7. Michael Roberts, *Sinhala Consciousness in the Kandyan Period, 1590s to 1815*, 50.

for smallpox had never before afflicted a king of the island. The king was taken to be akin to a divine being whose body was accorded the highest respect. The mark of smallpox was viewed, by Sri Vikrama and his court, as a sign of divine displeasure.[8] The moment of the king's illness was ill timed in political terms, given how the British were encircling his kingdom. Two years earlier, Sri Vikrama had defeated the forces of the British army. Yet by 1805, despite his victory, Sri Vikrama's position was increasingly unstable. Since the king was ill, Pilima Talauve, the chief minister, was presented with the opportunity of returning to the capital after being in the king's disfavor for two years. Instead of war with the British, the Kandyans turned at this point to seek peace.[9]

Before the arrival of the British, it was customary for the Dutch to include chests of medicine among the gifts of tribute to Kandy.[10] King Narendrasimha called in the Dutch to supply advice on medicine and disease at least three times; each time he did not submit to the medication and sometimes objected to examination.[11] The last visit, which came shortly before his death in 1739, was by a Dutch physician by the name of Dr. Danielsz.[12] "Doctor Danielsz went accompanied by his apprentice but all he could see of his royal patient was the ailing limb. Under such circumstances, it was impossible he could adopt any other course of treatment but what consisted of outward dressing."[13] Dr. Danielsz proceeded to prescribe a portion of tonics, which the king promptly spat out as undrinkable; he suggested that the doctor should combine the tonics with arrack, an alcohol brewed from toddy. This exchange eventually led to the doctor leaving Kandy at the instruction of court, unable to cope with the king's lack of cooperation. The humor of this anecdote aside, it points to how the sacrality of the king's body, evident in how Danielsz could only view the leg, dictated medical treatment in Kandy. Yet at the same time, the Kandyan court was open to learning from outsiders about cures. Indeed, the

8. Paul E. Pieris, *Tri Sinhala: The Last Phase, 1796–1815*, 82.

9. Colvin R. De Silva, *Ceylon under British Occupation, 1795–1833*, 1:119–20, 129–30; Pieris, *Tri Sinhala*, 82.

10. Uragoda, *A History of Medicine in Sri Lanka*, 69.

11. Ibid., 73.

12. "Diary kept by Dr. Danielsz of his journey to Kandy to cure the king," dated 10 May 1739, Lot 1/3289, SLNA.

13. This is the recollection of a commentator in 1872, cited in Adrian Senadhira, *History of Scientific Literature of Sri Lanka*, 49. See also Uragoda, *A History of Medicine in Sri Lanka*, 75ff.

Dutch doctors' visits only bear out how cosmopolitan the medical culture of Kandy was.

The point about cosmopolitanism is also important in making sense of the status of Muslim physicians in the court of Kandy. Arab migrants to the island from the thirteenth century onwards brought with them Yunani traditions; Alexander Johnston, the British orientalist, noted that Muslim priests and merchants on the island were well versed with Avicenna and had Arabic translations of Plato, Euclid, Galen, and Ptolemy.[14] There was a confluence and mixture of Yunani traditions with the island's ayurvedic medical practices, which came from India. The family of the Gopala Moors, who were Muslims, were accorded the rank of royal physicians right up to the end of the eighteenth century, and one tradition holds that they came to the island from Goa.[15] The Gopala Moors appear in the events of eighteenth century Kandy. In 1749, when Roman Catholic missionaries in Kandy sought to provide the sacraments to Catholics resident in the city, they had to bribe one of the Muslim physicians, who headed the King's *betge*, which is literally the house of medicines. In the 1760 conspiracy to dethrone the king, discussed earlier, it was again a Muslim physician who provided the king with information about the plot. During the reign of the last Kandyan king, there is evidence that a Muslim physician served as a judge. Muslim medicine therefore had a privileged place in Kandy, and physicians became trusted confidantes of the king, in addition to taking on other roles.[16]

In addition to Islamic Yunani traditions, the Kandyan court also turned to South Indian Tamil traditions of medicine, named Siddha. This was perfectly congruent with the ancestry of the last kings of Kandy. A line of South Indian *vaidyas* or medics played a role in the Kandyan court, and one of them, named Selendrasimha, became the Royal Physician to King Narendrasimha. Selendrasimha translated a Tamil medical text entitled the *Vaidyacintamani* into Sinhala and was also the author of the *Vattoru vedapota*.[17] The mixture of a variety of traditions of medicine in the island is exemplified in the language of some of the medical palm-

14. Uragoda, *A History of Medicine*, 14–15.

15. Jinadasa Liyanaratne, *Buddhism and Traditional Medicine in Sri Lanka*, 32.

16. This paragraph relies on Lorna Dewaraja, *The Muslims of Sri Lanka: One Thousand Years of Ethnic Harmony, 900–1915*, 123–28. See also, A.I.L. Marikar, *Glimpses from the Past of the Moors of Sri Lanka*.

17. Liyanaratne, *Buddhism and Traditional Medicine in Sri Lanka*, 3.

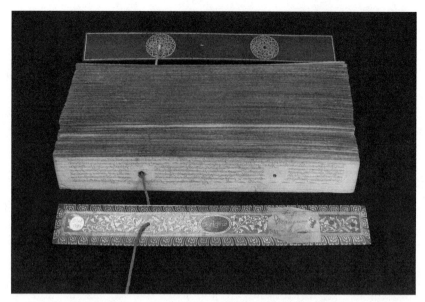

Fig. 7.1. *Bhessajjamanjusa-sanne.* From the Colombo Museum Library. A Sinhala
translation from the original Pali was completed by a Buddhist monk in the eighteenth
century. By permission of the Director of the Colombo Museum. Author's photograph.

leaf texts, which include Sinhala, Tamil, and Sanskrit words, in particular
names of plants and diseases.

Even as late-eighteenth-century Kandy was characterized by a medical pluralism, this period also saw a resurgence and popularization of ancient Pali ideas of medicine. The best evidence for this comes from a text
called the *Bhessajjamanjusa,* literally "The Casket of Medicine," which
was written primarily, but perhaps not exclusively, for the use of Buddhist
monks during the thirteenth century (fig. 7.1). This is arguably one of the
most important medical texts of the island. Valivita Saranamkara, who
reestablished higher ordination and sought to reform the Buddhist clergy,
completed a full translation of the *Bhessajjamanjusa* from Pali to Sinhala
at the request of Narendrasimha.[18] In fact, the first eighteen chapters of the
original Pali work had already been translated into Sinhala, and Saranamkara added the remainder with the help of his teacher Plakumbure Attha-
dassi and some of his pupils.[19] Saranamkara's full translation in prose has

18. C. E. Godakumbura, *Sinhalese Literature,* 332–33.
19. Jinadasa Liyanaratne, ed., *Bhessajjamanjusa,* introduction, 5.

the name *Bhessajjamanusa-sanne*.[20] Narendrasimha rewarded Saranamkara for the translation by presenting the monk with an elephant.[21] The
tradition of retranslating this work carried on into the later eighteenth
century; a verse translation appeared under the title *Yogapitakaya*, and another by the title of *Bhessajja Nidane* appeared in 1760 in Matara in the
south of the island.[22]

The *Bhessajjamanjusa-sanne* indicates the impact of Buddhism on
medical ideas. The return to this text in the eighteenth century can be
read as a reaction against the arrival of newer traditions of medicine on the
island, and this is in keeping with its placement within the resurgence of
learning initiated by Saranamkara. But despite the fact that the translation signals a reemphasis on an old Pali source, it is important to note that
the original text of the *Bhessajjamanjusa* was composed from a series of
Indian Sanskrit authorities.[23] Turns away from cosmopolitanism were difficult to orchestrate in a place such as Lanka. Just as European, Muslim,
and Indian traditions arrived in the island, the *Bhessajjamanjusa* traveled
to Southeast Asia from the island; copies have been found in Burma, Siam
and Cambodia.[24] It is vital to note that though the patronage of the king
was critical, the kingdom of Kandy did not have hegemony over the making of these kinds of medical treatises. Matara in the south was another
center of Sinhala medical interest in this period, and there were also other
medical texts which arose from areas outside Kandyan control.[25] Therefore,
the reassertion of what was deemed traditional learning within Kandy sat
alongside its diffusion and the operation of other centers of medicine.

Because the *Bhessajjamanjusa* was designed primarily for use by
monks, there is no discussion of diseases that afflict women or children
or reference to medical ideas connected with charms, incantations, or auspicious and inauspicious times.[26] Instead, Buddhist ideals have a bearing
on medical ideas: "The one who has constantly good food habits, and a
good conduct, who is prudent, not attached to worldly pleasures, generous, equally well disposed towards everybody, truthful, patient, and who

20. Ibid., 6.

21. Uragoda, *A History of Medicine*, 37.

22. Jinadasa Liyanaratne, "Studies in Medical Herbaria," 103–4.

23. Liyanaratne, ed., *Bhessajjamanjusa*, 22, and also Jinadasa Liyanaratne, "Indian
Medicine in Sri Lanka."

24. Liyanaratne, "Indian Medicine in Sri Lanka," 204.

25. Liyanaratne, *Buddhism and Traditional Medicine*, 9–10.

26. Godakumbura, *Sinhalese Literature*, 332.

seeks good company, will be free from disease."[27] Contentment, meritorious deeds, and the following of the truth of Buddha are also extolled. Diseases are divided into three categories: those arising from sinful deeds in this life; or from a previous life; or those caused by misdeeds in this and previous lives. The upkeep of the body in the meanwhile is put down to a humoral theory.[28]

While Dutch, Muslim, and ancient Pali traditions of medicine were attended to at the court, the medical landscape was even more heterogeneous at places away from the capital and stretching to the coastal belt. In rural areas, healing was dictated by an eclectic tradition of popular medicine. Rural medical-men had a variety of texts in their possession, including versions of those commissioned at court and translated into Sinhala. Of particular importance in the Kandyan period were two texts called *Prayogaratnavaliya* and *Yogarnavaya*. These were written by a Buddhist monk, Mayurapa Thera, and they contained exactly the popular information the *Bhessajjamanjusa-sanne* avoided. Material covered included pregnancy and children's illnesses in addition to mental disease.[29] In addition to this, village medicine-men used recitations and incantations, and these practices were tied in particular to a class of texts called yantra and mantra. Usually taking the form of ballads, the relevant yantra or mantra would be recited to cure illness and could also involve, for instance, the cutting of a certain number of limes above the head or other parts of the body, or the use of the teeth of a leopard, or the oil of the spleen of the bear for treatment. Gods and demons became attached to the imagination of disease. In the case of smallpox, there was an association between this disease and the goddess Pattini.

The view from eighteenth-century Kandy and the more widely dispersed coastal realms of the island is of a complex and variegated medical marketplace. The kingdom in the center of the island was accommodative of traditions from a variety of directions—European, Islamic, and Indian. Yet at the same time, Kandy sought to resurrect its Pali traditions and to guard the sanctity of the kingly body. The exchange and redefinition of medical traditions was tied to political need: Muslims moved inland, even as Europeans took over the coasts. South Indian traditions became popular, as the last kings of Kandy hailed from that region. Dutch medical assistance to the Kandyan kings came with the expectation of political

27. Liyanaratne, ed., *Bhessajjamanjusa*, 34.
28. Liyanaratne, *Buddhism and Traditional Medicine*, 18.
29. Godakumbura, *Sinhalese Literature*, 333–35.

cooperation. And a king's falling ill with smallpox had political significance for his rule. The ability to practice medicine and to control disease
brought political opportunities in this period of transition. Symmetrically,
political patronage was critical in determining the kinds of medical ideas
that took hold in Kandy. It was within this plural realm that British colonizers sought to consolidate their authority, and this also involved an attempt to make their medicine supreme over others.

JUNGLE FEVER?

Initially, the British struggled to come to terms with the Kandyan territories in medical terms, and this had direct consequences for the speed of
colonization.

In the battle of 1803, discussed in the previous chapter, the British forces marched on Kandy and took it relatively easily at first, since
it had been deserted, and then enthroned Muttusami, a puppet ruler, in
place of the king. Yet the victory was only temporary, for the Kandyan
forces had retreated from the city in a tactical move. The British were soon
surrounded and isolated. With the monsoon setting in, disease took hold
among them. Some Malay and Indian soldiers and assistants deserted to
the other side. In disarray, the British finally decided to evacuate from
Kandy, but were intercepted. Those who were on the way out as well as
those too sick to leave and who remained in Kandy were slaughtered, and
only three Britons survived. The official death figures were never made
public, but there were between 120 and 150 European soldiers in hospital
in Kandy, who were killed, and it is estimated that about three hundred of
the British 19th Regiment died in total. However, the number of Malays,
Indian, and Sinhala soldiers and helpers who were killed is likely to have
been higher.[30]

Though the British defeat of 1803 did not only result from disease,
disease is important for this story. The march to Kandy was an event requiring attention to the troops' health. Troops in Ceylon were usually
fed meat and rice. But in 1803, they struggled, being unable to get enough
meat.[31] This resulted in the onset of disease, which combined with physi-

---

30. Henry Marshall, *Ceylon: A General Description of the Island and its Inhabitants* (1846), 84. See also James Cordiner, *A Description of Ceylon: Containing an Account of the Country, Inhabitants and Natural Productions* (1807), 2:203–4.

31. Channa Wickremesekera, *Kandy at War: Indigenous Military Resistance to
European Expansion in Sri Lanka, 1594–1818*, 176.

cal lethargy and debility. A medical report for 1803 points to beriberi (vi-
tamin B1 deficiency) as one of the diseases that afflicted the troops; by
this time they even struggled to prepare their rice.[32] By 20 June 1803, upon
being holed up inside Kandy, the British were dying at the rate of six men
a day. Paddy, which is rice in husk, was the only means of subsistence for
the troops, and "and in their sickly state, they were unable to perform the
labor of clearing it of the husk."[33] The British failed to adapt their diet to
indigenous foods such as the grain used by the Kandyan troops. A widespread fever also appears to have affected the troops.

The lack of flexibility in British military practice is also clear with
respect to clothing, which in turn had medical consequences. While the
Kandyan troops hid themselves in the jungles and assailed their combat-
ants from afar, the British stood out in their uniforms. "The rays of the
sun . . . reflected from the bright arms and large brass plates in front of
the soldier's cap, together with his red jacket, white pantaloons, and white
belts."[34] This cumbersome attire prevented rapid movement. Moreover,
Major A. Johnston complained: "Surely the same dress which is adapted
to the snows of Canada would not answer in the burning plains of Hindostan."[35] The cap kept the head overheated, and its shape meant that the
lower head and neck were entirely exposed to the sun and rain. In heavy
rain the water dripped down the cap and down the soldier's back. Johnston
noted of the soldier: "He finds himself wet to the skin long before [the
rain] has penetrated his great coat. Thus circumstanced, he becomes cold
and chilly . . . and when on duty at night, or without the means of procuring dry clothes, it must lay the foundation of many disease, but particularly that known by the name of jungle fever."[36]

It was this mysterious "jungle fever" that was one of the causes of mor-
tality in 1803, and the name tells us something of how this sickness was

32. Cordiner, *Description of Ceylon*, 2:263. Cordiner wrote: "The diet of the men
at Candy consisted almost solely of beef and rice, without any admixture of fresh
vegetables, and but little addition of spice and other condiment. This is by no means
a healthy diet, and will readily be conceived to predispose to many disease, particularly Berry-berry, which has at different times, been so extremely fatal to the troops in
Ceylon" (265).

33. Marshall, *Ceylon: A General Description*, 72.

34. Arthur Johnston, *Narrative of the Operations of a Detachment in an Expedition
to Candy in the Island of Ceylon in the Year 1804* (1810), 110.

35. Ibid., 108.

36. Ibid., 98.

imagined.[37] It was seen to arise from the vegetation and the terrain. After the war, discussions of the causes of disease invariably tapped into a wider British imperial discourse of tropicality and climate and chimed with the established theories of the army physician John Pringle in India, who urged that tropical fevers arose from "putrid air."[38] Johnston provided the instance of a regiment of three officers and seventy-five men who marched into the interior, all aged between eighteen and twenty-three, and who had recently landed from the Cape of Good Hope. He noted that all of them fell "victims to the climate."[39] James Cordiner compared the complaint of this battalion to the yellow fever of the West Indies: "It baffled the skill of physicians and resisted the power of medicine."[40] A military man, Captain L. De Bussche, blamed the "noxious exhalations" from the swamp surrounding the city of Kandy for having occasioned the disease of the troops who took the city.[41]

The impact of jungle fever was spatialized in commentaries on the 1803 campaign. The official medical report on troops serving in Ceylon in 1803 provides the best support for this. This presented the case that the eastern side of the Kandyan territories was healthier than the western side, being less woody and marshy. Those who marched through the east therefore stayed healthier than those who marched through the west. This meant that the western slopes of the mountain kingdom abounded more with "foul and inflammable air."[42] It also held that closeness to riverbeds determined healthfulness, for the river water was usually full of vegetation and was thought to give rise to unhealthy airs.[43] The onset of the fever was even linked to the phases of the moon.[44]

THE MILITARY BIRTH OF CIVIL MEDICINE

After the defeat of 1803, the British knew that they had to reform their understanding of the diseases of Ceylon. In this quest, military needs were

37. Ibid., 113.

38. Harrison, "Disease and Medicine," 93.

39. Johnston, *Narrative*, 93.

40. Cordiner, *A Description of Ceylon*, 2:194.

41. L. De Bussche, *Letters on Ceylon, Particularly Relative to the Kingdom of Kandy* (1826), 70.

42. Cordiner, *A Description of Ceylon*, 2:266.

43. Ibid., 2:273.

44. Ibid., 2;270.

placed as the primary priority, and an ideological militarism had a bearing on the character and philosophy of the practice of medicine. Even as the army secured victory, the view of Ceylon's climate and diseases changed from this culture of fear evident in the campaign of 1803.

One of the clearest exemplifications of the dominance of a military ethos in the first decades of British rule of Ceylon is the controversial handling of an outbreak of smallpox in 1833–34, which resulted in three anonymous letters in the *Colombo Observer* and a printed pamphlet in reply by the Superintendent of Vaccination, J. Kinnis, and a long and defensive dispatch from Governor Robert Horton to London. The correspondent to the press described how "thousands of sufferers" had been left "under the merciless controul of minions in office."[45] Under government regulations, those affected had to be taken and placed in confinement in hospital. In criticizing these regulations, the anonymous writer cited the example of a man who was "mauled" and who died "not from small pox solely, but rather from the effects of the struggle to remove him, between the people employed for this purpose under the orders of the constable and others, and the relatives of the deceased."[46] The contention of these letters was that removing people to smallpox hospitals rather than letting them stay at home spread the disease more rapidly. Another case that was cited was "from the country" and was of a patient who was brought a considerable distance, followed by his crying relatives, through villages and even the town of Colombo. While on the journey, the bearers of the patient rested at every tavern and "the patient meanwhile permitted to avail himself of these intervals to creep out of the Dhooly and spread his mat in the open air to refresh his feverish body."[47]

The foundation of smallpox hospitals, such as those where these patients were later received, was a crucial initiative that sought to bring medicine to the public in the early nineteenth century. It showed an awareness that the health of those beyond the army was important, and these hospitals served as a way of shoring up an image of colonial benevolence. It is pertinent to trace their emergence, in order to make sense of the draconian legislative regime which came in their wake.

Four hospitals specifically for the military were set up soon after the

---

45. Cited in J. Kinnis, *A Report on the Small-Pox as it Appeared in Ceylon in 1833–34* (1835), 78.
   46. Cited in ibid., 81.
   47. Cited in ibid., 81.

British took the maritime provinces, in Colombo, Jaffna, and Trincomalee in 1800, and Galle shortly afterwards, and these are likely to have been hospitals already in use under the Dutch.[48] Robert Percival described the military hospital in Colombo: "It is very properly divided into distinct wards, so as to keep the sick of different disorders completely separate, and thus prevent infection from spreading."[49] Yet the restriction of medical care to the military, and the segregated nature of medical provision evident in Percival's description, raised a series of practical and moral problems for the rulers.

The outbreak of smallpox at the turn of the century raised the specter of infected islanders passing the disease on to the troops and necessitated a review. It was at this point that further hospitals were formed for receiving civilians who wished to be inoculated or who were infected with smallpox, partly to ensure the health of the troops. Each hospital had a British medical overseer and a Burgher purveyor.[50] Thomas Christie, who was placed in charge of these hospitals, described the epidemic at the turn of the century in graphic terms: in villages outside Batticaloa, on the east coast, soon after the healthy inhabitants deserted their abodes, elephants, cheetahs, and wild hogs came down from the jungle and "broke down the fences, destroyed and rooted up the trees, and ate the Paddy (unhusked rice) and other provisions, and what is still horrible carried off some of the sick, or at least consumed the bodies of the deceased."[51] Despite Christie's horror, this description may have been in keeping with the local custom of denying the rights of burial to those who died from smallpox.[52] But in describing a scene like this, Christie hoped to capture the moral high ground: by establishing smallpox hospitals the British showed that they were taking account of the situation and were concerned for those beyond the military.

Such a discourse of benevolence eventually had to come to terms with the obvious limitations in the colonial policy of attending to indigenous

48. Uragoda, *A History of Medicine*, 82.

49. Robert Percival, *An Account of the Island of Ceylon containing its History, Geography, Natural History, with the Manners and Customs of its Various Inhabitants* (1803), 105.

50. Thomas Christie, *An Account of the Ravages Committed in Ceylon by Smallpox, Previously to the Introduction of Vaccination* (1811), 8–9.

51. Ibid., 3–5.

52. See J. L. Vanderstraaten, *History of Medical Service in Ceylon*, 68, and "A Brief Sketch of the Medical History of Ceylon," 321.

peoples simply in the context of epidemics. In the first proposal to widen medical care and make it more regularly available to civilians, in 1817 Charles Farrell, the Principal Medical Officer of the island, in addressing Governor Robert Brownrigg, proposed the need for a Native Infirmary for the treatment of the poor. Farrell envisaged that the Native Infirmary could treat one hundred people, both male and female, and in proposing it he utilized a distinctly humanitarian tone:

> The wisdom and humanity that confined the Blessings of Vaccination on the population of this country could not remain insensible to the sufferings and distress of the poor and helpless labouring under diseases, which they had not the means to have properly treated, and which were their means ever so ample, the native Medical practitioners were incapable of treating.[53]

Farrell's memoir, which was sent on to London, stressed that this hospital would still be placed under military supervision, and it noted that military medical men would provide their services to it without remuneration as an act of charity. Farrell combined his humanitarian argument with the political economy of restoring the health of indigenous peoples so that they could return to labor; noted the need of "diminishing the infectious diseases, communicated by unfortunate females to our Soldiers, of which there are at this moment many instances"; and added that the healthfulness of the military would generally increase as a result of attending to the health of the inhabitants of the island. This reasoning points to how the discourse of humanity cannot be seen as straightforward, for it combined with other considerations, and civil medicine was conceived as an outworking of military medicine. There wasn't a simple turning away from the military origins of medicine to a more enlightened or humane age of care. Farrell's proposed hospital was opened in 1819.[54]

Farrell's philosophy of medicine was characteristic of British military engagement with Ceylon in the early nineteenth century after the taking of Kandy; he advocated an interventionist strategy and was full of optimism about the prospects of the British improvement of the colony. After Kandy had fallen, its climate and terrain were seen to be more hospitable than previously thought, and the incidence of disease was linked more

---

53. Letter dated Colombo, 26 May 1817, from C. Farrell MD to the Governor, CO 54/65, TNA.

54. Uragoda, *A History of Medicine*, 87.

readily to human habits and culture than to the environment.[55] Farrell noted: "The general healthiness of the country is much greater than we had any reason to expect." He added that the average rate of sickness in the troops in the year prior to May 1817 was one in twenty and that the rate of death was not more than one in two hundred fifty. He compared these rates favorably to those of Britain itself. Taking a popular stereotype of the time, namely that the station of Trincomalee on the east coast was especially unhealthy for troops stationed there, he argued for a more careful explanation. Though the soil and climate may have an impact on Trincomalee's healthfulness, Farrell urged that there were other factors, "most of them depending on ourselves and capable of being remedied." He pointed to the use of arrack by soldiers stationed in Trincomalee and their "long continued intemperance." Suitable clothing and good quarters, if adopted, would protect these men from the "sun and the weather."

The impact of the military birth of British medicine in Ceylon, evident in the early strategies for handling epidemics among civilians, continues to be evident into the middle of the nineteenth century in the role and distribution of medical practitioners. A running complaint in the Governor's letters to London was the scarcity of medical personnel in Ceylon. In 1821 Edward Barnes noted: "There is not in the Island any European medical practitioner except those in the Military Establishment." As a result of this, "all classes of the European population must and do look for medical aid" to the military practitioners.[56] The distribution of medical care in geographical terms matched the distribution of the army. Access to medicines was only possible via the military; by 1839, at last, there were proposals to set up "apothecaries" who would dispense medicine more freely.[57] The extent of militarization was in keeping with the situation in other Crown colonies.

The Colonial Office was reluctant to increase the number of medical men in the army in response to these complaints, and this raised strong feelings in Ceylon. Barnes's successor, Edward Paget, writing in 1822, stressed the economic consequences of this neglect.[58] The district of Man-

55. For a contrasting explanation based on climatic factors for the wars of 1803 and 1815, see L. De Bussche, *Letters on Ceylon.*

56. Dispatch dated 1 June 1821 from Governor Edward Barnes to London, CO 54/80, TNA.

57. See correspondence in Lot 10/169, from the Superintendent General of Vaccination and the Principal Civil Medical Officer, SLNA.

58. Dispatch dated 30 May 1822 from Governor Paget to London, CO 54/82, TNA.

nar was said to have lost a sixth of its population in a cholera epidemic in the early 1820s.[59] When cholera broke out in 1832, James Forbes, Inspector General of Army Hospitals, wrote in desperation to the Governor:

> All the posts of Badulla, Ratnapura, Kurnegalle and Chilaw, to which medical officers have always hitherto been attended, are now left to the care of native assistants. Between Colombo and Jaffna a distance of 230 miles, with numerous military posts, there is not a medical officer in the whole line. The working parties with the officers attached to them, on the different roads in the interior, are, in several places without even a Native Assistant & in the road now forming from Kandy to Trincomalee, through a very unhealthy country they are of necessity entirely left to the medical care of Native Assistants.[60]

London eventually urged the separation of military and civil medical spheres, and this became a more widespread aim of the reform of governance in the 1830s. In particular, the Colonial Office pushed for the training of "native sub-assistants," denoting usually those of mixed or Dutch or Portuguese descent, who could take on civil medical duties.[61] The Britons serving in Ceylon urged that the number of such "native" sub-assistants who were already performing medical duties was insufficient to meet the needs of the civil population, and skepticism about their ability is evident in Forbes's words above. In addition, "confusion, inconsistency, and collision of authority" had characterized their appointment so that there was a blurring of military and civil duties which was impossible to disentangle, and this is in keeping with the general chaos of governance in Ceylon in this period.[62]

To return to the controversial handling of evictions in the smallpox epidemic of 1833–34: this episode reveals yet again how civil and military cultures of medicine were inextricably linked.[63] It is clear from the Colonial Office papers that the press correspondent was not exaggerating the

59. Dispatch dated 1 June 1821 from Governor Edward Barnes to London, TNA.

60. Letter dated Kandy, 24 July 1832, from James Forbes, Inspector General of Hospitals, to the Governor, CO 54/117, TNA.

61. Letter dated Colombo, 27 July 1837, from A. Stewart MD to the Colonial Secretary, CO 54/161, TNA.

62. Letter dated Colombo, 22 November 1837, from A. Stewart MD to the Colonial Secretary, CO 54/161, TNA.

63. It was only much later, in 1858, that a separate Civil Medical Department was formed. See Uragoda, *A History of Medicine*, 93.

degree of force used in dispatching those with smallpox to hospitals. For instance, the official papers record how in 1837, in the context of a further epidemic, a District Judge issued a warrant for the removal of a woman with smallpox, but when her husband threatened to kill the Mudliyar of the district in question, the judge himself had to go to the house to have her removed, and her husband in turn petitioned the Governor.[64] In the same year, individuals were even appointed as "watchers" to spy out people who were infected by smallpox but lay concealed.[65] The controversy about forced admission to hospital raises a specific issue: the role of the colonial state and the way it took upon itself the duty of deciding the best interest of the public. In defending his handling of the outbreak in 1833–34, James Forbes made these rather telling comments about what should drive the colonial state, and how a victim of smallpox should not interfere with the higher good:

> The effects of government then are directed less to cure existing, than to prevent future cases, and if it be acknowledged that the measures adopted to this end (as might easily be shown) are more safely entrusted to its hands than to the hands of individuals themselves, who may be attacked, all his [the correspondent's] pathetic declamation on the cruelty and injustice of separating the infected few from the uninfected many falls to the ground.[66]

The call for civil medicine had been cloaked in a discourse of benefaction and charity, yet its outworking militated against individual needs. Indeed, the press correspondent blew the whistle on the discourse of humanitarianism with these words:

> Does it not raise a feeling of indignation to think nay to know that such things are? Are they permitted because this is not England? Are British subjects to be differently treated here from what they are there? Are a free people to be treated so anywhere? And let me ask, is this the way to raise the standard of native feeling now so much talked of by

64. Letter dated 29 April 1837 from R. Sillery, Superintendent of Vaccination, Lot 6/1399, SLNA.

65. Letter dated 23 June 1837 to G. Jones from unknown author referring to watchers being appointed in Jaffna, Lot 6/1400, SLNA.

66. "Answers to queries from the Right Honble Sir Robert Wilmot Horton to the Superintendent General of the Vaccine Department," CO 54/ 135, TNA.

exhibiting specimens of British humanity, such as have been described in my last letter?[67]

In devising rules about public interest in matters of health that circumvented the individual, and even the plight of the individual, the state created the space for ideas of class and other types of difference which cemented division in communities. One instance of this is the stream of complaints that arose from Cinnamon Gardens, which would become one of the prime residential districts for the British in Colombo, about the creation of a smallpox hospital in its environs. Because of the alarm of this neighborhood's elite residents about being infected by patients being taken through the streets, the hospital was moved away.[68]

The draconian measures to control smallpox illustrate how the military context of the colonial state's formation, with respect to medicine, lent a particular tenacity to its interest in intervening in matters of disease and in deciding public interest so to orchestrate evictions and confinements. Yet dissent to this militarization was evident, in London's call for a separation of the civil sphere from the military one, and also in liberal voices of protest and programs of reform.

THE CHANGING PLACE OF THE "NATIVE"

From the very beginning of the establishment of British medicine in Ceylon, it was envisaged that "natives" would be used to supplement the practice of British military medical men. On the one hand, the military medical establishment stood against the cosmopolitan landscape of cure and healing of the eighteenth century. However, on the other hand, within this establishment were signs that it also relied on extant traditions and islanders who could act as assistants. In particular, the use of people of Portuguese and Dutch descent in the medical establishment marked it as distinct from the mainland.

A Native Medical Establishment which operated under the army was founded at the start of the nineteenth century, and by the 1830s it included local peoples who were employed for both military and civil purposes. At the start of the nineteenth century, Alexander Johnston, the orientalist

67. Cited in Kinnis, *A Report on the Small-pox*, 87.
68. See dispatch dated Colombo, 2 January 1836, from Governor Horton to London, CO 54/146, TNA.

and Chief Justice, appointed a Muslim physician as the superintendent of
this medical department. Johnston noted of this Muslim physician:

> He was considered by the natives of the country as one of the best in-
> formed of the native physicians on the island, and possessed one of
> the best collections of native Medical Books, most of which had been
> in his family for seven and eight hundred years, during the whole of
> which period it had been customary for one member of his family at
> least to follow the Medical Profession. This same person made me a
> very detailed report of all the plants in Ceylon which have been used
> from time immemorial for Medical purposes by Mohammedan Native
> Physicians in the Island.[69]

This intermediary had the name of Miera Lebbe Mestriar Sekadie Marikkar, and, as in the case of Karatota Dhammarama, Johnston had a portrait
of the Muslim doctor drawn for him. Marikkar was Trustee of a mosque
at Maradana, close to Colombo, and also Head Moorman to the British,
and so the chief representative of his community, in 1818.[70] Here, again
as in the case of the founding of the first military hospitals, the British
inherited their networks of information from the Dutch, for Marikkar and
another Muslim named Sareek Lebbe had been appointed physicians by
the Dutch in 1791. Johnston wrote, in the letter just cited, that his own
suggestion that the British government found a botanical garden, in fact
emerged out of the information that followed the advice of Marikkar. This
fits into the story of the previous chapter of how this garden itself was
founded on the site of a garden kept by the king of Kandy. Putting all of
this together, the face of British medicine and natural history relied on established practices. Yet more than this, in turning to Muslim physicians,
the British turned to exactly the same class of individuals as did the Kandyan kings. Later in the century, the British adopted not only the class of
Muslim physicians but the very family of Muslim physicians who were
attached to the Kandyan kings. The Gopala Moors, who had served the
Kandyan kings, were themselves enrolled in the program of colonial medi-

69. From a letter to London of Alexander Johnston dated 3 February 1827, cited in
Mohammed Sameer Bin Haji Ismail Effendi, *Personages of the Past: Moors, Malays and
other Muslims of the Past of Sri Lanka*, 12–13.

70. The information in this paragraph is drawn from "Moorish Doctor" in Effendi,
Personages of the Past, 9–14, and Dewaraja, *Muslims of Sri Lanka*, 128–29.

cine: two Gopala Moors were appointed physicians by the British in 1828 and 1853.[71]

Despite the reliance on extant networks of indigenous practitioners, there was a persistent dismissal of "native medicine" in British records. The need to minister to local peoples, through a Native Infirmary, was stressed by Farrell to be important because of the "loathsome and aggravated form which disease assumes on most of them, from the ignorance of the native practitioners and the abject poverty of the wretched sufferers themselves."[72] Elsewhere he defended the need for a proper medical establishment in Ceylon on the grounds of the "utter ignorance of all branches of the medical profession" amongst the "native inhabitants."[73] Perhaps the strongest critique of the local medical system came from the pen of Charles Pridham in 1849. After a detailed survey, which documented how Sinhala medicine was intertwined with astrology and the reading of dreams, he wrote: "Every part of their system of medicine is deformed by such or greater absurdities; and little or no information is to be gained from it other than of a negative kind, such as the tendency of the human mind to fall into the most extraordinary errors and delusions."[74]

Yet the British disparagement of local traditions must be balanced by the fact of an attention on the part of other Britons to "native" medical traditions on palm leaf. This aspect of orientalism provides a broader context to explain why Johnston appointed Marikkar to the Native Medical Establishment. J. W. Bennett, writing as late as 1843, pointed out the considerable attention to medical botany on the part of the Sinhalese. He wrote of "Madung Appo, a native doctor of Galpiadde, near Galle, from whose skill in botany I derived much useful information." This medical man had sealed his fame by restoring the sight of a Portuguese-descent girl of seven years of age, after she was pronounced incurable by four separate European medical men:

> He began by ordering the child a milk diet; and during the six weeks that she was his patient, he employed no other medicine than a fine white

---

71. Liyanaratne, *Buddhism and Traditional Medicine*, 51–52. See also Marikar, *Glimpses from the Past*, 195–201.

72. "Remarks on a proposed Medical Establishment in the Pettah of Colombo for the relief of Sick and Indigent Natives of both sexes," CO 54/65, TNA.

73. Letter dated Colombo, 14 December 1821, from Charles Farrell, Deputy Inspector of Hospitals, to Governor Edward Barnes, CO 54/80, TNA.

74. Charles Pridham, *An Historical and Statistical Account of Ceylon and its Dependencies* (1849), 279.

powder, having all the appearance of quinine; this he gave in doses at
stated periods, and occasionally blew a similar powder, by means of a
quill having a piece of clear muslin at the end, into the child's eye.[75]

While crediting the skill of "Madung Appo," Bennett described him as an
"oculist."

Even as the cosmopolitan eighteenth-century medical marketplace in
the island was incorporated into British knowledge, the consistent attitude
of prejudice on the part of the British towards indigenous medicines soon
changed the terms of engagement. A desire to dispense European medicine
among local peoples was born. In the mid 1810s, the necessity of training
"young native men" so that they could be given a certificate prior to their
appointment as sub-assistants was mooted: "They are to attend the hospitals & Dispensary where they will have the means of learning the proper-
ties and uses of medicines and of acquiring some practical knowledge in
surgery & in the general treatment of diseases."[76] The need for a proper
training continued to be asserted into the 1830s. By this time the nature
of the education held to be important was specified quite clearly to be a
"European" one and one that extended beyond the practical experience of
witnessing others at work in hospitals.[77] Once this had been done, it was
envisaged that such pupils could be sent to Calcutta for further training,
and even to Europe.[78]

It was in the late 1830s that chemical reagents and anatomical apparatuses and books were requested for the Native Medical Establishment
for such a European education, and there was even the suggestion of Latin
classes for "native sub-assistants."[79] Among the texts listed as authorita-

---

75. J. W. Bennett, *Ceylon and its Capabilities: An Account of its Natural Resources, Indigenous Productions and Commercial Facilities* (1843), 107.

76. Undated memorandum, possibly from 1817, Lot 6/313, SLNA.

77. For the view that the early practice simply consisted of hands-on experience,
see letter dated Colombo, 24 July 1837, from A. Stewart to the Assistant Military Secretary, CO/54/161, TNA.

78. See letter dated Colombo, 19 July 1837, from T. Spencer, Medical Sub-Assistant,
to J. Kinnis, Superintendent of Vaccination, Lot 6/1400, SLNA. See also letter dated
Colombo, 27 July 1837, from A. Stewart to the Colonial Secretary, CO 54/161, TNA.

79. Letter dated Colombo, 18 December 1837, from A. Stewart to the Honble Colonial Secretary, Lot 6/1400, SLNA. On Latin classes, see letter dated Colombo, 3 January
1838, from A. Stewart to the Colonial Secretary, CO 54/161, TNA. See also J. L. Vanderstraaten, "A Brief Sketch of the Medical History of Ceylon," for details of how Governor
J. A. Stewart Mackenzie proposed a Medical School to the Legislative Council in 1839.

tive for "native students" were: R. W. Bampfield, *A Practical Treatise on Tropical Dysentery* (1819); J. Johnson, *The Influence of Tropical Climate on European Constitutions* (1827); and J. Annesley, *Researches into the Causes, Nature and Treatment of the More Prevalent Diseases of India* (1828).[80] The course of study was expected to cover Materia Medica, Botany, Practical and Clinical Medicine, Anatomy, Surgery, and Chemistry, and one timetable shows that teaching stretched across six days from 10 am to 5 pm.[81] Yet the training and use of "native sub-assistants" for civil duties, especially among Europeans, was a controversial decision because of the prejudice of the British at having to be seen by a "native."[82] In addition to the scruple of Europeans, this category of practitioner labored under poor wages. The wage of sub-assistant was considerably lower than that of clerks in other parts of the bureaucracy, which meant that some of them transferred to other roles in the colonial state structure.[83]

The increasing emphasis on the need for education, and a European education directed at the pliable minds of young men, was complicated by an interest in utilizing established "native doctors" and other "intelligent natives" as vaccinators in the period from the 1810s onwards. The reason behind this conflicting proposal was that islanders were more likely to agree to the novel and cutting-edge method of vaccination by a known and respected individual who could convince them to "overcome native prejudice." It was proposed that this class of helpers would not simply substitute European medicine for traditional medicine, but that they would be "invited and encouraged to add this branch of knowledge [vaccination] to their craft."[84] L. De Liveira, the Mudliyar of Attapatto, in writing to J. Kinnis, Superintendent of Vaccination, in 1837 suggested that the use of people in whom the inhabitants of the island placed confidence as vaccinators would serve as a cost-effective measure. He sketched out how such "native doctors" could walk from village to village in search of subjects

80. Letter dated Colombo, 18 December 1837, from A. Stewart to the Honble Colonial Secretary.

81. "Brief suggestions as a basis upon which to commence the improvement of the native medical establishment of Ceylon," CO 54/161, TNA; also Lot 10/128, SLNA.

82. "Memorandum for the Right Honble the Governor in the present Establishment of the Medical Staff of the Island," CO 54/114, TNA.

83. Dispatch dated Colombo, 3 February 1838, from Governor J. A. Stewart-Mackenzie to London, CO 54/161, TNA.

84. Letter dated Colombo, 17 August 1837, from the Vaccine Department to the Honble Colonial Secretary, Lot 6/1400, SLNA.

and produce a list once a week or a fortnight or month of all the people they had vaccinated.[85] In this way they would become part of the colonial bureaucracy, despite combining a practice of traditional medicine with vaccination.

While British colonial medicine at the start of the nineteenth century had relied on personal contacts such as those of Alexander Johnston, as the decades rolled past and the colonial state was extended, a hierarchy became evident in the medical establishment. British surgeons were at the top, with "native sub-assistants" toward the bottom, and vaccinators, who were sometimes "native doctors," beneath them. However, it is important to note that by the 1830s "native sub-assistants" referred primarily to descendants of the Portuguese and Dutch or to mixed-race individuals rather than to the Buddhist monks or Muslim physicians who had provided information to the orientalists at the start of the nineteenth century. A return of the civil branch of the Native Medical Establishment in 1837 included a list of names arranged as "medical sub-assistants" in the first, second, and third classes, and "medical pupils" in the first, second, volunteer, and extra vaccinator classes. The names amounted to forty, and the most senior included Hoedt, Toussaint, Loftus, Heyn, Woutersz, Spencer, and Jansen.[86] One Dutch Burgher who was praised in particular was E. F. Kelaart, who was a member of the first class to take lessons in medicine from a British doctor, Dr. Kevett, in Ceylon.[87] Kelaart later went to London take a medical degree and published work on natural history. Before taking his classes in medicine, he worked in the cinnamon distillery as Native Superintendent Clerk. Here he was observed to have "proved himself a most zealous and useful servant in the sorting department. He was said to understand the language of the sorters, and kept their accounts with the Inspector."[88] The change from Marikkar to Kelaart is striking—for here we have a shift from Muslim physic to European medicine dressed in indigenous guise.

The colonial archive contains several petitions from medical sub-assistants, with Dutch and Portuguese names, complaining about their

85. Letter dated Colombo, 9 October 1837, from L. De Liveira to J. Kinnis, Lot 6/1400, SLNA.

86. "Revised schedule of the Civil Branch of the Native Medical Establishment," 30 December 1837, CO 54/161, TNA.

87. See Vanderstraaten, "A Brief Sketch of the Medical History of Ceylon," 326.

88. Letter dated Colombo, 16 November 1826, from Charles Farrell, Lot 6/1022, SLNA.

wages, the conditions of their work, which necessitated arduous travel, and also the illnesses to which they were exposed. P. E. De Zilwa entered the service in 1814 and was sent to the interior straight after the British took Kandy, where he had to undertake medical duties on a very extensive scale in a province that had just been conquered. Eventually he ended up as attendant in the hospital that Farrell had formed for the poor in Colombo. In a petition, De Zilwa described himself in these words:

> From his not having been stationary but ordered on duty to different parts of the interior, he has had repeated attacks of fever and dysentery and latterly from the same effects lost his wife and children. . . . Increased debility, domestic calamity and embarrassed circumstances in a pecuniary point of view (the latter unavoidable to a medical man whose life is a roving one) at last rendered his removal to a Maritime Province necessary—accordingly in 1829 he was removed to the Pauper Hospital in Colombo.[89]

J. B. Misso wrote in 1834 of how he had hazarded his life for the service, but that his pay was only £4.10s per month, and that it had not been increased since 1819.[90] These complaints tell us enough of how the British utilized this category of helper.

The expansion of the colonial state and its bureaucracy was the changing factor in the engagement and placement of rival medical ideas. These increasingly rigid structures placed the heaviest demands on those on the bottom of the rung. Dutch and Portuguese and mixed-descent individuals found a place for themselves in the difficult middle ground which opened up between the colonial masters and the colonized. Meanwhile, British army surgeons made a career of moving across the empire, forging a very different trajectory of service. A. Stewart, who was Superintendent General of Vaccination in 1837, is a case in point: he began his career in 1805 with an expedition to Hanover, then went to Latin America, Spain, and Portugal, was involved with the army of occupation in France in 1816–18,

---

89. Petition dated Colombo, 14 January 1834, From P. E. De Zilwa, Medical Attendant of the Pettah Pauper Hospital, to the Governor, Lot 6/1024, SLNA.

90. Petition dated Colombo, 14 January 1834 from J. B. Misso, Medical Subassistant in charge of the Pioneer Hospital, Lot 6/1024, SLNA. See also petition of J. C. Vansanden, dated 19 June 1836, to G. Jones, the Principal Medical Officer, Lot 6/1321, SLNA.

and served in Canada from 1828–32 and Mauritius from 1834–37 before be-ing appointed to Ceylon.[91]

AN INCISION INTO THE BODY

The story of the use of "natives" in the British establishment raises the difficulty of delineating the "native." This account shows the confluence and changing status of different European cultures, in how the British followed Dutch precedent, and then turned to the descendants of previ-ous colonizers. It also shows the continued importance, rather than the marginalization, of island traditions. Beyond the confines of the medical establishment, vaccination was perhaps the most potent topic of medi-cal practice that necessitated relations between the state and its peoples, and relations between the British and the islanders, to be articulated and contested. The new technique of Jennerian vaccination—innovated by the English physician Edward Jenner—was first introduced into the island via Constantinople at the same time as it was to India, in 1802, only a few years after Jenner developed it. Though Lankans had used methods of variolation prior to this, British reports suggest that this was neither wide-spread nor popular.

In an unusual initiative, J. Kinnis wrote a public letter to the inhabit-ants of Ceylon in 1837, in the form of a printed pamphlet, urging them to undergo vaccination. The letter contained copious use of statistics to prove that the rate of fatality from smallpox was high, and that it was arrested by the adoption of vaccination. The form of Kinnis's defense of vaccination reveals the ways in which inhabitants of the island resisted the procedure. For instance, Kinnis provided his readers with a detailed explanation of how vaccination impacted the skin, so as to dispense with their fears of needles and blisters. He spoke of vaccination as the "mere act of inserting with a lancet a little vaccine lymph," which then set in motion a whole series of other changes in the skin, which were essential for successful vaccination. He described the result of the incision into the skin as "a little, hard, inflamed swelling, or pimple . . . terminating in a blister or vesicle, with a depressed centre, and full, swollen margin." This blister would be encircled by a discoloration of the skin, which would be a "natural" or "florid red" in Europeans and a "darker shade" in "natives."

91. Letter dated Colombo, 17 August 1837, from T. Stewart to the Honble Colonial Secretary, Lot 6/1400, SLNA.

He added that when this vesicle was not opened or broken, it "gradually dries, hardens and falls off, with a slight, though permanent mark." This was what Kinnis wanted his readers to look for and which he termed a "successful" mark.[92] In providing graphic description, Kinnis sought to dispel the magic of vaccination.

Kinnis also replied to some of the excuses presented by the inhabitants of Ceylon. First, there was the excuse of providence, where local peoples apparently answered the vaccinator that it was their lot to die of smallpox and that the British could not redirect the will of God. Second there was the accusation that the vaccinators were interested not in the inhabitants of the island but in their pay:

> If vaccination is so very beneficial, keep it to yourselves; we want none of it: sure we are, that, were it worth a single pice, you would not be so liberal. If you really are, as you say anxious to do us good, bring and distribute among us some rix-dollars and thus prove the sincerity of your professions.[93]

In these accusations it is possible to discern how local peoples saw a misfit between the political economy of colonization and the humanitarian pretensions of medicine, which opened up a site of resistance to medicine. In responding to this last excuse, Kinnis took on an allegorical style, and told the story of King William IV, who wished to have "idle rogues and vagabonds" as his subjects in Ceylon, and so asked for money to be given freely to all. This king's subjects were described as the opposite of loyal, industrious, and "good and true.[94] The making of the vaccinating regime therefore involved a good dose of statistics, combined with measures of moral and economic pleading. Personal choice was circumscribed by the will of the state and the rhetoric of good labor and responsible obedience to the Crown.

The debate over vaccination, as this pamphlet illustrates, was a public one, and one of the first public debates of the island that touched on the nature of colonial intent. Governor Brownrigg wrote to London that he was even considering legislative measures to force people to be vaccinated.[95] It

92. J. Kinnis, *A Letter to the Inhabitants of Ceylon on the Advantages of Vaccination* (1837), 20.

93. Ibid., 24.

94. Ibid.

95. Dispatch dated Colombo, 2 November 1819, from Governor Robert Brownrigg to London, CO/54/74, TNA.

was this public context that explains isolated attacks on vaccinators. One incident, described in the correspondence of Governor Brownrigg to London, came shortly after the rebellion of 1817–18, thus causing further concern of insurrection.[96] Assistant Surgeon Tonnere and a Malay soldier accompanying him as translator were assaulted by "a riotous and disorderly" gang armed with sticks and clubs in the Kandyan Provinces, close to Kurunegala.[97] The transcripts of interviews from this episode reveal the fear that Sinhala people expressed towards vaccinators, and how the sight of a man with a needle necessitated the hiding of children and women. In this instance, Tonnere first sat himself down on the rock of a Buddhist temple, amid about one hundred people assembled there, including those keeping their vows and participating in ceremonies, and this was an unwise decision.

Upon entering the garden of a nearby house with a view to vaccinating, he got into an altercation with a group of men, who turned violent, lifted up their clothes and showed their posteriors to the vaccinator, and then attacked him. At the center of this fight was a misunderstanding: one of the men thought the vaccinator was about to forcibly insert a needle into his arm. In giving his testimony, this man, named Ambakerria, noted that one of his children had run into the jungle at the sight of Tonnere, while he passed the other into the care of another adult. Tonnere was accused of seizing him by the arms and "by signs" making clear that he wished to vaccinate Ambakerria. Ambakerria accused Tonnere and the Malay of dragging him for some distance, until he was able to dislodge their grip. In his own testimony, Tonnere explained that he gripped Ambakerria in order to take him back to the temple and place him in charge of a headman, since he suspected his intentions. Tonnere's testimony included these words:

> I defended myself as well as I could, but finding they were beginning to cut me off from the path, I made off as fast as I could, defending myself with the Stick [he had seized from one of them]. I am certain that Ambakerria, Appoo Nainde, the 2nd Prisoner a Person who has escaped from Goal and others unknown to me made many blows at me as I was running along, and I have no doubt but that had they knocked me down I should have fallen a sacrifice.[98]

96. Dispatch dated Colombo, 6 January 1820, from Governor Robert Brownrigg to London, and enclosures, CO 54/76, TNA.

97. Ibid.

98. Testimonies before James Gay, Judicial Commissioner for the Kandyan Provinces, 23 December 1819, CO 54/76, TNA.

Concern surrounding vaccination arose in part from the common fact, evident here, that the vaccinator was a military man or attached to the military in some way. In fact in 1819, John D'Oyly, the Resident at Kandy, specifically wrote to the Agents of Government in the Kandyan Territories setting out that military medical officers would take up the task of vaccination.[99] It is noteworthy that Tonnere himself was aware of the symbolism of his red jacket, and removed it "lest it may occasion fear" before putting it on again in order to speak to a Buddhist priest, who appeared from the temple.[100] Directly in the aftermath of this episode, a military operation was conducted to find the men involved, and one innocent passerby complained about how he had been struck and seized by the military party.[101] As punishment Ambakerria was thrown into prison for twelve months of hard labor, and was also given a hundred lashes on his bare back in public view in the vicinity of the incident. The manner of resistance evident in Ambakerria's actions was not atypical, though quite unusual. In another incident "native sub-assistant" Bernard De Zilwa complained that the Mudliyar of Mahagampattu had instigated an attack on him by a group of people in the southern district of Hambantotte in 1831.[102]

Even if violent resistance was isolated, it spoke also of a more general fear of intrusion. In particular, there was a good amount of commentary on how Muslims found vaccination to be unacceptable, as it necessitated a Christian man coming into contact with women and girls. Opposition to vaccination was found in Barbaryn, the very village from which the ancestors of Marikkar, the Muslim physician, were said to hail, so much so that the villagers were asked to select two or more from among their number to act as vaccinators.[103] In the Galle district in the south, a Buddhist priest was said to have taken up vaccination in the early nineteenth century.[104]

Jonathan Forbes wrote of how belief in the goddess Pattini impacted on daily life when infection let loose. Villagers offered gifts to Pattini and

99. Circular letter from John D'Oyly dated Kandy, 31 October 1819, Lot 21/52, SLNAK.

100. Testimonies before James Gay, CO 54/76, TNA.

101. Ibid.

102. Letter dated 19 November 1831 from the Medical Sub-assistant to the Staff Assistant Surgeon at Galle, Lot 6/1023, SLNA.

103. Letter dated Colombo, 29 January 1816, from Andrew High to James Gay, Lot 6/313, SLNA.

104. Christie, *An Account of the Ravages Committed in Ceylon by Small-pox*, 57.

then wore shields, bangles, and other indicators as a sign of Pattini's protection against illness.[105] The extension of the vaccinating regime did not erase the character of the medical culture of the eighteenth-century island; vaccination became entangled here with established cultural practices. These preexistent traditions could serve as a basis for resistance. But it is also suggestive to consider whether the military state's attempt to take over the ground of the "indigenous" through links with Buddhist monks and Muslim physicians provoked a counter-culture of resistance based on the same ground.

The spread of vaccination was reported regularly and confidently in the colonial archive, often in the form of tables, with columns denoting the name of the district and the number of people vaccinated. These charts were important attempts on the part of the state to count up the people of the island and make the point that the provision of healthcare served as a means of conceiving of the populace of the new state. The arithmetic of vaccination spoke also of the extension of the state, and the way it came to terms with more and more people, and also of the demography of the island. Though vaccination was urged as a humane practice in the interests of the individual, it saw the state conceptualizing populations rather than individuals.

AN ISLANDED MEDICAL STATE

In addition to utilizing numerical charts, the reportage of smallpox and other diseases such as cholera was a spatialized narrative, which gave energy to the boundary-making propensities of the colonial island state.

In controlling smallpox particular concern was directed towards Moors and Malabars, the latter later termed Tamils, who arrived in the island from India, and also to Malays. Writing in 1835, J. Kinnis noted that all occurrences of smallpox on the island were "imported from the continent of India" by the arrival of people from India at multiple points on the coast. In the outbreak of 1833–34, he argued that smallpox arrived on the island through two cases, both of Muslim migrants: first, a "Moorman who affirmed from first to last that he had landed at Galle, in a Nagore dhoney from Madras" and who had been taken in by the Muslims of the mosque in New Street in the fort of Colombo, and second, a twelve-year-old boy

105. Jonathan Forbes, *Eleven Years in Ceylon: Comprising Sketches of the Field Sports and Natural History of that Colony and an Account of its History and Antiquities* (1840), 2:355.

who was "in the service of a Mahomedan priest."[106] Kinnis noted the implications of these discoveries:

> The government agent of the Western Province and the Superintendent of Police were instructed to use all their vigilance in detecting any others that might occur, and the Collector of customs and master attendant to cause a diligent search to be made in all vessels, then in the harbour, or that might afterwards arrive and to place in quarantine any vessel on board of which a case of small-pox might be discovered.

Two points are evident here: the idea that smallpox was linked with migrating peoples rather than indigenes, and that it could be controlled by policing maritime movement.

In 1837, a hired coolie who was working on a canal, and who according to one report had arrived from India and according to another from the coast at Arippu, was taken with smallpox, creating panic among the three hundred other workers. The medical assistant who examined him wrote: "His name is Mutoo Lingram, a Malabar."[107] He added that this man was taken ill with an eruption all over his head and face, body, and extremities.[108] Linked with the fear of diseases arriving from India was the particular concern that the pearl fishery in the north of the island, based in Arippu, one of the places where this laborer might have come from, was a bed of disease, as it saw peoples from both Ceylon and India working together in large numbers.[109] Suspicion was also directed towards merchant communities: in 1816 a "Chitty merchant" was blamed for the spread of smallpox from the south to the Mannar district. Restrictions were also on occasion directed towards wandering holy men, including Roman Catholic priests.[110]

The quarantining of incoming vessels became a regular feature of the

---

106. Kinnis, *A Letter to the Inhabitants of Ceylon on the Advantages of Vaccination*, 1–2.

107. Letter dated Chilaw, 12 April 1837, from L. Kelly, Medical Sub-Assistant, to G. Jones, Lot 6/1399, SLNA.

108. A Moorman was detected with smallpox on board a vessel which arrived in Colombo from India in 1825; see letter dated Colombo, 25 February 1825, from C. T. Whitfield, Lot 6/1223, SLNA.

109. For comments on the pearl fishery, see memorandum dated 12 February 1830 from the Medical Department, Lot 6/1023, SLNA.

110. See, for instance, letter dated Mannar, 24 August 1816, from the Vaccinator at Mannar to the Medical Superintendent, Lot 6/312, SLNA.

handling of outbreaks. In 1831, when smallpox spread across the Coromandel Coast of India, the Inspector of Hospitals in Colombo recommended that dhonies, or small boats arriving from India, be thoroughly examined.[111] In 1832, when cholera reappeared in the island, there was the need to quarantine vessels from the west, and the "Persian Gulf and the Red Sea" were noted in particular.[112] In 1833, "the fisher class" close to Jaffna, in the north, was said to be particularly susceptible to cholera. Their diet was said to be poorer than other island inhabitants, and their position on the coast was particularly favorable to the spread of a disease.[113] The children of a discharged Malay soldier were found to be infected with smallpox while on board a vessel outside Galle, in the south of the island, in 1837. The vessel was thoroughly fumigated and the clothes of every person on board were washed in sea water and hung up on the riggings to dry.[114] For a colonial regime which was deeply suspicious of the movement of peoples and of what was called vagrancy, migrants became easy targets. Later this discourse expanded when Indian indentured laborers were brought to the island in greater numbers and succumbed to disease.[115]

The importance of controlling disease also allowed the state to regulate housing and urban space more generally. Note this directive with respect to cholera:

> Under any circumstances the best way to prepare and guard against the effects of the invasion of cholera is to observe cleanliness in every respect and proper ventilation; to remove all nuisances, the source of foul or noxious exhalation from about houses, streets, by-lanes etc, to live temperately and regularly, and to avoid so far as may be, too great bodily or mental exertion—With these precautions and a state of mind free from apprehension, little comparatively is to be dreaded from this disease.[116]

111. Letter dated Colombo, 10 December 1831, from James Forbes to P. Anstruther, Lot 6/1023, SLNA.

112. Dispatch dated Colombo, 24 October 1832, from Governor Horton to London, CO 54/118, TNA, SLNA.

113. Letter dated Colombo, 11 April 1830, from A. Stewart to the Honble Colonial Secretary, Lot 6/1401, SLNA.

114. Letter dated Point de Galle, 24 April 1837, from R. Sillery to G. Jones, Lot 6/1399, SLNA.

115. Uragoda, A History of Medicine, 97ff.

116. Letter dated 24 April 1837 from A. Stewart to the Honble Colonial Secretary, Lot 6/1401, SLNA.

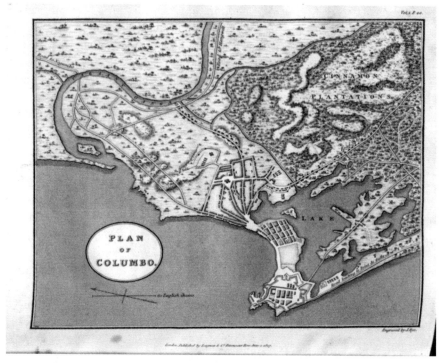

Fig. 7.2. "Plan of Columbo" showing the city in the early nineteenth century. From James Cordiner, *A Description of Ceylon* (London, 1807), vol. 1, facing page 40. Reproduced by kind permission of the Syndics of Cambridge University Library.

Given that Colombo was the heart of British government and settlement, it came under increasing surveillance and spatial documentation in the context of the spread of disease (fig. 7.2). One rather good example of this is the lengthy report on the spread of cholera in Colombo fort in 1832 prepared by James Forbes and forwarded to London.[117] Forbes began by sketching the urban geography of the fort. In the context of drought, the water of the fort was impure; the canal had evaporated and left "stagnant pools, covered with a green putrescent matter," and the lake had also dried out in many parts. The breeze in the fort, just before the cholera hit in March, had rarely any "freshness or force." Forbes also described the pettah, or area inhabited by "natives" outside the fort:

117. All quotations that follow are from dispatch dated Colombo, 24 October 1832, from Horton to London, CO 54/118.

The Pettah is a regular Dutch built town, the walls of the houses are built of Cabook (the ordinary building material of the Island), the floors of Brick flags raised two or three feet above the streets, the roofs and verandahs tiled, the bungalows glazed, venetian or latticed. It consists of five principal streets. . . .

Forbes then went into a detailed explanation of the layout of these streets and their general construction, and the workings of the sewers, the facility of the drainage, and the locations of the slaughter house, burial grounds, and cattle sheds, noting in particular the "densely populated and ill ventilated stalls" of the bazaar, and the "fisher's quarter" where the cholera first appeared. In dealing with the barracks in the fort, he calculated that each man should be allowed a "breadth of 4 feet against the wall." This led to the assertion that there was severe overcrowding. Having made such a comprehensive survey of the lay of the land, Forbes was ready to trace the epidemic episode by episode as it spread across the pettah and the fort. Forbes disputed the widespread belief that the disease came from Madras by sea, pointing to how it spread to the harbor of Colombo from the pettah.

He also appended a detailed return showing how the disease spread numerically and "topographically," using information from the hospital, the jail, and the collector of customs and master attendant. His "topographical return" listed the streets one by one and then documented the first and last case, the numbers infected on each street, and those who had died. Forbes noted from his computations: "The Disease had prevailed in the Pettah 17 days before it attacked the Troops, in the Fort 24 days before it attacked the Dhonies, in harbour; and 28 days before it attacked the Convicts in Goal." In addition to providing a sort of aerial view of the disease, Forbes's report contains detailed information about the living conditions of families in the pettah and the fort. He used a list of one hundred and ten families in this area to calculate how the cholera affected each of them, according to the size of the family, and also to determine whether there was a disparity according to gender and age. Using the number of people in particular houses, the construction and ventilation of the houses, and their proximity to stagnant water, he sought to explain the precise trajectory of the epidemic. In detailing the impact of the disease on European troops, he divided them as follows: "1st Single and Married, 2. Sober and Drunken, 3. Those who came out with, and those who have arrived subsequently to their Regiments." The result in this case was rather extra-

ordinary: it was against matrimony, and for sobriety. As an example of
the degree of scrutiny this survey necessitated, it is pertinent to note his
description of the small vessels in the harbor. They were said to amount to
"343 dhonies" with a crew of twelve each. "From so many small vessels, a
great deal of floating rubbish such as old mats, cadjans &c were drifted on
shore. It appears also that a quantity of beche de mere [bêche-de-mer, sea
cucumber] imported on the 21st March remained about a fortnight before
it was removed from the Custom House." A "nauseous smell" was said
to have arisen from stacked coral waiting to be turned into lime. He also
noted:

> Attached to the Port Department are 20 government Boatmen, 40
> Moors and Malabars employed in licensed Boats and a small but vari-
> able number of carpenters and Blacksmiths. One of the Carpenters,
> who resided in Land Street, died on the 20th March, another was at-
> tacked and recovered. A licensed Boatman was attacked on the 1st of
> April . . . [etc.]

Forbes concluded that the streets and quarters in which the cholera epidemic began were particularly marked by imperfect ventilation, overcrowding, and "vegetable matter undergoing the process of decomposi-
tion." The epidemic apparently respected class, "as no individual above
the rank of Non-commissioned officer was attacked."

The scrutiny of maritime relations and the urban planning of Colombo
are two different cases of the state's reach in terms of disease control. Yet
both of these speak of the spatial vector of the control of disease, and how
outbreaks provided the state with the reason and incentive to police forms
of livelihood and mobility. In this sense the control of epidemics and island state-making were intertwined quite neatly, and statistical projects
such as that undertaken by Forbes saw the state documenting and thus
making more rigid the demarcations of race, gender, and class.

CONCLUSIONS

There was a series of steps, which denoted a rather dynamic picture of
convolutions rather than a strict rupture or singular moment of transition
in the medical history of the island, between the patronage of the kings
of Kandy and the wider political possibilities of the eighteenth-century
culture of medicine and the making of the British colonial state. The
first was a phase of history where disease could make and break kings

and their collaborators, and medicine followed political changes and migrants from India, Asia, and the Islamic world. Then came a point where the British sought to take over these networks by working alongside Muslim physicians and Buddhist monks, as well as by opening hospitals where the Dutch had instituted theirs. From here, the medical establishment branched out as the state's apparatuses filled up: a Native Medical Establishment was formed, and even "native doctors" were used as vaccinators. However, the increasing bureaucratization of medicine meant a hierarchy of practitioners and the inscription of different categories of the "native" and "European." A military stamp was evident throughout this period, for medicine was dispensed by those attached to the military, and the civil wing of British colonial medicine came out of and bore the marks of the military context. But the last step which this chapter has sketched encompasses the seeds of dissent to militarism, as a civil sphere linked to the press contested the authoritarianism of medical provision.

In bringing medicine into the center of discussion, the chapter has focused rather less on the discourses and practices of medicine and instead pointed to the process of state-making. By showing how medicine was critical to the way states could think of themselves as states, it is possible to carve out a more vibrant place for medical histories. The statishness of medicine might include but should not restrict itself to the domain of institutions.[118] For the state is better seen, following Michel Foucault, as a form of governmentality, encompassing institutions alongside procedures and reflections. In this frame it becomes the means of the radiation of power, and a central agent in the enaction of violence. This is why attending to statishness brings to view and makes sense of the character of resistance, the form it takes, and the object to which it is directed. Vaccination was seen as a militarized system and one connected to the state's program of political and spiritual economy, and it was attacked as such rather than for its specific medical implications. Staying with Foucault, it is possible to find the resources to place medicine at the center of governmentality. For as he says, the problem of population unfroze the art of government, making it more than a juridical body of thought.[119] This metaphorical way of thinking of the origins of the state is in keeping with the argument of this chapter: state formation is best seen as a dynamic and mutating process rather than as a gestalt switch from an earlier age of monarchism.

118. For more on this see, S. Kaviraj, "On the Construction of Colonial Power: Structure, Discourse, Hegemony."

119. Michel Foucault, "Governmentality," 215.

There are some historical specificities to this story that should not be forgotten in a broader contextualization in theory. The chapter has presented an account of how an island was made into a state in part through medicine. Islandedness matters here, for this island had a distinct path to colonization, which included Portuguese and Dutch control, and this had a legacy for British medicine which is quite noticeable in relation to its assistants. At the same time, Muslim medicine continued to play a role, alongside Buddhist traditions. The colonial state's power lay in the way it placed these elements within itself, rather than displacing them in an outright fashion. As Frederick Cooper writes, empires find it very difficult to balance the making of new "bureaucratic authority" with the exercise of power which links them to "patronage structures, networks, and idioms of authority in conquered territories."[120] This is certainly the case in Sri Lanka's imperial history. At the same time islandedness matters for the particular discourses that attended colonial medicine. An attention to migration and geography in medicine allowed the boundaries of the state and the island to coalesce, and the unity of these forms created a turbulent site for articulation of racial and class-based anxieties to define the political space that was born. These anxieties were part and parcel of the context for an age of reforming liberalism, which comes into view in the next chapter.

120. Frederick Cooper, *Colonialism in Question: Theory, Knowledge, History,* 201.

CHAPTER EIGHT

Publics

Whatthe British sought to accomplish in Ceylon arguably became a model of what should constitute a colonial state, particularly among Crown colonies, where the island has rightly been seen as pioneering in constitutional terms.[1] Yet the mechanisms of reform that the British adopted are too often cast in the role of causal forces in the creation of modern Sri Lanka: a set of interventions from outside which, on the one hand, organized the island into a rational grid of provinces, which in turn allowed the integration of the whole island as a unit, and, on the other, put in place a modernized system of laws, free trade, and new norms for the use of labor.

Instead, the age of reforms needs to be seen as another convolution in the dynamism of colonial state-making and islanding. The reforms were neither a start for the nation, the bureaucracy, and liberalism, nor an end for an old system of patronage, mercantilism, and monarchism. Because of the continuous political experimentation that characterized British rule of the island, and which predated the reforms, this new set of programs generated a further arena of competition in the effort to take control of the schools and the press. If there is a single thread of importance which might be pinned on the reforms, it was the institutionalization of extant notions of difference pertaining to ethnicity, language, and European descent and status.

Writing of the nature of liberal thought in South Asia, C. A. Bayly has sought to recover a history of "speech acts" in the region, which were transnational while at the same time deeply located and formed in South

1. K. M. De Silva, "The Colebrooke-Cameron Reforms," 256.

Asian social and intellectual contexts.[2] He argues that a form of "counter-preaching" had roots in the intellectual resourcefulness of South Asia. Writing specifically about Ceylon, Bayly urges that elements of a liberal agenda took shape earlier in the island than in India itself. Following on from Bayly's work, it is possible to stress a different chronology and spatial framework for reforming tendencies in Ceylon. The age of reforms of the 1830s needs to be located in relation to earlier traditions of liberal orientalism and their intellectual lineage in turn among assistants, informants, and local political structures. In addition to this, liberal reforms should not be seen as a package that somehow came to Ceylon through the agency of the two men who led the Commission of Inquiry: Colebrooke and Cameron. Liberal political thought and what it meant in practice was debated within Ceylon. Governors took an active role in dispensing with much of the real force of the recommendations of liberal reformers. At the same time the press and public men from multiple backgrounds redefined liberal thought in new and more radical ways, supporting the extension of representation and urging protest against government and even feeding rebellion, so as to remake liberalism to fit Lankan aspirations and agendas. There were different styles of liberalism, some more anticolonial than others, and different manifestations of liberalism within the government and beyond it. The liberalism of Ceylon bears a further sign of local ancestry, in opting to make ethnicity, language, and nation a positive cornerstone of reform, set against more "feudal traditions" of hierarchy, in particular caste. These meanings were molded with an awareness of the international context of debate, as is evident in the events of the 1848 rebellion in Ceylon, where European news was covered extensively by the press.

In sketching the instruments of liberal debate, this chapter explores the extension of the press and the school system in the island, the second of which incorporated the possibilities of vernacular language teaching much earlier than in India. These instruments bear out the consolidation of a public sphere in Ceylon in the 1830s. Before this argument is misunderstood, it is important to qualify it. First, it is not the case that public spheres arose in complete form out of colonial reforming activities. Second, these public spheres did not allow equal communication and exchange of ideas and information among even the elite constituency of the colony. Third, this is not to argue that there wasn't a sphere of exchange in

2. C. A. Bayly, *Recovering Liberties: Indian Thought in the Age of Liberalism and Empire*.

Lanka prior to this period; the preexistence of a tradition of criticism and discussion was evident for instance in the account of the literary productions of Matara, in a previous chapter. By the 1830s, these were being taken up by new centers of literary production and monastic education in Galle and then Colombo.[3] Rather, it is possible to witness the polarizations that surrounded early British rule. What took shape was more a debate to take hold of the new sense of "the public" as a sphere connected by modern institutions. The rise of the public in Ceylon even in this period did not lead to it acting as a unitary or horizontal space of imagination, attachment, and representation.

In making these claims, I call for a more nuanced history of the emergence of the public sphere in South Asia. From its very definition as a concept in Lanka, the public was not one but many, not colonial alone or simply dominated by English-speaking commentators and readers. On all sides, however, civil servants, journalists, educated elites, conservatives, and radicals sought to define a public and to conceive of its rights or traditions and the nature of equality or hierarchy within it. As in the previous chapter, it is possible to see how the state sought to circumvent the individual by seeking for close relations with the press and schools. Even at their birth, Lanka's publics had a contested relationship to governmental authority. The publics in view here—linked to new schools and the press—represent a tiny minority of the island's population, and yet their discussion of how to conceive of society—or control it—came to have a lasting significance for traditions of governance and ethnicity. They ignored and sought to overtake the role of monastic centers of education and set the context for the divisive polemics between Christians and Buddhists in the later nineteenth century in Sri Lanka.[4]

In identifying the manner in which the public was envisioned in Ceylon, it is important to take account of the smallness of the island, for this generated a particularly fragmented political and social scene. The size of the territory also allowed modes of British-styled education to make deeper inroads. Because of these points, it is possible to hold that publics are forged not in imitation or through diffusion but through local circumstances and political necessities, and this provides another reason why they took so many and divergent forms in the nineteenth-century British empire. In taking this view, there is a good amount of evidence to support Ritu Birla's suggestion that the public in South Asia was deeply impacted

---

3. Malalgoda, *Buddhism in Sinhalese Society*, 187–88.
4. For a brief discussion of these polemics, see the Conclusion.

by the terrain of the market—so much so that forms of public engagement, organization, and association took on features of capitalism.[5] In Lanka, the interests of established landed elites and new European merchant classes and planters were at the fore in rival definitions of the public.

Too often explanations of the momentum of the public sphere in South Asia have centered on issues of class and collaboration. In returning instead to ideas of ethnicity and nationality and their relation to British colonization, this last chapter circles back to the first one of the book and to the theme of "partitioning." If the territorial and ideological elements of ethnicity were in place as the island was being partitioned with the advent of the British, the period that followed, which is the subject here, added an institutional format. The new Legislative Council was organized along ethnic and national lines, and the schools had a view of the different alignments of language and ethnicity, as did appointments to the civil service. While it is difficult to make the case that a consciousness of ethnicity pervaded the public spheres of Ceylon in the age of reform, the key point is more that ethnicized institutional formats, which would be filled out much later, had been outlined. A more guarded version of this argument has already been made by John Rogers, who argues that a historicist form of ethnicity, in a theory of nations, was popularized by the British in the 1840s in their search for the Sinhalese history of the island.[6] The current argument locates the ancestry of these discourses within the institutional politics of Ceylon more widely.

THE COMMISSION AND THE GOVERNORS

The age of reform in Ceylon is usually dated rather precisely to 1833, when some of the recommendations of a Commission of Inquiry which visited the island after being appointed by the House of Commons were brought into effect. G. C. Mendis is the foremost advocate of the radically transformative effects of these reforms. He calls them "the dividing line in Ceylon History."[7]

5. Ritu Birla, *Stages of Capital: Law, Culture and Market Governance in Late Colonial India.*

6. John D. Rogers, "Early British Rule and Social Classification in Lanka."

7. G. C. Mendis, *Ceylon, Today and Yesterday: Main Currents of Ceylon History,* 80. Mendis continues: "From them we can look back to the past, to the ancient Sinhalese and Tamil systems, from them we can also look forward to the future development of Ceylon on modern lines." Colvin R. De Silva ends his history of the arrival of British rule with a description of the effect of these reforms: "The year 1833 constitutes a definite and impor-

Yet as Kingsley De Silva notes, in contradiction to G. C. Mendis, the Commission served better as a demolition squad than a construction crew. They abolished compulsory labor or rajakariya, mercantilist monopolies such as that for cinnamon, and discriminatory regulations; however, their constructive recommendations were less successful in reaching the stage of implementation.[8] In developing this argument, we should note that the Commission marked a point of transition. But this was a point of transition which followed many others which have already been discussed. Before its arrival, the island was already being conceived of as a unit. A program of road-building was unlocking the Kandyan territories and spreading European objects of commerce into the interior before the 1830s. There was also prior dissent against "military rule" and the authoritarian position of the Governor. Further, a liberal sentiment of reform, which sought to incorporate indigenous peoples within the fabric of rule and as providers of information, existed prior to the 1830s and was best embodied in the figure of Alexander Johnston. The Commission's recommendations could not be the steering mechanism for a radical shift of direction.

Upon arrival, the Commission consisted at first only of one member, the pragmatic W.M.G. Colebrooke, who had served in the army both in India and Ceylon and also in the British expedition to Java in 1811. He later became a governor in the Caribbean. After his arrival, Colebrooke immediately published, in the English, Sinhala, and Tamil languages, the terms of his commission, and gathered petitions from across the island. He then issued a questionnaire to all of Ceylon's civil servants. The Commission's second member, who arrived a year later, was the more doctrinaire Charles Hay Cameron. Cameron was a lawyer and the offspring of a governor of the Bahamas. Colonial service and Crown government ran in his veins. He came under the influence of Adam Smith, Jeremy Bentham, and James Mill.[9] The radical intention of Colebrooke and Cameron was underpinned by a commitment to the Utilitarianism that came from these sources.

Yet in real terms it was not these men who decided the fate of the age

---

tant landmark in the history of Ceylon. . . . The [British] could now turn unhindered, with a reorganized and modernized administration, to the task of opening up the Island and developing her resources. . . . Ceylon was firmly set on the highway of modern development. A new era of her history had begun." (De Silva, *Ceylon under British Occupation, 1795–1833*, 2:594). For another account of the Colebrooke-Cameron reforms see David Scott, "Colonial Governmentality," and discussion of this work in the Conclusion, below.

8. K. M. De Silva, "Colebrooke-Cameron Reforms," 248–49.

9. Summary of details relies on G. C. Mendis, ed., *The Colebrooke-Cameron Papers: Documents on British Colonial Policy in Ceylon, 1796–1833*, 1:xxx–xxxv.

of reforms. Within the island, resistance to reform was led and orchestrated by successive governors, whose terms overlapped across the Commission's period of work and reporting: namely, Edward Barnes, the enthusiast of roads, and Robert Wilmot Horton. Barnes was openly hostile, while Horton's opposition was curious as he was involved in the Commission's appointment when in London as under-secretary of the Colonial Office. When Horton became Governor of Ceylon in 1832, he took the view that the Commission was guided too heavily by theory and played up his on-the-ground experience. Horton also sought to lobby opposition against the reforms, most potently through the *Colombo Journal*, the island's first English-language newspaper, which was effectively a state organ.[10]

At the heart of the government's opposition was the sore point that the Commission sought to interfere with the symbolic and real power of the governor. One rather small example of this was the 1831 instruction of the Colonial Office to the Ceylon government to abolish the attendants on the governor. Horton responded defensively that these consisted of forty-two individuals, of whom only one interpreter, an *aratchy*, a *kangani*, and twelve *lascorins* were specifically attending on him in person. He described the role of these fifteen people as "messengers" and not "private servants." He explained that the remainder of the forty-two individuals held honorary titles: "The Governor by conferring on a Native the Title of Modliar [Mudliyar] of his Gate and Guard grants him the highest and most valued distinction in his power."[11] However, by 1833 Horton's hand had been forced, and he sent the Colonial Office a list of the "messengers" whom he had struck off.[12]

These years therefore had a direct impact on the symbolic status of the Governor. Horton wrote to London, defending the old style of engagement between the Governor and his subjects:

> I think it in the highest degree inexpedient that a Governor should present himself in Public here without the facility of interpretation being afforded to him and above all without having accredited Mes-

10. For Horton's views of the reforms, see Vijaya Samaraweera, "Governor Sir Robert Wilmot Horton and the Reforms of 1833 in Ceylon."

11. Dispatch dated Colombo, 21 November 1831, from Governor Horton to London, CO 54/113, TNA. See also dispatch dated Colombo, 1 February 1832, from Governor Horton to London, CO 54/117, TNA.

12. Dispatch dated Colombo, 23 February 1833, from Governor Horton to London, CO 54/127, TNA. Civil servants were also instructed that they could not keep attendants for private purposes.

sengers with him, into whose hands the people will venture to entrust their Petitions.

In this way Horton painted a picture of the governor as someone assailed on all sides by the public and requiring guarding and tending, which was rather consistent with the monarchical image of the Kandyan kings.[13] In a private letter, which he insisted was "perfectly serious," Horton wrote:

> I have always concluded that it was deemed politically expedient to give a sort of semi-military character to a Governor, by dressing him up with cocked Hat, having plumes, a steel scabbard, a gold sash and spears a foot long, a dress utterly choking in this Climate, and which I never keep on one instant longer than I must. If on the contrary that is not the case, and it is deemed desirable to extinguish his portion of military adjunct—then it can be done without prejudice to the Service. Revoke my letters Patent and "civilise" me as I have suggested.[14]

The relevance of this exchange lies in how it points to the slow birthing of a different style of governance and to a different relationship between the state and the "public." In Horton's words, bereft of his attendants he would appear in public as a "private individual."[15] Until this point, the government had been personified by a military Governor. Horton in his own person provides evidence of how this was being reconfigured, as he was the first civilian Governor of the island.

Related to this, one of the central outcomes of the Colebrooke and Cameron Commission was the creation of an Executive Council consisting of the senior civil servants, and a Legislative Council, which minimized, at least in appearance, the constitutional power of the Governor and included nonofficial members, namely local people and representatives of the European community. Originally the Commission conceived that the Governor himself would be excluded from the fifteen-member Legislative Council. This was overturned by London. The Colonial Secretary, the Viscount Goderich, wrote to Horton: "To this body will be entrusted the power of enacting all such laws as may be required for the welfare of the Island."

13. Dispatch dated Colombo, 1 February 1832, from Governor Horton to London, CO 54/117, TNA, and dispatch dated Colombo, 23 February 1833, CO 54/127, TNA.

14. Private letter dated 31 August 1833 from Horton to R. W. Hay, CO 54/129, TNA.

15. Dispatch dated Colombo, 1 February 1832, from Governor Horton to London, CO 54/117, TNA.

The nonofficial "natives" and Europeans were expected to provide accurate "local information" while "convincing the population that the laws made for their Government, proceed from persons participating in their own interests and general opinion."[16]

However, the way in which the Legislative Council came to function in its first decade was rather different to the way in which it was conceived, which bears out the importance of attending to the local history of political debate. Horton noted that the inclusion of three "natives" and three Europeans in the Legislative Council was premature and worried that he would become a "political puppet."[17] Given the Governor's right to choose nonofficial members, loyalty to the local British officers became a key qualification for a seat in the Legislative Council. This meant that it was convenient to turn to the small group of English-speaking Christians from among the established elite. Successive governors continued to labor under the view that "heathens" could not sit on the Legislative Council.[18]

Alongside the recommendation of making the Legislative Council a means of engaging with the views of "natives" was the Commission's recommendation to open the civil service to the Ceylonese. Before this

16. Mendis, ed., *The Colebrooke-Cameron Papers*, 1:250.

17. P. D. Kannangara, *The History of the Ceylon Civil Service, 1802–1833: A Study of Administrative Change in Ceylon*, 234. At first Horton imagined that these proposals would mean that "natives" would consist of those holding traditional titles, like his attendant Mudliyars, who had worked for the colonists, and critiqued this idea saying that the "natives" looked to the Europeans to prevent the encroachment of the higher classes of their number upon them rather than to exalt such established elites to new positions of power. Later when it became clear that the recommendation was that "natives" on the Legislative Council were supposed to be untitled, he presented the opposite criticism that such individuals would not command any respect among their countrymen and urged that the seats in the Legislative Council should be salaried to encourage candidates to resign from titled positions. Private letter dated 23 January 1833 from Governor Horton to London, CO 54/127, TNA. Barnes was decidedly opposed to persons "of colour" being placed either in government or in such a council. In a private missive to London, he endorsed the view of Horton that such individuals would be placed in a "very embarrassing situation" and that no civil servants would wish to see "such a person" come among them. Private letter from Edward Barnes, c. 1827, CO 54/97, TNA.

18. In 1844, Colin Campbell still worked under this presumption, and London had to correct him on this point, calling such an idea "irrational" and tracing its origin to Horton's mean view of native "capacity" and "character." Dispatch dated Colombo, 20 November 1844, and note following it, from Governor Campbell to London, CO 54/213, TNA.

period, "natives" were restricted to traditional titles, the Kaccheri and Korale Mudilyars were the most senior, while titles such as Muhandiram, Aratchy, and Vidane were held by lesser individuals.[19] Johnston had already made the suggestion of reform of this system. But successive governors were stridently opposed to "leveling" the ground. Barnes held the view that admitting "natives" would create the momentum for the loss of the colony: "Admitted to one situation they would have an equal claim to another, so that unless you contemplate the supercession [sic] of all the European authorities, not excepting the Governor, I could not see where you could stop."[20]

After succeeding Barnes as Governor, Horton sought to open the civil service in a primarily ceremonial rather than real manner. In 1837, he wrote about the appointment of Frederick De Livera, who had been educated by the Baptists of Calcutta, as Acting Assistant Government Agent of Hambanotte; but his formal appointment as District Judge of Tangalle and Matara was held back for seven years, despite his acting in these capacities. This was chiefly as a result of the objections of the Colonial Secretary Philip Anstruther.[21] In the same year, Horton also alleged that he had appointed another "native," who he specifically noted was a "Malabar," as the partner of the "Singhalese" De Livera, as District Judge at Keyts, the island off the northern coast. These appointments gave him the opportunity to circulate the view of the relative liberality of his government in Ceylon, compared with how "natives" were placed in the civil service in India.[22] J. A. Stewart Mackenzie in turn, in his time as governor, wrote to London endorsing the view of Anstruther, and also of George Turnour, that local people should not be admitted to the civil service. His dispatch on the subject included this line by Turnour: "The colonial service should not be sacrificed for the maintenance of a principle inapplicable to Ceylon."[23]

This bears out how battle between the governors and the reformers was won by the governors, who redefined the premises of liberal thought.

19. Kannangara, *The History of the Ceylon Civil Service*, 196–97.

20. Ibid., 208.

21. Patrick Peebles, "Unchartered Justice: Revising the Ceylon Charter, 1833–1848," paper presented to the workshop, "Early Modernity in Sri Lanka, South Asia and Southeast Asia," at the Annual Conference on South Asia, University of Wisconsin-Madison, 2010. See also private letter dated 12 July 1837 from Horton to London, CO 54/155, TNA.

22. Dispatch dated Colombo, 2 September 1837, from Horton to London, CO 54/155, TNA.

23. Letter dated 5 June 1839 from Governor Mackenzie to London, CO 54/171, TNA.

Horton in particular was able to use the reforms to authorize and publicize his tenure, and rendered these changes ceremonial rather than real. Nevertheless, in the midst of this tussle the relations between the state and its subjects were being reconfigured.

INSTITUTIONALIZING LANGUAGE, ETHNICITY, AND COMMUNITY IDENTITY

One striking feature of the new structures of administration was their organization along communal lines, thereby providing an institutional basis for notions of difference. By convention the Legislative Council was supposed to include one Sinhalese member, one Tamil member, and a Burgher. Once this arrangement was made, members of the three communities guarded their right of representation. There was protest, for instance, when the governor sought to fill a vacancy for a Tamil member with a Sinhala member in 1848.[24] The Council's later history also became the site for caste competition. The age of reforms therefore created a public sphere which was not one but many.

One of the best-known early members of the Legislative Council was the scholar Simon Casie Chetty, who we have encountered before and who replaced a deceased "Malabar" member, A. Commaraswamy Pulle, and held his post until 1845.[25] Casie Chetty exemplifies the new style of representative. He was the author of the *Ceylon Gazetteer* (1834) and *Tamil Plutarch* (1859).[26] He was steeped in orientalist research and published in the first issue of the journal of the Ceylon branch of the Asiatic Society in 1845.[27] This trajectory of work came out of the maturing relationship he enjoyed in the 1830s with the colonial elite. Casie Chetty tried very

24. Letter from "An Occular," *The Ceylon Times*, 28 March 1848.

25. Casie Chetty was the Tamil voice on the Council for its first phase of operation, as Commaraswamy Pulle only held office for a few months. For the names of the representatives on the Legislative Council, see K. M. De Silva, *A History of Sri Lanka*, 248.

26. For some details of his biography, see the foreword to Simon Casie Chetty, ed., *Ceylon Gazetteer*. Within weeks of his arrival, Mackenzie wrote of Casie Chetty to London, proposing that he be one of the "natives" on the Legislative Council, to replace a deceased "Malabar" member. Casie Chetty served in this position until 1845, when he became one of the country's first local-born civil servants. Mackenzie urged the need to pay him a salary while on the Council to make the seat of interest to him. See private letter dated 3 January 1838 from Governor Mackenzie to Lord Glenelg, CO 54/160, TNA.

27. Casie Chetty is said to have been an original member of the Royal Asiatic Soci-

hard to emulate their intellectual activities. In 1839 he wrote to Governor
Stewart Mackenzie about the discovery of the "ruins of a city supposed
to be Tammanoua Nuwera of Vijaya Raja," the legendary first capital of
the island, prior to Anuradhapura. "I visited the place on the 8th ulto and
have since engaged a surveyor to make a survey of it for me, to accompany the account of the ruins which I propose to send to the Royal Asiatic
Society."[28] Mackenzie proceeded in turn to visit these ruins. The critical
difference between Casie Chetty and the Buddhist monks and Muslim
physicians who had served as informants in a previous age was that Casie
Chetty sought to imitate as well as aid the British. He took a leading part
in his local church, where he read the service in Tamil every Sunday. He
wrote to Mackenzie:

> I rejoice exceedingly to find that our little church at Calpentyn has
> attracted your kind notice. Since your Excellency saw it, I have got all
> the doors and windows nearly furnished with shutters, and if your Ex-
> cellency will do me the favour to direct 2000 square bricks. . . . The
> majority of the Protestants in this part of the island being Tamils, a
> Tamil clergyman is of course required at Calpentyn and it would af-
> ford us much satisfaction if Mr. Christian David were ordered to estab-
> lish himself amongst us.[29]

Rather tellingly, this was the David who had also served as an informant
to Alexander Johnston. Casie Chetty's story gestures to the evolutionary
nature of colonial transitions: contacts and forms of work were carried
on, even as the packaging, in the form of Casie Chetty himself and his
conduct, generated a new style of "native." When Mackenzie left, Casie
Chetty wrote in 1841 that his heart was filled "with inexplicable sorrow,"
for he could "never expect to experience from another that kindness"
which Mackenzie had shown him.[30]

Casie Chetty's political views appear in a series of letters he wrote
to the short-lived *Colombo Journal*, under the pseudonym "Native," just

---

ety and yet his name does not appear in the first "list of members." See *Journal of the
Royal Asiatic Society Ceylon Branch* 1 (1845): i–iii.

28. Letter dated 27 April 1839 from Casie Chetty to Governor Mackenzie, Lot 10/98,
SLNA.

29. Letter dated Calpentyn, 31 October 1839, from Casie Chetty to Governor Mackenzie, Lot 10/98, SLNA.

30. Letter dated Calpentyn, 13 February 1841, from Casie Chetty to Governor
Mackenzie, Lot 10/98, SLNA.

prior to the publication of the Commission's report. These letters were deferential in tone. The first began by contrasting "the tyrannical rule of the Kandyan despots, the religious bigotry of the Portuguese and the sordid policy of the Dutch" with the "mild and benignant way of Britain." Despite his view of the desirability of British rule, Casie Chetty ventured to assert that there was "one thing needful": the admission of "natives" into positions of real rather than symbolic power in the government of the country. The punchiest lines took the form of three questions: *"Are we too black? . . .* Is it because we wear petticoats? . . . Perhaps it may be argued that we have not yet attained the state of civilization which capacities us for offices of trust—but how can this be rationally expected when there is no inducement held out, to incite us to exertion?"[31] Yet it is important not to read these letters as the sign of unbounded liberalism. For Casie Chetty was pressing the claims of a particular class of "native" who had a "competent knowledge of the English language, laws and customs." The sentence which generated the greatest controversy was an offhand remark about the descendants of the Dutch. He wrote that all the positions that might be occupied by those he called "Tamuls and Cingalese" were taken by the descendants of the Dutch.

Casie Chetty's letters brought on a war of words between advocates of "native" advancement and those defending the rights of the Burghers. This exchange highlights again the mutating shape of intermediation and definitions of indigeneity in the specific context of Lanka. Were the Burghers to be treated as natives? Were the English-speaking natives proper representatives of the natives? Related to this, Casie Chetty self-identified as a Tamil and ran a Tamil newspaper called *Udayaditiya* for a short period, despite being a member of a Chetty community that now distances itself from the category of "Tamil." Casie Chetty's views on indigeneity were strikingly clear-cut. He held that those who were recent settlers did not count as "natives": "The descendants of a nation, who have only been settled about 200 years in the Island, are not so fairly entitled to what may be considered important offices, as the despised 'indigenous don.'"[32] In advocating the need to open the lower civil service in particular to those he saw as true natives, Casie Chetty used the example of the East India Company as a point of support, where he said the measure had been adopted "in all their territories."

The batteries of words in the *Colombo Journal* which followed the let-

31. Letter to the editor, *Colombo Journal*, 29 February 1832.
32. Letter to the editor, *Colombo Journal*, 11 April 1832.

ters of this "Native" reveal the many different perspectives and positions from which people were able to make claims and counter-claims of their right of access to the fabric of colonial government and political debate in the 1830s. One correspondent, "Neuter," sought to expose what he called "that rather arrogant race—the Dutch descendants." "Who shall dare to bar the progress of native merit up the steep heights of Fame? If there be any so presumptuous it is the *Dutch descendants*." In stridently racist language this correspondent accused the Burghers of being unable to arrange words or compose original letters in English. "Neuter" held that "natives" were better at the English language than the Dutch, who were "mere *copyists*."[33] In response one Dutch correspondent held that rather than occupying numerous posts in governance, various appointments were barred to the community; the pinnacle of possibility for a Dutch descendant was said to be a "petty Magistracy." He asked the natives to "stick to their farms and train their Children to follow the plough" rather than to dream of colonial appointments. From another perspective, "Philo" held that natives should not press for more offices as they have more opportunities now than in the age when they were under the tyranny of oppressive chiefs and headmen.[34] Another correspondent sought to show that Tamils were not fit for offices in government, using his experience from Jaffna, where he said Tamil headmen have been dismissed for fraud.[35] One Sinhalese correspondent wrote asking why low-country Sinhalese, from the coastal belt, were being treated better than those from up-country, from the old Kandyan territories.[36] These racial concerns and debates burst onto the political scene of Ceylon with some suddenness and force, as was evident at the time too.[37] They may be in keeping with a period when there was a particular attempt on the part of the Burghers to present themselves as Europeans.[38]

The letters in the *Colombo Journal* mesh well with the sentiments expressed by Governors Barnes and Horton to London about the plight of European descendants. In fact, a later cynical reading of the controversy between Burghers and natives held that it was orchestrated by Barnes him-

33. Letter to the editor, *Colombo Journal*, 24 March 1832.

34. Letter to the editor, *Colombo Journal*, 28 March 1832,.

35. Letter to the editor, *Colombo Journal*, 11 April 1832.

36. Letter to the editor, *Colombo Journal*, 14 April 1832.

37. Letter to the editor, *Colombo Journal*, 18 April 1832, asking for the war between Burghers and natives to stop.

38. See, for instance, Michael Roberts, Ismeth Raheem, and Percy Colin-Thome, *People in Between: The Burghers and the Middle Class in the Transformations within Sri Lanka, 1790s-1960s*, chaps. 4 and 5.

self.[39] In 1829 Barnes had written that the "use of the Dutch language is declining fast" and that Dutch Christians were gradually conforming to the practices of the English church.[40] Horton sketched a much bleaker picture. Burghers were said to be in the "most pitiable situation in the World," for there was now no use for their services. In a prescient proposal, given one of the directions of flight of this community in the twentieth century, Horton asked whether they might be sent off to New South Wales.[41] This hardening view of the status of the Burghers was a feature of these years of British rule.[42] The widespread enmity between the Burghers and the natives may well have been linked to issues of birth and caste. The "native elites" were disinclined to mix with Burghers for fear that the latter had an element of low-caste ancestry within them. On one occasion, for instance, Horton drew attention to how the Second Maha Mudliyar, Abraham De Saram, had declined to serve on a grand jury because of not wanting to be tainted by the "low-caste" blood of a Burgher.[43] This is ironic given how often the De Saram family had married Burghers in this period; yet it gestures once again to how mutable racial relations could be in successive decades.[44]

Moving from the eve of the Colebrooke-Cameron Commission, when individuals like Casie Chetty were dreaming of positions, to the years after it, the emergence of a Legislative Council composed of merchants and natives also signaled something quite different from the rise of a unitary public sphere. The Council also exacerbated tensions between various established and novel categories of individuals, linked to attachments to caste, nation, and ethnicity. Horton's dispatch following his receipt of instructions to establish a Legislative Council is illustrative. It contains unusual commentary on caste. He notes that the progressive relaxation of caste differences in the island which was already underway would be undercut by the appointment of "natives" to the Legislative Council, who would necessarily be members of higher castes. Interestingly, the Commission itself had received numerous petitions which denoted caste anxieties. They had received complaints on the irregularity of observing caste

39. "The Impartiality of the Times," *The Observer*, 7 November 1848.

40. Dispatch dated Colombo, 11 March 1829, from Barnes to London, CO 54/101, TNA.

41. Private letter dated 1 September 1833 from Horton to London, CO 54/129, TNA.

42. See for instance Roberts et al., *People in Between*.

43. Kannangara, *The History of the Ceylon Civil Service*, 209.

44. From Roberts et al., *People in Between*, 53.

duties on the part of lower castes, and the government's preference for higher castes in making appointments to traditional titles.[45]

Alongside his reflections on caste, Horton provided London with a theory of nations which is rather striking:

> The native inhabitants of this Colony consist of three distinct and separate nations, if they be so called, having different religions and different customs and habits, the Singhalese, Malabars and Moors, all of whom may be considered to have an equal claim to be represented on the Legislative Council.[46]

In addition to presenting a venue for competition between castes and ethnicities, the Legislative Council also pitted the European community against itself. Some of the earliest correspondence relating to nonofficial members of the Legislative Council dwells at copious length on how to select merchants, and also contains the complaints of the merchants at the way in which appointments to the Council had been handled. Horton derided the merchants as men with no deep interest in the colony, transient residents unsuited for such posts. In a petition, the merchants complained that Horton had delayed their appointment to the Council; they also held the view that the Legislative Council would be firmly under Horton's control as he had the right to nominate members to it.[47] They later accused Horton of filling the first Legislative Council with two pensioned natives who had held titles, and a proctor, and then exercising a "hostility" to the merchants by swearing in the natives before the merchants. The reason why this was an insult to the merchants was because of the principle of seniority: the prior swearing-in of "natives" meant that the merchants were obliged to sit below "natives." In a huff, the merchants boycotted the Legislative Council. They were of the view that there was a definite effort to "lower the character of the British members of the Council (not in the service of Government) in the eyes of the Native community."[48] However, after the Colonial Office instructed Cey-

45. Kannangara, *The History of the Ceylon Civil Service*, 243–44.

46. Dispatch dated Colombo, 23 November 1833, from Governor Horton to London, CO 54/131, TNA.

47. Dispatches dated Nuwera Eliya, 17 April 1814, from Governor Horton to London, and Kandy, 5 September 1834, from Governor Horton to London, CO 54/135, TNA.

48. Dispatches dated Colombo, 8 January 1836, and Colombo, 18 January 1836, from Governor Horton to London and enclosure of letter dated Colombo, 29 December 1835, from merchants to the Colonial Secretary, CO 54/146, TNA.

lon to look to other classes to fill these vacancies, the merchants decided to waive their objections.[49]

While Casie Chetty's story gestures to the Tamil perspective, the Sinhala elites were also united in their loyalty to the Crown and their delight at advancement to positions of governance. They organized what Horton called the "first native meeting" to draft and sign a memorial of thanks for their representative's appointment to the Council. The first Sinhalese member was J. G. Philipsz Panditharatne, who held the post until 1843. E. De Saram, the Interpreter Mudliyar, whose English Horton described "as a better hand" than either his own or the Colonial Secretary's, wrote to the Governor telling him that he expected 20,000 signatures to this memorial. The meeting was held in a home of the De Saram family. The chairman of the meeting, J. L. Perera, called its members to "perpetual loyalty" to the British, because of how local-born people had attained "perfect equality with His Majesty's English subjects." The meeting chanted together, "Long live our great King." They thanked Horton for the dispatch with which London's recommendations had been implemented.

It is important to highlight that the meeting's resolution was addressed from "His Majesty's Singhalese and other native subjects." The Committee to gather signatures was composed wholly of Sinhalese, even though the subordinate regional committees included a Muslim member from Trincomalee and F. R. Mutoo Christna (Muthu Krishna) and J. R. Moothiah (Muttiah) from Jaffna in the north of the island. This was a fully loyalist gesture of liberal sentiment, tempered by a growing political sensibility among the Sinhala elites. The most radical undertone lay buried within the words of the memorial. The appointment of "native" members to the Council was said to be an "earnest given" of "many future privileges, and what we prize above all—as a public and lasting recognition of our political existence." Curiously these members sought also to take hold of the old tradition of palm-leaf learning. The meeting included the recitation of a Sanskrit blessing on George IV composed for the occasion, called the *Sardulavikridila*. This blessing adopted the traditional idiom of the glorification of the island's kings and began: "He who is adorned with a glory whose widespread lustre is as auspiciously bright as the Milk ocean billows. . . ."[50]

---

49. Note from Lord Glenelg to Governor Horton dated 24 July 1834, CO 54/154, TNA. For a later history of this period of the Council, see "Suggestions—Legislative Council," *The Colombo Observer*, 21 June 1847.

50. Letter dated Colombo, 29 September 1835, from E. De Saram to the Governor, CO 54/141, TNA.

Therefore in seeking to use indigenous peoples in governance, this age saw an attempt to move away from the old style of indigenous informant and often reconstructed the "native" as the English-speaking Christian, and identified him as Sinhalese or Tamil rather than Burgher. Yet the English-speaking elite who were within the British colonial state came from the ranks of the old titled classes. This was not a radical change as much as a mutation of the idea of what counted as indigenous, and an evolution in how indigenous elites needed to present themselves to make themselves suitable assistants. The age of reforms did not reform as much as create and accelerate the development of paths of pressure.

SCHOOLS AND THE PRESS

The divisiveness of the age of reforms with respect to the nascent public spheres is evident also in two other areas: the schools and the press.

European-style schools had been established informally in the island by various missionary groups—including the Baptists, the Wesleyans, the Anglicans, and the Americans—in the years before the arrival of the Commission.[51] On the eve of the reforming Commission's arrival, the state also officially supported ninety-four "government schools" run by chaplains, and these were based primarily in the southern provinces and catered to the Sinhalese. The government also patronized a "seminary" in Colombo which allowed youths to acquire English so that they could serve as assistants in governance.[52] In 1832 Horton wrote of his visit to the annual missionary examinations in Kotte and Jaffna, which consisted of public questions put to "native youths" educated in missionary schools. He was delighted by their performance and immediately sought to appoint some of these youths to the medical establishment as sub-assistants and also to the surveyor's department.[53] The relationship between missionaries and the state concerning the running of schools was thus a fluid one; divisions between "government" and "missionary" schools did not indicate much.

The intervention of the Colebrooke and Cameron recommendations in this picture has been described as a "liquidation of the old order." In fact the Commission deepened some of the trends which were already

51. L. A. Wickremaratne, "Education and Social Change, 1832 to c. 1900."

52. Dispatch dated Colombo, 11 March 1829, CO 54/104, TNA.

53. Dispatch dated Colombo, 15 October 1832, from Horton to London, CO 54/118, TNA.

evident.[54] The Commission members' Utilitarianism provided an ideological foundation to the already emerging enthusiasm for English-language education. Practically, it gave rise to a School Commission in 1834, which operated until 1869. At first the School Commission was led by Anglicans, but a reorganization in 1841 brought in nonconformist and Roman Catholic representatives. The School Commission took under its umbrella the already existent "government schools" while quickly establishing a raft of new ones to diffuse English.[55] It made the Colombo Academy the leading school for dispensing education to the Ceylonese elite, including Burghers; this Academy grew out of a "seminary" run by Rev. J. Marsh, who continued as headmaster in the new institution. It was later renamed Royal College and continues to be the leading state school of the island and has given rise to a host of politicians.

It is also tempting to see the educational policy which followed the Colebrooke-Cameron Commission as a uniform one, when contrasted with the fragmented interests of missions and the state which preceded it.[56] This is a fallacy. The School Commission was characterized by entrenched denominational wars and personality clashes, which were in keeping with the character of this age of reform. Governor Mackenzie bemoaned the power of the established clergy over the School Commission, and he sought in a private letter to get London's approval to support dissenting missionaries in their educational endeavors instead.[57] He noted that "bitter squabbling" had made the School Commission nearly dormant. He also wrote of how the Anglicans wished to take the schools run by the School Commission solely under their own purview, and he insisted that government schools should be open to all Christian denominations, and also to the children of Buddhists, Hindus, and Muslims.[58]

The liberally minded Mackenzie clashed with the Tory High Church Archdeacon who ran the School Commission, Ven. J. Glenie, and his son

54. See Wickremaratne, "Education and Social Change," 174.

55. See Dispatch and enclosures dated 4 October 1837, CO 54/156, TNA.

56. K. M. De Silva, *Social Policy and Missionary Organizations in Ceylon, 1840–1855*, 142: "Before 1833 State activity in education in Ceylon was both slight and sporadic, but thereafter with the implementation of Colebrooke's recommendations it assumed a more regular and definite form."

57. Private letter dated 1 June 1838 from Governor Mackenzie to Lord Glenelg, CO 54/163, TNA.

58. Dispatch dated Colombo, 3 May 1839, from Governor Mackenzie to London, CO 54/170, and private letter dated 4 June 1839 from Governor Mackenzie to Lord Glenelg, CO 54/171, TNA.

Rev. Owen Glenie. Owen ran the newspaper *The Ceylon Chronicle* (later *Ceylon Herald*), which launched a personal attack on Mackenzie's connection to dissenting missionary schools.[59] This eventually gave rise to a libel case surrounding the newspaper's comments about the Governor's visit to the northeast of the island in an attempt at converting the Vaddah or forest-dwelling aboriginal peoples, an ambition it described as contrived rather than real.[60] Schooling was a critical issue in this clash between the Anglicans and the Governor. *The Ceylon Herald* gave this advice to parents:

> If you require any assistance from government in educating your offspring, unite with the English Dissenting Missionaries and you shall have everything; if you unite with the clergyman appointed and paid by Government to take care of you and your children's spiritual interests, you shall have nothing.[61]

When Mackenzie sought to remove the younger Glenie from Colombo, there arose another set-piece debate about the Governor's authority, typical of this anxious period. Glenie raised an uncomfortable question for London: did a colonial governor have authority over a colonial church? The Colonial Office dodged the issue and referred it to the Bishop of Madras. By 1845, Mackenzie's successor Colin Campbell managed to pass an ordinance which effectively severed the privileged status of the Anglicans in Ceylon.

The Chief Justice, speaking at the libel case against *The Ceylon Herald*, described the public sphere in Ceylon in graphic terms, picking up the argument of this chapter about the need for theorists of the public sphere in Asia to attend to local conditions. He spoke of the "inhabitants cooped up in this Island," who "like insects of mantis species" specialize "in tearing each other to pieces." "The newspapers were made conduit pipes for the discharge of hatred and malice against one another."[62] Following this

59. De Silva, *Social Policy and Missionary Organizations in Ceylon*, 32–33; also dispatch dated Colombo, 3 May 1839, from Governor Mackenzie to London.

60. Dispatch dated Colombo, 17 December 1839, from Governor Mackenzie to London, CO 54/173, TNA. See also H. C. Selby, *Report of the Trial of John Mackenzie Ross for Libel* (1839).

61. Enclosed in dispatch dated Colombo, 3 May 1839, from Governor Mackenzie to London, CO 54/170, TNA.

62. Dispatch dated Kandy, 10 February 1840, from Governor Mackenzie to London, CO 54/178, TNA.

description, the public sphere emerged not whole but polarized by different brands of politicized Christianity, which sought to seize the resources
of access to it. Mackenzie was also aware of the necessity of taking control
of public opinion. He wrote of the "deep injury to the tone both of thinking and of public discussion—in a small colonial society" caused by these
newspaper reports, and also of the "influence" which the Glenies wielded
over the respectable classes and "half-educated natives" by virtue of their
position in the church and in schools.[63] He justified his turn to the courts,
in a dispatch to London, by invoking the "mind of the Public." He considered "the Public at large" to require protection from the press, because
of both the public's and the editors' lack of education.[64] To Mackenzie's
astonishment, the jury acquitted the editor.

The divisiveness of the system of schooling in Ceylon in the years
before 1850 also centered on another issue. Mackenzie, unlike the Commission, took the view that education should be conducted primarily in
vernacular languages, so as to cater for the masses rather than the elite,
and as preparatory to English-language education. In this he was heavily
influenced by the Wesleyan minister and scholar Rev. D. J. Gogerly and
cited the examples of the "Highlands, of Wales and the whole West of
Ireland," where the need for native languages had been established.[65] He
wrote to London of the "very great" poverty of most of the inhabitants of
the island, which called for a program of "general instruction."[66] He also
noted that there were six hundred schools in operation run by natives,
where they learned how to write on palm leaves. Though this is probably
a reference to temple schools, Mackenzie shrewdly avoided a mention of
Buddhism in his dispatch, given the controversy surrounding the colonial
state's connection to Buddhism in this period. His proposal amounted to
a means through which the tradition of palm-leaf writing could be modernized, so that the masses could have access to published books and to
schoolmasters trained by the colonial state. "I believe" he wrote, in pointing to the ultimate aim of this scheme of vernacular education, "no work
translated into Singhalese has been more sought after by the Natives than

63. Private letter dated 4 June 1839 from Governor Mackenzie to London, CO
54/171, TNA.

64. Dispatch dated Colombo, 17 December 1839, CO 54/173, TNA.

65. Governor's minute conveyed to London, dated Colombo, 14 December 1839,
CO 54/173, TNA.

66. Dispatch dated Kandy, 11 March 1840, from Governor Mackenzie to London,
CO 54/179, TNA.

our book of Common Prayer." Mackenzie hoped to remodel the realm of
the Buddhist priests and informal village tutors, and the sphere of information which had characterized the island since Saranamkara's reforms
in the eighteenth century.

Gogerly described the education given by monks and other tutors in
the village setting in a detailed report which was submitted to London:

> The chief object aimed at is to obtain the ability to combine the letters
> and give their sound without the intellect being in any way exercised.
> Their chief educational works are in Sanscrit and although a Singha-
> lese interpretation is appended, the style of language is so remote from
> that in general use that only a very small proportion of it is intelligible
> and numbers do not so much learn to read as commit to memory.[67]

Without the Colonial Office's sanction, Mackenzie instituted a Translation Committee to set in motion his scheme of bringing Sinhala books
to the rural public. Gogerly laid out plans for publishing a series of school
books through this committee, which ranged across the subjects of science, history, and general literature, and he noted the importance of a periodical for use in schools.[68] The periodical *The Ceylon Examiner* came out
in favor and compared the rural schoolmaster's task to that of the road-
builder: "Whilst the latter does but bring village in contact with village,
or town with town, the schoolmaster connects the past with the present—
the most remote corners of the earth with each other."[69]

Yet London's preference was for English-language education for all,
and this followed the line taken by the Colebrooke and Cameron Commission. Behind the scenes, however, Mackenzie and his successor were
able to push through with limited success the place of vernacular education. This became possible through the medium of the remodeled School
Commission, which was dominated by Gogerly. In 1845, the first Native
Normal School was founded, where education was given in the Sinhala
language to forty young men and ten young women who were Sinha-

67. Letter dated Galle, 5 August 1840, from D. J. Gogerly to the Governor,
CO 54/181, TNA. For a similar proposal advocating vernacular education, see "The
School Commission and Education," *The Examiner*, 15 January 1848.

68. Letter dated Galle, 5 August 1840, from Gogerly to Governor, CO 54/181, TNA.

69. "Government Education in Ceylon," *The Ceylon Examiner*, 11 August 1846, and
"Report of the School Commission," *The Ceylon Examiner*, 15 December 1846; quotation from "Schools and Roads," *The Ceylon Examiner*, 21 October 1848.

lese, who could then act as teachers; by 1848 there were twenty-four such schools.[70] Gogerly's plans were somewhat derailed by the financial trouble which hit the coffee plantations in the middle years of the 1840s. Yet the argument can be made that this incorporation of vernaculars accentuated a phenomenon which was already evident with the expansion of schools in the 1830s. Vernacular education, like the government schools before it, separated the Sinhalese population from the Tamil.

Mackenzie's initial proposal asked for London's approval for one Normal School for the Sinhalese and another for the Tamils, and separate superintendents for schools in Sinhala and Tamil districts.[71] The census of schools under government supervision, which he sent to London with this request in 1840, conflated nationality with language. In the government record, each school was listed against the schoolmasters teaching within it, and then followed a column with the heading "of what nation or language?" Each schoolmaster was assigned a categorical label, and these included, "Dutch descendant," "English," "Scotch," a single "German," and "Singhalese" or "Tamul." In no school among the three dozen listed were Sinhalese schoolmasters mixed with Tamil ones.[72] In Gogerly's report, the northern and eastern districts were denoted as primarily Tamil-speaking areas, while the rest of the island was seen as Sinhala-speaking. Gogerly wrote that there were one million speakers of the "Singhalese" language on the island, of whom about 150,000 should be under a course of instruction. He argued that they were in greater need of printed books than the Tamils, as the Madras Presidency had published various works in Tamil. "I do not mean to intimate that the Committee should not direct their attention to that [Tamil] language, but the destitute state of the Singhalese gives them a priority."[73]

In this way the state sanctioned a connection between literacy and na-

70. De Silva, *Social Policy*, 181. For the numbers in the First Native School, see "Government Education in Ceylon," *The Ceylon Examiner*, 11 August 1846.

71. Dispatch dated 11 March 1840 from Governor Mackenzie to London, CO 54/179, TNA. Even before the arrival of the Colebrooke-Cameron Commission, education for the Tamil elite was conducted separately in Colombo in a government school run by Rev. Ondatjee, while the Colombo "seminary" had lessons in Sinhala but not in Tamil. See dispatch dated 4 October 1837 from Governor Horton to London, CO 54/156, TNA.

72. "Return of schools and scholars under the Government School Commission," CO 54/179, TNA.

73. Letter dated Galle, 5 August 1840, from Gogerly to Governor Mackenzie, CO 54/181, TNA.

tionality or ethnicity, which had long-term consequences for forms of attachment and belonging. Even today education in the island is divided between the Sinhala and Tamil "mediums of instruction." The way in which education became demarcated along communal lines is also evident in some other types of evidence. In the colonial civil service, increasingly, areas were being denoted as "Sinhala" speaking or "Tamil" speaking. For instance *The Examiner* of 1848 called on the government to promote civil servants who spoke Sinhala only in Sinhala areas, and those who spoke Tamil in Tamil areas.[74] The growing alignment of education and community identity was not seen only in the sphere of the Sinhala and Tamil. In a petition delivered to Governor Mackenzie by the Roman Catholic Burghers, J. B. Misso, the president of the Roman Catholic School Commission, wrote that in the absence of system government support, Roman Catholics had to support themselves in educational endeavors. It had become "impossible for them to give their offspring a liberal education, and bring them up in the principles of religion of their fathers, but with extreme self-denial and insuperable expence."[75] Misso could seek support from the state at this point because this period saw the government supporting the idea of separate education for separate communities.[76]

From another perspective a correspondent who called himself "A Native," writing in *The Observer*, critiqued the government, asserting that the state sought to give natives a "commercial" education, which was "a flat, plain, mechanical education" rather than a profoundly intellectual one. This "Native" asserted that such a course of teaching prevented his number from keeping pace with Europeans and from entering fully into the temple of the sciences and arts. Perceptively and eloquently, this correspondent nailed the wrongs of a utilitarian philosophy of learning:

Such is the principle of their education,—they leave nothing of mental responsibility and action to their subjects. They expect to exact a soulless obedience. A down trodden people become indifferent to the

---

74. "Native Languages," *The Examiner*, 12 August 1848.

75. Memorial of J. B. Misso, President of the Roman Catholic School Commission to Governor Stewart, dated Colombo, 16 October 1840, CO 54/182, TNA.

76. Dispatch dated Kandy, 17 December 1840, from Governor Mackenzie to London, CO 54/182, TNA. Mackenzie made a plea to London to provide more systematic support to Roman Catholics, writing of how large the community was, and also how poor and ignorant it was.

higher engagements of life, but are anxious merely for the wants and lusts of life. It is then called happy. Nothing breathes and stirs, self reliance is destroyed, the song of liberty is forgotten, of that liberty and freedom of thought, by which alone we think, feel, and act.

This correspondent asserted that this education sank the "natives" into a lethargic state, making them compliant to the colonists. It directed attention away from political questions such as the need for representation; it made its citizens silent. In the words of the author:

No questions about the interests of a nation, about finance, taxes, post office or internal improvement to decide or discuss. We will not be asked how a road shall be laid, or how a bridge shall be built. Although in one case we have to perform the labour and in the other to supply the materials. The tax gatherer will tell us how we have to pay.[77]

How were letters like this publishable in the press? The first newspaper, *The Colombo Journal*, where Casie Chetty had sent his own pleas for "native advancement," was instituted in response to the arrival of the Colebrooke-Cameron Commission. Governor Horton wrote to it under various pseudonyms, and his private secretary was the assistant editor.[78] Horton's interest in the press was hailed as novel; even Alexander Johnston, from his office at the metropolitan Asiatic Society, congratulated Horton on the "change [he] introduced into the press of India."[79] Yet Horton's close involvement with the press was motivated by a desire to control it before it could take a dangerous and independent form. In 1835 he wrote that he had instituted *The Colombo Journal* to prevent the take-off of a rival "factious newspaper." The first issue of the paper acknowledged that it was controlled by "high powers," but that it would "stand in the place of a free press," and this sentiment prevailed even when it announced its impending demise.[80]

According to Horton, the instruction from London to shut down the *Colombo Journal* led to the birth of another paper, *The Observer*, which

77. "A Native on Education," *The Colombo Observer*, 18 September 1847.
78. F. Beven, "The Press," 301.
79. Letter dated 12 September 1833 from Alexander Johnston to Horton, CO 54/134, TNA.
80. See the first issue of *The Colombo Journal*, 11 January 1832. And for an announcement of the cessation of the journal, *The Colombo Journal*, 3 April 1833.

became a "vehicle of slander" against Horton and was monopolized by the merchants.[81] In this context, Horton pushed through a rival periodical to counter the abuses of *The Observer*. This was *The Chronicle*, which was founded in 1837 with the assistance of Owen Glenie. But as we have already seen, *The Chronicle* (later renamed *The Herald*) turned against the governors of Ceylon. This paper was—in its short period of existence—the organ of the Anglicans. By this time, the press had thus been set upon by three rival groups from within the colonial elite: the government, the merchants, and the established church. These early newspapers, together with some other smaller short-lived ones, were joined in 1846 by two organs which would live long, *The Times* and *The Examiner*. Both of these were founded by merchants as biweeklies; the first took a conservative political position by taking a particular interest in the civil and political rights of "Englishmen" and the second a middling one.[82] *The Examiner* was later taken over by a group of Ceylonese men, including the liberal C. A. Lorenz.[83] By the late 1840s the relations between *The Times*, *The Examiner*, and *The Observer* became especially hostile around the events that led up to the 1848 rebellion. *The Times* accused its now stridently liberal-minded rival, *The Observer*, of taking up "Rampant Radicalism" and the banner of "independence" by seeking for justice for all classes and ranks of people. This accusation, penned before the rebellion, was accompanied with this warning to its rival: *"not to be a net in which to entrap the unwary, the unthinking and the discontented."*[84] The fractious relations between these papers mapped on to relations of class and ethnicity. One letter published in *The Times* before the 1848 rebellion held that the Sinhalese race was inferior in intelligence and capability, as evident in the greatness of the arts, sciences, and manufactures of Europe, making it impossible to extend the privileges of government to local people. It noted that "semi-barbarians" needed to be governed with force and restraint rather than with equal rights.[85]

81. Private letter dated 27 June 1835, from Governor Horton to R. W. Hay, CO 54/140, TNA. See also private letter dated 2 May 1834 from Governor Horton to Lord Stanley, CO 54/135, TNA.

82. For the view of *The Times* as especially interested in the rights of the English, see "Our Trumpet," *The Ceylon Times*, 13 July 1847.

83. See Roberts et al., *People in Between*, and Beven, "The Press," 304.

84. "Independence," *The Ceylon Times*, 11 June 1847.

85. Letter to the editor, *The Ceylon Times*, 7 September 1847. See also the defense of the compulsory Road Ordinances as fit for "half-savages" in *The Ceylon Times*, 13 June 1848.

From its very inception therefore the Ceylonese press took up the views of particular communities, bearing out the point that the press divided rather than unified an emerging public sphere. Just as it divided the Europeans according to interest groups, a couple of decades later it would divide the islanders by language and ethnicity. Though newspapers in Sinhala and Tamil are usually dated to the 1860s, in fact there were some in circulation from the 1840s. The *Ceylon Times* of 1846 carried news of plans to start a "publication of a Periodical in the Singhalese language, for the general information of that race," under the auspices of Gogerly, the Wesleyan missionary. It reported:

> Nearly every Singhalese can both read and write his own language, but to him (beyond the few wretched compositions on the ethical writings of the Priesthood, and a few Poems written on apochryphical [*sic*] subjects), there is not one work to afford him any information either of the races of the world, their arts or their sciences. At present his mind is a chaos at the present moment as it was two centuries ago. The object of the proposed Periodical is to enlighten his mind.[86]

The Examiner of 1848 also carried news of a periodical called *Lanka Nidhana*, which had drawn the "jealousy of Buddhists" because of its religious contents. It noted that its successor, *The Illuminator*, which may well be the paper described above, would follow more in the steps of the English *Penny Magazine* and would encompass practical knowledge.[87]

With the expansion of the press in a myriad of controversial and conflicted directions, which mapped on to the rival sentiments of communities, it became necessary to bring in tougher measures of surveillance. Prior to the expansion of the number of titles in the 1840s, Mackenzie brought a Press Ordinance into force to regulate the printing and publishing of newspapers in Ceylon. This ordinance made it necessary for every copy of a newspaper published in Ceylon to be deposited with government, and for every new newspaper to be registered with the Colonial Secretary in Colombo.[88] Yet though the government stepped up attempts of surveillance, the year 1848 brought another episode of contestation involving the politicians of Ceylon and its tense and rival public spheres.

86. "Gratifying," *The Ceylon Times*, 12 November 1847.
87. "The Illuminator," *The Examiner*, 25 November 1848.
88. Ordinance to Regulate the Printing and Publishing of Newspapers, CO 54/178, TNA.

THE *OBSERVER'S* REBELLION?

The year 1848 marked the second major rebellion against British rule in Ceylon, at a time of global revolutions. Within the island, this rebellion was taken in part to indicate the dangers of the press and an open public sphere. At the heart of it was a different and more radical strand of liberalism than that taken up by the Ceylonese elite, including those who sat on the Legislative Council as loyalists. This was a liberalism writ large, and beyond the confines of the government.

The rebellion shook both the Kandyan provinces and the urban center of British colonialism in Colombo and began as a set of grievances about a series of new taxes; the worst of these was the proposal for a Road Ordinance which made it obligatory for every male to perform six days of labor on the roads or pay a tax in commutation. This was seen as akin to the rajakariya system abolished by the Colebrooke-Cameron Commission a decade earlier. The newspapers carried a vigorous debate on what the increase of taxation would cost: *The Observer* newspaper calculated that each Ceylonese already paid 6s 3d in taxes annually, while the *Times* claimed that it was no more than 2s each.[89] As the newspapers reported, the Kandyan villagers were incensed by the expansion of the culture of coffee into the remote quarters of their territories, and also by the arrival of low-country Sinhalese and worst of all of Indian indentured laborers to work on these plantations. Another bone of contention was the way in which the colonial state, in the wake of the influence of evangelicalism, was seeking to distance itself from the promises it had made to govern with accord to Buddhist religious principles. A group of leaders, including a pretender to the Kandyan throne, Gongalegoda Banda, who was crowned as such by supporters, tapped into this rising tide of anger about British governance. The pretender was described by *The Times* as "a fat, half stupid boy."[90] In Colombo the chief advocate of resistance to new taxes was the Irishman and physician Dr. Charles Elliott, who sought to infuse European revolutionary thought into the island and urged meetings, petitions, and marches of protest. Governor Lord Torrington (the seventh Viscount Torrington) panicked when rebellion took off a year after his arrival: he responded with great force and declared martial law. The rebels ransacked

89. "Taxation of the Natives of Ceylon," *The Examiner*, 27 September 1848; "The 'Observer' and the 'Times,'" *The Examiner*, 30 September 1848.

90. "Insurrection of the Kandians," *The Ceylon Times*, 31 July 1848; also "Another King Captured," *The Ceylon Times*, 26 September 1848.

various government buildings but were not highly organized. In fact, the degree of force the British brought to bear was rather unnecessary: just one British soldier was wounded, while at least two hundred rebels were killed.[91] After an inquiry by the British House of Commons all new taxes except for the Road Ordinance were repealed.

The nature of modern and liberal governance was critical to debates surrounding the rebellion, and these drew on the global context of 1848. At the eve of the rebellion, events surrounding the French Revolution of 1848 and other revolts and disturbances across Europe were widely reported in the Ceylonese press, alongside discussions of the proposal for new taxes on the island. "King-craft and priest-craft" were some of the topics which dominated the periodicals. A cartoon published in *The Observer* showed a man atop an elephant with missile in hand seeking to combat "the last protectionist"; above his head were the periodicals in support of free trade, "Colombo Observer, Ceylon Examiner, Bengal Hurkaru, Calcutta English-man, Madras Athenaeum and Madras Crescent"[92] (fig. 8.1). Debates about protectionism veered into the specific question of taxation, and discussions of the merits of governing by local headmen expanded into arguments for and against feudalism. The press also divided rather sharply on the issue of republicanism.[93]

The rebellion can be seen as one which gathered momentum out of the work of *The Observer* newspaper, which until 1847 had Charles Elliott as its editor. The newspaper was quick to defend itself from accusations that it had "fathered" the rebellion, claiming that this notion was circulated rather mischievously by its rival *The Times*. In its first reports of the rebellion *The Times* pointed to the effects of its rival periodical and some low-country Sinhalese in spreading insurrectionist feelings in the highlands, among "excited savages" who were planning to spring "like the prowling tiger in the dark" on defenseless victims.[94] Later it insinuated—without naming him—that Elliott was the "arch traitor," "the man who ha[d] sown the seeds from which this 'monster' rebellion had sprung."[95] Elliott had the support of a group of Burgher lawyers, who shared liberal views of na-

---

91. This paragraph summarizes K. M. De Silva, *Letters on Ceylon, 1846–1850: The Administration of Viscount Torrington and the 'Rebellion' of 1848*, introduction.

92. Cartoon appearing in *The Colombo Observer*, 8 May 1848.

93. See for instance the articles in *The Examiner*, April-June 1848. Also "Radical Times," *The Colombo Observer*, 4 May 1848.

94. *The Ceylon Times*, 11 July 1848, from "The Mob in Kandy," the first report of the rebellion.

95. "Who is the Real Traitor?" *The Ceylon Times*, 8 August 1848.

Fig. 8.1. Cartoon appearing in *The Observer*, 8 May 1848. Author's photograph.

tive rights, quite unlike those who had resisted the pleas of Simon Casie Chetty in the *Colombo Journal* a decade earlier.[96]

Prior to the outbreak of rebellion there was a series of complaints which the *Observer* took up, which amounted to attempts to expose the interventionism of the colonial state. These were later channeled into the rebellion itself. First, Elliott's paper criticized the "centralisation" of govern-

96. In fact, one rather curious change between the mid 1830s and the mid 1840s is the more amicable relationship between Burghers and other local people. On this see a letter commenting on the change in *The Colombo Observer*, 7 November 1848, article entitled "The Impartiality of the Times."

ment, noting that it was exploiting rural areas for "the improvement of the metropolis and its neighbourhood" and to support the officers of government.[97] Taking this critique further, *The Observer* compared the relationship between the Colonial Secretary's office and the rural Ceylonese to a medical practitioner who prescribes potions by the post without seeing the patient.[98] The newspaper proposed, first, that each Government Agent sit at the head of a council composed of government functionaries as well as local people, which could have the responsibility of raising revenue as well as coming to decisions on public works. It compared this proposal to the function of municipalities in Britain. Second, *The Observer* sought after a radical reorganization of the Legislative Council. It proposed an "elective franchise" to replace the system whereby the governor nominated members to the council: "It becomes the duty of England to prepare as speedily as may be, her Eastern possessions for the enjoyment of privileges like her own, by encouraging and granting, and thereby qualifying them for a participation in self-government." The periodical justified this proposal by asserting that the proportion of "intelligent European descendants" and industrious and peaceable "Natives" in Ceylon was greater than in India.[99] Third, *The Observer* took a leading role in what was called "the verandah question," an issue which pertained to the rights of the Burgher inhabitants of Colombo and which occupied large columns of print in the months before the rebellion. The government sought to take over the legal title of ownership of "verandahs," which stretched in front of old Dutch houses in the pettah of Colombo. The government justified this action by asserting that they encroached on the public roads of the city. Elliott himself, together with others including J. B. Misso and James De Alwis, represented the rights of Burgher residents before the newly arrived Governor Lord Torrington and spoke of the "immemorial possession" of this land by Burghers. To the governor himself, Elliott claimed that the colonial state was the "greatest encroacher" in the way it impinged on individual rights.[100] Torrington would have none of it and declared that all verandahs that would not be purchased by their owners would be pulled down.

The broader argument that structures of reform generated points of discord is well illustrated again, for *The Observer's* invectives about re-

97. "Centralisation," *The Colombo Observer*, 3 May 1847.

98. "Centralisation," *The Colombo Observer*, 6 May 1847.

99. "Suggestions Legislative Council, Concluded," *The Colombo Observer*, 28 June 1847.

100. "The Verandah Question," *The Colombo Observer*, 8 July 1847.

form fed into rebellion. On 3 July 1848 *The Observer* published a letter from "An Englishman" living in the "outstations" who had access to the feelings of the rural populace, and who called upon the people of Ceylon to refuse to pay taxes and to group together to build a more equitable society. This letter was then printed in Sinhalese, along with some other documents. According to a correspondent in *The Times*, the Sinhala type for such a pamphlet was provided by a "Roman Catholic press," indicating Burgher involvement; twenty thousand copies were "circulated with the greatest industry through every nook and corner, every shady glen of the Kandian country, as well as through the entire of the maritime provinces."[101] *The Times* printed excerpts from the pamphlet translated into English, though these translations were disputed by *The Observer*. One translated letter from "An Englishman" read as follows:

> The Government of Ceylon is DOING INJUSTICE *like the Government of* RUSSIA. I SEE NO DIFFERENCE BETWEEN THESE TWO GOVERNMENTS EXCEPT IN NAME. . . . IN PROPORTION AS OTHER RACES ARE DELIVERED FROM INJUSTICE MORE AND MORE INJUSTICE IS COMING UPON THE INHABITANTS OF THIS COUNTRY. NOW I SAY, IS IT PROPER THAT THE SINGALESE PEOPLE SHOULD SUBMIT TO SUCH SEVERE INJUSTICE? WILL THEY DO SO? . . . I THINK THE SINGHALESE PEOPLE WILL SHOW THEY ARE NOT A RACE OF SLAVES without doing (not doing) such severe things as Europeans lately did in order to be delivered from injustice.[102]

Elliott was also directly involved in a large gathering of locals who met to sign a petition addressed to Torrington protesting the new taxes. Eventually twenty-two hundred signatures were collected for this petition. [103] This meeting took place in Borella, in Colombo, and people assembled from other regions including Kotte and Negombo, and rumors circulated that they were threatening to march on the fort of Colombo. The meeting degenerated into a violent clash with the police, and Torrington appeared there himself.[104] *The Times* described these protesters as having Elliott literally on their shoulders and as their chief inspiration. It called for him

101. "From the Ceylon Times Extra," *The Ceylon Times*, 28 July 1848.
102. "Our Local Lycurgus," *The Ceylon Times*, 20 October 1848.
103. "Presentation of the Sinhalese Petition," *The Examiner*, 29 July 1848.
104. "From the Ceylon Times Extra," *The Ceylon Times*, 28 July 1848.

to be "put down."[105] The rival journal satirized the politics of the *Observer* by drawing attention to the thin line evident between organizing demonstrations and condoning violence. It also saw Elliott's comparison of the French and Sinhalese to be rather absurd:

> To be sure, the said Editor has a great horror of bloodshed, "don't fight, oh no, pray don't resist the law," only look at the condition of France, and other countries where people *know* how to demand their *rights*, look at (what think you!!!) Pondicherry—look at the Punjaub, and above all—what!! why—at the Editor of the *Observer*—but don't fight![106]

What is pertinent to note, for our purposes in this book, is how the rebellion gave rise to questions about the role of the press in Ceylon and the nature of the freedom of expression. In the run-up to the rebellion, in debates about "verandahs," *The Examiner* pronounced on the role of the press: "The press is powerful for good but powerless for evil."[107] According to *The Times*, the press was "free" only so far as it did not spread abuse and discontent between the rulers and the ruled.[108] "Away with the miserable jargon of 'Liberty of the Press,'" *The Times* held, "when we see a journal circulated amongst a handful of Europeans and [its principles amongst] thousands upon thousands of half civilized natives . . . and when the mainful object of that journal is to make the latter dissatisfied and discontented with the former—their Rulers."[109] It took the view that the rebellion had arisen partly out of the policy of education adopted by the government—for the rebels had benefited from the new education but turned the fruits of this education against those that had provided it to them.[110] For its part, Elliott's periodical urged that the role of print had been critical in a different sense, by allowing recourse to the legal right of petitioning and public protestation, and by keeping violence at bay. In a letter to Henry Grey, 3rd Earl Grey, Secretary of State for War and the Colonies, in London, he urged that had *The Observer's* Sinhala addresses been circulated before the rebellion, the course of protest would not have turned vio-

---

105. "Liberty of the Press and its Prostitution," *The Ceylon Times*, 15 August 1848.
106. "The Observer's Public," *The Ceylon Times*, 8 August 1848.
107. "The Influence of the Press," *The Examiner*, 3 July 1847.
108. "Liberty of the Press and its Prostitution," *The Ceylon Times*, 15 August 1848.
109. Ibid.
110. "Our Local Lycurgus," *The Ceylon Times*, 20 October 1848.

lent.[111] Either way, the expansion of the printed word was entangled with resistant histories; the rebels clearly benefited from Elliott's outpourings, even though the entire rebellion cannot be traced back to them as *The Times* accused. From within the very means that the British introduced to Ceylon in order to "civilize the natives," and which were first dominated by the English-speaking elites, arose some of the means of resistance.

It would be an exaggeration to see the role of print to be only an organ of resistance, for it was also a bastion of conservatism. It was the divisiveness of this age of reforms, in relation to public campaigns around the rebellion, which is noteworthy. Greeted by the petitions and public meetings against the regime of taxation, counter petitions and public meetings were organized by European settlers in support of the colonial government. One of these was a public meeting of "Roman Catholics" who wished to declare their support of the government; another involved a public meeting of ninety "gentlemen" who wished to provide the government with public proof of their loyalty and support.[112] The Chamber of Commerce organized a delegation of merchants to wait on the governor with a resolution in order to convey the merchants' good feeling and support.[113] *The Times* printed a petition in support of the government by a set of "natives," and at the termination of the rebellion a similar statement came from Sinhalese residents of Kandy.[114] There was also a "manifesto" signed by some British residents of Kandy, calling for the government to suppress *The Observer*, which caused Elliott to threaten legal action for libel.[115] Governor Torrington himself issued a printed letter to the Sinhala inhabitants of the island calling them to obey the law and to recall the benefits that British rule had brought to them.[116]

In keeping with this torrent of public protestation and pronouncements, *The Examiner* picked up on the precise point which this argument wishes to establish: it said that "memorial mania" characterized the island's society after the rebellion. All classes were said to be involved in organizing public meetings, petitions, and memorials to the governor and

---

111. "Letter to Earl Grey," *The Ceylon Times*, 22 August 1848.

112. "Intended Public Meeting" and "Another Intended Public Meeting," *The Ceylon Times*, 4 August 1848.

113. "Resolution of the Chamber of Commerce," *The Ceylon Times*, 4 August 1848.

114. *The Ceylon Times*, 15 August 1848; 22 August 1848.

115. "The Kandy Manifesto," *The Colombo Observer*, 11 December 1848.

116. "By His Excellency the Governor," *The Ceylon Times*, 4 August 1848.

the Colonial Office: missives from the merchants, planters, loyal head-men, and the inhabitants of the town of Galle were mentioned.[117] The roll-ing out of the public sphere opened up conflicting possibilities—and one such possibility was that taken up by Elliott of forging a radical liberalism that took the influence of the printed word away from the urban English-speaking elite and into the rural provinces. Yet his was not the only path through which the impact of print and public association was felt.

The story of British colonialism and those who responded to and re-sisted it was already evolving apace. Shortly after the period in the 1830s, when the English-speaking elite emerged as the choice intermediaries, the 1840s saw resistance to this definition of indigeneity. One critical factor in changing the engagement of the British government with the islanders was the extension of schools and the press, which raised up a new class of voices who sought to engage with the structures of governance. Yet those categories of people were themselves divided by the very structures they confronted. Yet again it is difficult to trace a linear history of the local people who were caught on the sidelines of the information regime either as collaborators or rebels, for their lineage was ever changing.

CONCLUSION

The topics of nativeness and lineage, which have been running themes of the book, have returned again to this discussion. The critical addition made by the 1830s and 1840s concerned the issue of representation. The question "Who belongs?" turned into "Who represents?" The politics of representation generated rapidly changeable relations between the urban and rural sphere, between English-speaking elites and the Sinhala-speakers who turned to rebellion, between locally born public men and European-descent liberals and conservatives. All of these factions sought to present a voice, and a manner of response to the newly liberal age of reforms, ei-ther by taking a more radical stance or by seeking for counter-reform. The variety of different definitions of indigeneity in evidence here is astound-ing. Indegeneity could be taken to indicate longevity of residence and fam-ily ancestry; but to qualify as a truly representative indigene one could be required to be a Protestant or an English-speaker or to be educated. At the same time, the lineage of a different sense of nativeness was still alive and well in the notion of kingship which was at the heart of the 1848 rebellion. Those who adopted the identity of representatives were not that different

117. "Memorial Mania," *The Examiner*, 23 August 1848.

from the old ranks of titled elites. Notions of the indigenous were also dominated by capital, embodied in merchants and planters, who intruded with great energy into this story, often with a desire to keep the natives in place or to see them as "half-savages." The rise of a vocabulary of descent was also evident in the response of minorities such as Dutch Burghers and Roman Catholics, who sought to make a claim for their rights in response to those accorded to other communities. In these contesting definitions of the native one thing cut across: indegeneity was increasingly indicated by language, nation, and race.

It cannot be doubted that there were signs of interest among the publics of Ceylon in questions of ethnicity. More broadly, this suggests evidence from Lanka to supplement an argument that is now being made in the historiography of India, that race mattered earlier, and that it should not be seen as an ideology which took a concretized form only after the 1857–58 rebellion in India.[118] The critical claim of this argument is that this interest in ethnicity was related to the formation of the Legislative Council, the opening of the civil service, and the School Commission's policies, even if the representative width of these institutions should not be exaggerated in order to make the reforms appear revolutionary. Discourse and feeling were forged in line with structures of governance.

The burden of this chapter has not been to show how developments in government in Sri Lanka preceded those elsewhere, though such a case may indeed be made. Nor indeed is this an argument for the early origin of the "public sphere" in Ceylon—though the British appear to have used the term "public" here at a comparatively early date. Instead the transition that was enacted by the age of reforms followed many others. It accentuated trends which were already under way, and engagements around reform carried through the complicated social positioning of elites and patriotic ideology over the longue durée. If approached in this way, what matters is not an assessment of whether colonialism was the agent of change or continuity, but rather the idea of transition as a process which was everchanging and which lay in the varying and uneven interactions of people across the axis of colonizer-colonized, even at the moment when the state and its publics were being forged out of early modern conceptions of political sovereignty and lineage.

118. Shruti Kapila, "Race Matters: Orientalism and Religion, India and Beyond, c. 1770–1880."

CONCLUSION

Convolutions of Space and Time

The frontispiece of Jonathan Forbes's *Eleven Years in Ceylon* (1840) is an engraving titled "Exhibition of Buddha's Tooth at Kandy, May 1828" (fig. 9.1). In the midst of tempestuous weather, the casket of the Tooth Relic emerges from the Temple of the Tooth, atop an elephant, with crowds of people and elephants paying homage to it. In his book, Forbes wrote that such an exhibition had not been undertaken since the days of King Kirti Sri Rajasimha, fifty-three years prior to this. The Tooth Relic was exhibited for three days, under the patronage of Governor Edward Barnes, and "night and day, without intermission, during the continuance of this festival, there was kept up a continual din of tom-toms, and sounding of Kandian pipes and chanque shells."[1] In 1844 in an official dispatch to London, Governor Colin Campbell made specific reference to this engraving in the broader context of making the point that the British colonial state's patronage of Buddhism had changed: "No British officer, Civil or Military, either has been in recent times, or now is in the practice of assisting at any heathen rites," he wrote, "nor does the Government interfere in the appointment of any Priests or other attendants of the Temple of Budhu."[2] Campbell noted how his government had received a request from the Kandyan monks, upon the arrival of a delegation of priests from Siam, to make an exhibition of the relic again. The reply that was sent back was that "the Government would hold no communication on the subject, either by au-

1. Jonathan Forbes, *Eleven Years in Ceylon: Comprising Sketches of the Field Sports and Natural History of that Colony and an Account of its History and Antiquities* (1840), 1:294.

2. Dispatch dated Colombo, 24 January 1844, from Colin Campbell to Lord Stanley, CO 54/120, TNA.

Fig. 9.1. "Exhibition of Buddha's Tooth at Kandy, May 1828." From
Jonathan Forbes, *Eleven Years in Ceylon* (London, 1841). Reproduced by
kind permission of the Syndics of Cambridge University Library.

thorising or prohibiting the exhibition," for the Lankan priests were said
to have the absolute legal right to exhibit the Tooth Relic themselves.

At least three historical moments are interwoven in these two sources:
Kirti Sri's spectacular festivals are invoked in telling of the exhibition of
the Tooth Relic patronized by Barnes, and reference to that later memory
in turn is utilized by Campbell to distance his own practice. Within this
lineage of rule from Rajasimha to Campbell the possibilities of kingly patronage were still open. For, in an ironic twist, Campbell attended the exhibition organized by the priests for the Siamese delegation, together with
"a great number of people, Europeans as well as Natives of all religions."
London took note of this and underlined the throw-away phrase in the
dispatch which read: "and I [Campbell] went." Yet, despite Campbell's return to the practice of a king, within this span of time the colonial state's
entanglement with the kingdom was being unravelled. At the instigation
of the Colonial Office in London, the British government was aiming to
withdraw its involvement in the patronage of Buddhist rites, including the
appointment of chief priests. This was the ground that was so close to the
heart of John D'Oyly, who had seen the Tooth Relic to be in his personal
oversight. Yet the severance of the state's connection to Buddhism could

not be absolute, for another issue, the ownership of lands connected to temples, was still unresolved.[3] Campbell wrote: "The Governor . . . continues to appoint the day managers of Lands belonging to Temples, which in many instances comprehend considerable districts; it having hitherto have found impossible to find any body in whom this species of patronage may be vested."[4] The coeval histories of the kingdom and colonial state are in full evidence here. Yet the colonial state was by the late 1840s starting to run loose at last from its precedents.

By looking out of the overlooked space of Sri Lanka, it has been possible to put together a new narrative in the theory of colonial transition in South Asia. The consolidation of British rule in the subcontinent needs to be interpreted alongside its decolonisation, for Sri Lanka's relationship to India was being partitioned before 1830. In the island, the British sought to create a separate unit for colonial governance and trade, in keeping with understandings of ethnology and religion. The British also islanded this space, within a discourse of romanticism, tied to a view of Sri Lanka's highlands and the lowlands as opposites, which were bound together, and consistent with how islands were used to test schemes of colonialism, ranging from programs of reform to medical control. In this view the whole territory of the island was a meaningful ontological unit of colonization; this did not necessarily homogenize it in practice, but it certainly united it as an object of bureaucratic empire. *Partitioning* and *islanding* did not orchestrate a radical rupture but rather narrowed the pathways already extant in the island, and one important pathway was that of the interior kingdom of Kandy. In this sense the kingdom and the colonial state were enmeshed. But this is not an argument for continuity or change, but for the constant *movement* and *recycling* that characterized colonial takeover. The "native" or indigenous was defined changeably by the British in time and space, taking up the legacy of the Dutch and Portuguese regimes that preceded it.

This book has been an attempt to rework the chronology of Sri Lankan history in the eighteenth and nineteenth centuries by moving beyond the traditional divides of "Kandyan period," "Dutch period," and "British period" histories, with which those of us who were schooled in Lanka are too familiar. Time was neither linear nor static in the process of colonial

3. For the eventual outcome in relation to temple lands and to how Kandyan monks as individuals were vested with absolute rights, see Malalgoda, *Buddhism in Sinhalese Society*, 177–78.

4. Dispatch dated Colombo, 24 January 1844, CO 54/120, TNA.

transition. The idea of kingly rule could return to the scene just when we might have expected its disappearance. Practices of pilgrimage, travel, restoration, and natural improvement connected the kingdom to the colonial state and could operate in the same moment. The kingly could also give meaning and context to colonial forms of knowledge-making. The legacy of the Dutch and other European nations continued as descendants from among these communities took up posts in greater numbers in the newly expanded colonial state, and this was the exact moment when the erasure of past legacies would have been anticipated.[5] The time of colonial transition differed greatly according to locality; the question of whether Nuwara Kalawiya, the region in which Anuradhapura lies, or Sabaragamuwa, the region in which Sri Pada lies, is in focus, rather than Kandy itself or Colombo, matters to this story. An argument could be made that the peripheries of the Kandyan kingdom sustained notions of kingship much longer than elsewhere. Meanwhile, the value of kingly ideas was also felt along the coastal belt, which had already been colonized by other European powers. The historicization of the land continued apace, consolidating a linear history for the island, through the publication of texts such as the *Mahavamsa*. Yet even here, the linearity of these histories should be interpreted as a recontextualization of forms of text-making, linguistic study, and patronage which were being revived in the eighteenth century within the Buddhist *sasana* of Kandy and by its patrons, the kings. The modernity of historical time in British colonialism in Lanka was thus determined by a different sense of time, which merged the governance of the island with the mythic past.

Moving from time to space, in islanding Lanka the British sought to make this small territory a unit in the Indian Ocean. Here too, there were contradictory impulses. For the island discourse was resisted by local conceptions of heritage and space. Despite the forms of spatial attachment that sought to break the island apart, and to find the temperate within it, the dissonances served to bolster the islanding project. The island took a more discrete character even as its different parts fell into place within the greater whole constituted by Crown colonialism. The centers of political power, be they Kandyan or British, sought to bind the island together, and increasingly so as the British took control. The parallelism between the island space and the colonial state was deeply useful for purposes of trade, for dictates over migration and belonging, and for the extension of

5. This extends the argument recently made by Alicia Schrikker, *Dutch and British Colonial Intervention in Sri Lanka, 1780–1815: Expansion and Reform.*

the powers of the state in the control of epidemics or in experiments about reform. In the light of the history of British islanding, it is imperative to move beyond the island as a fundamental unit of historical study.

If we look back in the historiography of Sri Lanka, one essay by Senake Bandaranayake may be seen as prophetic about the present moment of transnational Lankan studies. He wrote of "three spheres" which have been critical for the creation of modern Sri Lanka: internal processes, its location as an offshore island a few miles off the South Asian continent, and its position at the center of the Indian Ocean. In conclusion, however, Bandaranayake insisted on the hegemony of what he called the internal over the external: "Sri Lanka's external connections are governed, conditioned and filtered at each juncture by the strength of its local traditions and its internal dynamics."[6] His was a critique of an Indo-centric view of Lanka's past. Bandaranayake's approach was picked up in later writings concerned with Sri Lanka's international relations over the longue durée. The present book is not an international history, for it posits a separate series of concentric circles to those proposed by Bandaranayake: the local history of the periphery, the metropolitan status of Colombo or Kandy, the island as a unit, the island in the sea, and the island as British. In addition to the spatial spheres covered here, one would need to add views out from the coastal belt, from the east, the southwest, and the north. Shifting emphases and powers were vested in each of these circles at particular points of time, and the distinction between one or the other cannot be classed as an opposition between the inherently internal and the alienable external. Nevertheless, the particularity of the local, rather than the national, needs to be kept in view, so that long-distance connections do not totally transform it. Even as such particularity is stressed, it is important to appreciate the wider geographical context of Lanka's past, lest our scholarship marginalize the island once again.

A sense of the changing space of political life was part and parcel of the string of rebellions that ran through the first decades of British rule. Rebellion is an important theme of this period's history, and the movement from kingdom to colonial state was key to its occurrence. Kumari Jayawardena has rightly sought to bring the revolts of the eighteenth and nineteenth centuries together into the same frame of analysis, and asks: "Why were [they] based on a feeling of deep loss over the monarchy and the old order, with a marked devotion not only to its restoration but also

6. Senake Bandaranayake, "The External Factor in Sri Lanka's Historical Formation," 84.

the continuation of the religious culture, structures and rituals of the *ancient regime!*"[7] This is an important observation, which fits the argument that is offered here. Reports that reached London invariably commented on how pretenders to the throne were central to insurrection. In early 1842 for instance, under Campbell's governance, an investigation revealed that there had been "nightly meetings in the jungle " of the Kandyan provinces, where an individual in "an enclosure of white cloth was introduced as a new king about to drive the English out of the country." The "king" cited a dream as verification that he would be successful in this bid.[8] The following year, Campbell noted again that a pretender to the throne was going about "exciting the people and representing himself as a descendant of the late king of Kandy." A rising of the people was said to be planned to coincide with the "festival of the Perahera," and "it was their intention to attack Kandy."[9] A few months later, a lengthy account of interviews from the District Court revealed a well-established stratagem: insurgents had sought to make and hide gunpowder, and the pretender had traveled through the country holding meetings, while sitting on an old log to receive homage.[10] This information provides another gloss on why the British sought to officiate at Buddhist festivals; Campbell noted that he had made a point of being present at the exhibition of the Tooth to make sure that there was no uprising. The mantle of kingship was thus not taken on board unwittingly; there was an element of opportunism about the claim of the British to rule like kings. Such a claim prevented the spread of resistance and supported loyalist sentiment. This political posturing meant that violence could arise at the interface of different forms of rule, tradition, and knowledge. Rebellion was one mode of expressing opposition. But it is possible to expand the compass of resistance to take in opposition to public works and vaccination, in addition to attempts to evade trade monopolies; all of these have been documented through the course of this work.

Jayawardena provides evidence of how Mauritius started to play a part in Ceylon's revolts, as much as news of British wars in India, and this

---

7. Kumari Jayawardena, *Perpetual Ferment: Popular Revolts in Sri Lanka in the Eighteenth and Nineteenth Centuries,* 4.

8. Letter dated Kandy, 5 July 1842, from the Deputy Queens' Advocate to the Colonial Secretary, CO 54/199, TNA.

9. Private letter dated 18 August 1843 from Colin Campbell to the Secretary of State, CO 54/204, TNA.

10. Dispatch and enclosures dated Colombo, 21 October 1843, from Colin Campbell to Lord Stanley, CO 54/205, TNA.

brings us back to new senses of space. Mauritius's connection to Lanka lay in the fact that rebels were transported and exiled there. One of them, a monk by the name of Ihagama, who was involved in the 1817–18 rebellion, returned to the island in the early 1830s, spoke some French, and was taken to have French contacts who could be invited to help free the island from British control.[11] One plan was to send the monk to Mauritius to solicit help, and to send ivory with him to tempt the French. This bears out the new resonance of the Crown colonial web on the character of resistance. At the same time plans for appeal to the Siamese king could also be entertained in 1834, keeping alive older translocal attachments. The discussion of the 1848 rebellion in the last chapter also bears out the way in which resistance was being shaped by the character of colonialism, in the spread of the press and in the new notion of the public. Rebels and resisters were working against a moving target, and this is why they deployed evolving modes of operation. From the 1840s on the pretenders started to be people from the lowlands, claiming to be Nayakkar, rather than Tamil speakers. In fact a pretender in the 1840s had a rather colorful past, having at various times been a monk from the lowlands as well as a baptized Christian and a distiller of the alcohol, arrack.[12] From another direction of response to colonialism, from its very inception to the middle of the century there was no singular public controlled as such by the colonial state, but rather bickering and contentious battles to take control of the emerging public sphere. In all of these ways, the eclectic and cosmopolitan past of the island, or its place in an ocean, was not totally ironed out into uniformity by the islanding project. However, resistance and reaction to colonialism were also shaped in the changing forms and context of British rule, as much as by the Theravada ecumene.

Despite the continuous upheaval of British colonialism, there were some interventions which would have long-lasting consequences, and those of ethnicity, religiosity, and language literacy have been highlighted. By the end of the period covered in this book these three social markers had been interlinked in colonial sociology and programs of education, representative government, and racial classification, and in turn had been defended by recourse to the separable status of the island. For instance, as early as 1807, the colonial chaplain James Cordiner in one of the popular works on the island wrote: "The great body of the inhabitants of Ceylon is divided into three general classes, Cingalese, Candians, and Malabars.

11. Jayawardena, *Perpetual Ferment*, 91–92, on the conspiracy of 1834.
12. Ibid., 124.

The first and second are descended from the aborigines of the island; the third are the offspring of colonies, which have emigrated from the Indian peninsula."[13] Yet this discourse emerged directly out of the state-making processes of colonialism. In this description, the "Cingalese " and "Candians " are separable natural categories given the independence of the kingdom in the interior and the British rule of the coasts. Inhabitants of the former are labeled as "Candians," while the latter are termed "Cingalese." In the same vein "Malabars" denotes inhabitants that have come from the Company's territories, and so the distinction between the two types of British rule matters. Moving into the nineteenth century, attempts on the part of the state to differentiate maritime spaces, to document population, and to survey urban settlement in Colombo point to the undoubted activity of what Indian scholars describe as the ethnographic state. Yet even here, while state-making is in view, the powers of British colonialism should not be exaggerated.

Take one of Bernard Cohn's key instruments of "objectification," namely the census.[14] A palm-leaf poem on a census—which was presumably a regional tally—composed in 1841 by Gonigoda Mudiyanselage Kalubanda, a resident of Alujjoma at Walpaluva in Hat Korale, is revelatory. The author wrote in folk idiom, utilizing a form of poetry reminiscent of the boundary ballads discussed in this work, and successive verses tell of how headmen and village elders provided the numbers for the census:

> The village headman of Pallegoma, counted the men as sixty-six.
> The number of men entered by the headman of Lel Oya was fifty-two.
> Was it not one hundred and three, the number the headman of Akara
>   Embiya gave?
> The exact number given by the Aarccila of Mela Dule was one hundred
>   and eight.[15]

While verses such as this speak with some excitement about the act of counting, elsewhere the poet tells of how it was unclear whether animals should be counted, and how one headman included the dead in the count. In one location, the numbers were recorded in one place in a "white

---

13. James Cordiner, *Description of Ceylon: Containing an Account of the Country, Inhabitants and Natural Productions* (1807), 90.

14. Bernard Cohn, *An Anthropologist among the Historians and Other Essays.*

15. *Humorous Verses on a Census in the Up Country*, palm-leaf manuscript from the Wellcome Trust Library.

palm-leaf book." The poem ended by reverting to the question of taxation, while simultaneously documenting the income due from temple lands, for instance to the Temple of the Tooth. A suspicion of the survey was also in evidence; one headman feared to add his children to the number. Elsewhere, the Tamil population of one area was left out of the count. Regardless of its veracity, this account speaks of a different engagement with the census and the state more broadly. The state's new powers were being enrolled within extant notions of narration, recording, tribute, and difference. The modern and its past merge in this source. To be clear, the British state was a violent interjection into Lanka, even when placed alongside the undoubted violence of Kandy. But it is also important to attend to the context of its operation, and how it was always in the making.

This palm-leaf ballad is a reminder that the sources for this book have been varied, and they have been interpreted side-by-side without too much concern to their genres, or for determining an authentic or reliable voice. At one level it is impossible to distance the colonial archive from palm-leaf texts too greatly, for British officials became patrons of palm-leaf writing and appear in them. The palm-leaf books differ according to their origin: in court, their production was patronized by kings, while those from rural spheres such as the boundary ballads are written in a different idiom. At the heart of this book is the idea that the indigenous is a mutating concept that is used for political purchase on all sides. Keeping with this claim, it would be wrong to isolate local ballads and traditions from European sources. Indeed, in some of the palm-leaf sources considered in this work there is an awareness of translocal relations for trade, pilgrimage, and ordination. In linguistic terms these sources employ a rich spectrum of possibilities ranging from Arabic and Tamil to Sanskrit and Sinhala and drawing in some European words too.

This has been an attempt to read a select few palm-leaf texts within and alongside the canonical sources of colonial history, namely the colonial archive. By the end point of the period covered in this book, with the expansion of print the tradition of learning epitomized by palm-leaf texts was enrolled as a subject of attention in the debates about representative government and education. Meanwhile, there is some evidence that the norms of palm-leaf writing were transferred onto paper.[16] Just as the kingdom and the colonial state have a complex relation, palm-leaf sources and the colonial archive should be placed together at the heart of Lanka's his-

16. See, for instance, Sirancee Gunawardana, *Palm-leaf Manuscripts of Sri Lanka*, 31.

tory. One scholar who has attempted this task with great aplomb already is Michael Roberts, who has written of the oral and associational culture of the Sinhala people, using these sources to show how the notion of the king as god allowed "the galactic polity" of the kingdom to coalesce.[17] This work adds to his scholarship while offering the argument that the possibilities inherent in these texts were more variable in relation to loyalty and attachment. The oral and ritual culture of the Lankan people did not simply cast itself as Sinhala against an alien other; within it was the possibility of encompassing British rule and of speaking in loyalist terms of the colonisers; of distinguishing Tamils from Sinhalese, as well as casting the last kings of Lanka as Sinhala, despite their ancestry.

In picking up the point about loyalty to the British, and in the spirit of the current work as an Indian Ocean history, it is useful to note the terms of a late sandesa poem, the *Mayura Sandesa*, written in the longstanding genre of a bird in flight. While one of the oldest sandesa poems takes the same title, the one in question possibly dates to around 1850.[18] What is striking about this poem is the unproblematic depiction of colonial capitalism, norms of gender, and social life. The poet was sending a prayer to the shrine at Kataragama, and the route of the bird—a peacock—takes in the coastal region of the south of the island. The poet instructs the bird: "When you fly by the ocean, enjoying the breeze, you will see steam boats sailing far away." The poem refers to the noise of traders and customers at Tangalla, and mentions places to eat, "full of people from ships." There is delight at the sight of British women: "Lovely white women along the streets decorate their long, blue hair with flowers." Tangalla is said to be prosperous and wonderful, like the city of Sakra. If colonialism reforged Lanka in the Indian Ocean economy, this is narrated here in keeping with established traditions. The "pleasures" and opportunities of colonialism are contextualized within the norms of piety, prayer, and travel of another time. Straight after Tangalla, the bird is instructed to watch out for people paying homage at a Buddhist temple at Henakaduva. If violence and resistance were noticeable aspects of the island's response to the rolling out of the state and its discourses, that opposition could be melded into alliance, friendship, and collaboration. Colonial transition kept this spectrum of responses open by its uncertain variability.

Looking beyond the middle of the nineteenth century, it is possible to sketch in outline the later history of the processes of becoming described

17. Roberts, *Sinhala Consciousness in the Kandyan Period, 1590s to 1815.*
18. C. E. Godakumbura, *Sinhalese Literature,* 207.

here. Yet such an excursus must begin with the reminder that partitioning and islanding do not have linear developmental narratives, nor are these accounts of closure where the partitioned island becomes a final object. After 1860, the colonial state continued to see the island as distinct from the mainland, while encouraging links between Lanka and separate territories in the imperial web, and there continued to be an interplay between cosmopolitanism and indegeneity on the island. One issue that marks a divergence of policy between Lanka and the mainland is the matter of caste.[19] In the aftermath of the Colebrooke and Cameron Commission, caste was downplayed in colonial sociology, in part because Lanka was perceived to be a Buddhist land, and Buddhism was said to be a religion without caste. As John Rogers writes: "As the nineteenth century progressed, the greater importance of history, nation and race—and the lesser importance of religion and caste—was taken by many British observers and some middle class Lankans as indicative of Lanka's civilizational superiority compared to India."[20] Caste was absent in a whole array of colonial policies—from the census to the selection of jurors. Yet in practice, the state could not fully partition Lanka from the mainland, for caste carried on being discussed in the press, and the organization of the Buddhist *sangha* reflected caste competition.

As Lanka entered the later nineteenth century, state policies drew from imperial models elsewhere. After the East India Company lost its control of India in the aftermath of the Indian Rebellion of 1857–58, the structural irritant between Crown and Company was removed. Policy could at times be shared between the mainland, the wider world, and the island. A classic example of the connections in imperial policy lies in the regime of Arthur Gordon, who arrived to govern Ceylon in 1883 from Trinidad, Mauritius, Fiji, and New Zealand. In Fiji, Gordon had taken up the reins in the immediate aftermath of the British annexation of 1874, and this was justified in lieu of the protection afforded to Fijian tradition. Gordon cultivated an elaborate style of romantic politics, governing through a narrow group of Fijian chiefs and institutions. This style of indirect rule reemerged in Lanka when Gordon moved to the island: Kandyan elites and the aristocratic families of the southwest had a newly exalted role in propping up British governance. If this points to the genesis of connected policies across the greater Indian Ocean, Henry McCallum, who arrived as gover-

19. This material that follows arises from John D. Rogers, "Caste as a Social Category and Identity in Colonial Lanka."

20. Ibid., 66.

nor in 1907, adopted the Indian idea of the "durbar," or mass ceremonial assembly, and there were annual durbars of Kandyans and Jaffna chiefs, dressed in traditional garb.[21] McCallum believed the village to be the basic unit of Ceylon's society. At a fundamental level, even as models were borrowed, the essential difference of the island was upheld. John Ferguson, in his directory of Ceylonese affairs for 1883, picked up the islanding theme by writing of Lanka's relationship to the mainland: "To that continent it is related as Great Britain is to Europe, or Madagascar to Africa."[22]

Whilst these redefinitions of indigenous tradition were afoot, the cosmopolitan was also reconstituted in the later nineteenth century. According to Mark Frost, the Anglicized Colombo elite, who had emerged from privileged schools set up by the British as a result of the educational reforms in the 1840s charted here, were able to critique attempts to communalize politics along the lines of rival classes.[23] The dominant notion of citizenship among this class was that of the "Ceylonese"; the Burgher-run *Ceylon Examiner* of 1868 proposed dropping "Sinhalese" and "Tamil" and replacing these terms with "Ceylonese." This elite drew the energy of their new style of politics from the mechanical vigor of the British Indian Ocean, linked by new steamers, correspondence, and the circulation of the press and news beyond particular localities. Colombo's isolation—which appeared for instance in India's hold over it in matters of trade discussed earlier—had changed. From 1890 to 1921, with half of the city's population non-Sinhalese, it became a critical port of call in the circulation of people and capital over water. Ananda Coomaraswamy (1877–1947), the historian of art, could lecture the Colombo elite on the need of a political union of Ceylon and India. W. A. De Silva, the Buddhist educationalist, could see India as the island's mother country. Both Coomaraswamy and De Silva came together to form the Ceylon Social Reform Society in 1905, to promote understanding between different communities, and this would later feed into the Ceylon National Congress, founded in 1919. The Ceylon National Congress had limited appeal, but it is interesting in the light of

21. See ibid., 72.

22. John Ferguson, *Ceylon in 1883: Describing the Progress of the Island since 1803* (1893), 8. I thank John Rogers for pointing me to this quotation.

23. This paragraph summarizes information in Mark Frost, "'Wider Opportunities': Religious Revival, Nationalist Awakening and the Global Dimension in Colombo, 1870–1920," and Frost, "Cosmopolitan Fragments from a Splintered Isle: Colombo and 'Ceylonese' Nationalism."

a history of the cosmopolitan to note that its first president was a Tamil, Ponnambalam Arunachalam. The point is not that this sort of cosmopolitanism opposed conceptions of the indigenous or the distinctive nature of the "Ceylonese"; rather, it articulated the indigenous in light of new connections wrought by print and association across the Indian Ocean.

Frost asserts that after 1920 there was a new wave of state initiatives which overturned this culture of elite cosmopolitanism. Interestingly, many of these state initiatives were driven by Ceylonese who came to take up responsibilities within state structures. At the same time, this later period should be seen from the perspective of the colonial state's activity over the course of the longer nineteenth century; cycles of state-making had already sought to create a unit of territory in the sea, and to tie that unit to notions of the indigenous, and such a program had already wrestled with a variety of different kinds of cosmopolitanism. One can think back to how the 1820s British state sought to come to terms with migration or trade, as covered in this book. The 1920s marked another wave of that process of partitioning. One significant figure was Governor William Manning, who sought to weaken the Congress lest it become more populist. Manning's policy was "divide and rule": pitting the Kandyans against lowland Sinhalese and Tamils against the Sinhala. The Donoughmore Commission of 1926–27, which came to the island to review constitutional matters, saw an explosion of identity politics, and this has been seen as a turning point in rival assertions of communal belonging. According to Nira Wickramasinge: "At the Donoughmore sittings authenticity became a pliable instrument used by all small communities that feared they would lose their place in society if the law of numbers were to prevail."[24] Following the Commission, Sri Lanka became the first British colony to receive universal franchise, which was exercised in the election of 1931. If "Who belongs?" had given way to "Who represents?" in the nineteenth century, this period saw another mutation with the take-off of the question "Who has rights?" A sensitive issue was the rights of Indian migrant communities in the island, and this became acute in the light of the Depression of the 1930s. The Donoughmore Commission only granted voting rights to those Indians who could prove their long residence on the island as well as their desire to become permanent residents, and anti-Indian legislative measures such as that targeting the rice trade from India, whereby traders

24. Nira Wickramasinghe, *Sri Lanka in the Modern Age: A History of Contested Identities*, 103.

were obliged to buy local produce, saw the state creating new contexts for definitions of attachment to the island.[25]

The year 1915 was the hundredth anniversary of the fall of the Kandyan kingdom. It was a millennial moment for some of the island's Buddhists. *The Buddhist* periodical reminded its readers that this was the anniversary of when "we lost . . . our independent existence as nation." It called the adherents of the faith to forge a new empire, "far vaster and much more enduring than the mightiest temporal kingdom the world has ever seen."[26] The riots that occurred that year occurred firstly in historic Kandy, and there is some evidence that the Sinhala press was playing up the significance of the fall of Kandy by republishing the Convention that marked its takeover and by highlighting how the British had not protected Buddhism. But there is also evidence, from police reports, to suggest that the British were on the lookout for unrest in this year, given the anniversary.[27] Varying meanings given to the memory of the fall of Kandy partly generated the context for violence directed in different directions—on the part of Buddhists against allegedly alien Muslims, termed "Moors," who were seen to have recently arrived from South India, and on the part of the British against rioters and suspected sympathizers. From a global view, in the context of the First World War, the rioting Sinhalese may have seen Muslims as fair game, because they believed that they were at one with the Turks, who had sided with the Germans; and the economic consequences of global war also had an impact on the involvement of the urban working classes in this riot. The British state responded by bearing down on the temperance movement, which was suspected to be a political front, and which at this stage stood against the state's excise policies with respect to alcohol. In the aftermath of the riots the British had charged hundreds and killed dozens. It is important not to overdo the significance of 1915. Yet in the context of this work it gestures to the rather muted but continuing afterlife of the Kandyan kingdom. Nineteen-fifteen served to consolidate a vocabulary of self-rule which had been circulating among some Ceylonese since the turn of the century. The period that followed was thus not

25. This material is drawn from ibid., chaps. 3 and 4.

26. Cited in Frost, "Wider Opportunities," 963.

27. The material of this paragraph relies on Charles S. Blackton, "The Action Phase of 1915"; Kumari Jayawardena, "Economic and Political Factors in the 1915 Riots"; and A. P. Kannangara, "The Riots of 1915 in Sri Lanka: A Study in the Roots of Communal Violence."

only characterized by state attempts to demarcate the authentic but also by newly consolidated discourses of the indigenous popularized by certain local elites.

One individual who was caught out by the state, after his pronouncements against the "Coast Moors " as foreigners and enemies, was Anagarika Dharmapala (1864–1933). Born as Don David Hewavitarane to a family involved in the timber trade, Dharmapala had some Christian education in Colombo, including at the Colombo Academy; he then became a key early Sinhala nationalist. He was a child of the revival of Lankan Buddhism in the late nineteenth century and had witnessed the famous public debate in 1873, in Panadura, south of Colombo, which took place before an audience numbering around six thousand, between a Christian missionary and a Buddhist monk. The Panadura debate was won by the monk by public acclaim and was emblematic of the growing intellectual confidence and public status of Buddhism. At the same time, Dharmapala embraced theosophical writings in the context of the visit of the American President of the Theosophical Society, Henry Steel Olcott, to Lanka in 1880, and this may be taken as characteristic of Colombo cosmopolitanism. Yet his views evolved from that of a theosophical interest in the universality of religion to a narrower conception of the greatness of Buddhism, and this move coincided with a critical engagement with Hinduism. His vision was for Buddhism to overtake the Hinduism of India, and so to recover the heartland of its past. This became particularly urgent after he set up the Maha Bodhi Society in Calcutta to reclaim the Bodh-Gaya, the site of Buddha's enlightenment, from the control of Hindus. The British were suspicious of Dharmapala's connections to India and even thought that he was a German agent. Dharmapala sought to defend the Sinhala Buddhists in the context of the riots of 1915; writing to the British of what differentiated them from others, he argued that the purest form of Buddhism existed in Ceylon, "and the people for 2,222 years have never forgotten their faith, although the Portuguese, Dutch and British have made every effort to stamp it out."[28] He did not hold himself back in his criticisms of the British, as "demons," "vampires," and "sea-wolves," among other descriptions.[29] He was interned in Calcutta for five years in the immediate after-

28. Ananda Guruge, *Return to Righteousness: A Collection of Speeches, Essays and Letters of the Anagarika Dharmapala,* 538.

29. See for instance Michael Roberts, "For Humanity. For the Sinhalese. Dharmapala as Crusading Bosat."

math of 1915. Dharmapala marks the growing visibility of a harder-edged Sinhala Buddhist nationalism, which sat cheek-by-jowl with a culture of cosmopolitan connection, which was also evident in his own biography.

The distinction between India and Sri Lanka appears to have been a sensitive one in the first year of the Donoughmore Commission's work in Sri Lanka. Gandhi visited Lanka in this year, and he trod a careful path in talking of the island's future. Writing in his periodical *Young India*, he noted: "I have no imperial ambitions for India of my imagination. I should be content to regard Ceylon as an absolutely independent State; but I should not hesitate to accept Ceylon as part of free India if the Island-ers express their wish to be so in an unmistakable language."[30] Given the anti-migrant feeling in Lanka, Gandhi's advice to Indian resident commu-nities in the island was not to assert kinship with the Ceylonese as a mat-ter of rights but rather to take kinship up as a matter of service. The issue which drew a more activist line of politics from Gandhi, especially in later years, was the status of Indian laborers in the island. By 1945, he called for all such Indians to be given full citizenship. He said that he knew that the Ceylonese had long hoped to be separated from India, although "the culture and problems of both were similar." Yet "India and Lanka were one" and Lanka was "a pendant of a long chain which was India."[31] Ad-ditionally, he saw his philosophy of self-help and village uplift as being one which could cut across the divisions of the sea and across the various regions of South Asia in this period. The ambiguity in Gandhian thought about Ceylon is reflective of the changing structures of nationalism, both in the island and in India, and of Gandhi's double sense of himself as an insider and outsider to the island.

From the later 1920s onwards Tamil and Sinhalese politics was bifur-cated at the elite level, especially over power at the State Councils, elected in 1931 and 1936.[32] The identification of the Indian laborer and migrant as a threat, which arose in sharper form with the emerging prospect of self-rule, quickly led in turn to another shift in the idea of the indigenous. Minorities within the island—for instance, European descendants and

30. Mahatma Gandhi, "Burma and Ceylon," *Young India*, 10 March 1927, repro-duced in Gopalkrishna Gandhi, ed., *Gandhi and Sri Lanka, 1905–1947*, 22–23. This paragraph relies on reprinted letters and articles in this volume.

31. Mahatma Gandhi, "Message to People of Ceylon," *Bombay Chronicle*, 28 March 1945, reproduced in Gandhi, ed., *Gandhi and Sri Lanka*, 320.

32. For the material of this paragraph see Nira Wickramasinghe, *Sri Lanka in the Modern Age*, chaps. 3 and 4.

Muslims—found their status increasingly precarious from the late 1930s. In particular, the extra-territorialisation of Tamils was entangled with the extra-territorialisation of Indians. For instance, when in 1944 Ceylon Tamils and Ceylon Indians grouped together to form the All Ceylon Tamil Congress, the line between Indians and Tamils had been blurred for strategic purpose. In 1938, a Muslim League and a Ceylon Moors Association were also formed. In intellectual terms there were even attempts on the part of linguistic purists like Munidasa Cumaratunga to turn away from India's cultural legacy to the island by separating the Sinhala language from both Pali and Sanskrit.[33] According to the Citizenship Acts of 1948–49, Indian laborers without papers were denied rights, and the higher class of citizenship on the island was reserved for those of long residence. Yet all of this did not lead to a violent moment of decolonisation.

There is one practical reason why Sri Lanka has been misplaced from histories of partition in South Asia: the period of 1947–48 was thoroughly uneventful. All that occurred on the date of independence was the swearing in of a new British Governor General for the Dominion of Ceylon; republican status came much later in 1972. The decade that followed 1948 in virtue of this has been dubbed one of "fake independence," characterized by the continuation of imperial traditions and elite power. Yet when seen in the light of this work, the lack of violence at independence may be interpreted not only in relation to what was to come in the 1950s with the emergence of populist politics led by S.W.R.D. Bandaranaike, who was prime minister from 1956 to 1959, but also as reflective of the long character of partitioning between Sri Lanka and India, and Asia. Here, partitioning, because of its cyclical trajectory, did not necessitate a radical event or events such as that which marked the partition of Pakistan and India. Indeed the islanding and unislanding of Lanka and its connection to programs of state-making has carried on, both on the part of islanders and outside powers. A potent recent period was the presence of the Indian Peace-Keeping Force, from 1987 to 1990, followed by the assassination of the Indian premier Rajiv Gandhi by a female Tamil Tiger suicide bomber. Lanka's relation to the mainland—and its becoming as an island—continues to follow an unfinished trajectory.

Sri Lanka does not often serve as a site of meditation in postcolonial theory, yet David Scott's work is one instance in which it takes central ground. For Scott, the modern colonial state was concerned with "disabling

33. Sandagomi Coperahewa, "Purifying the Sinhala Language: The Hela Movement of Munidasa Cumaratunga (1930s to 1940s)."

non-modern forms of life by dismantling their conditions."[34] Discussing
the nineteenth-century age of reforms in Ceylon, Scott argues that they reconfigured colonial power by creating new targets and fields of operation,
which changed conduct in keeping with the form of governmentality. To
really understand the making of the modern state, it is necessary, following Scott, to focus not on the point of application of that power but rather
on what it does to change the conditions in which life is lived and defined. While I agree with Scott that there was a reconfiguration of the field
in which life was lived in Ceylon, I would wish to state his overall argument rather differently. For the point is not that there was a wholesale dismantling of what came prior to colonialism—or indeed the construction
of something new from nothing. Rather the power of colonialism came
through the recontextualization and redefinition of its antecedents, which
were not non-modern but rather alternative pathways of becoming modern.
This recontextualization might be seen as a narrowing of possibilities, and
as an intrusive intervention, which worked out its implications and intentions over time. It was certainly not a process which was confined to the
decades covered in this work. This is not the argument Scott caricatures,
quoting Partha Chatterjee, as a spiriting away of the colonial intrusion and
making all its features the innate property of indigenous history. Rather it
is an attempt to see the constantly mutating forms of colonialism, which
makes it difficult to pin the colonial state down.

It is important to underscore that this has not been a comparative history that has considered the colonial transition of Lanka in the late eighteenth and early nineteenth centuries alongside those of India, Mauritius,
and the Cape. Nor has it been a traditional connected history. It has kept
the specificity of the island alive, by taking it seriously for itself and by
looking at it from multiple points of view from within and without. In
this sense this is an Indian Ocean history, but one which is grounded and
which pulls back from grand generalizations over water, so as not to lose
the manner in which small stretches of land could remake and recontex-
tualize structures of empire and practices of knowledge-making. Only by
deep investigation of particular places set in a wider seascape can claims
of connectivity be borne. Discussions of connectivity in turn may lead
to observations of how Lanka could become dislodged—or partitioned
from the mainland—even as it looked to a new horizon and took a new
place upon this great sea. After all, a connection is a disconnection, when
viewed from another direction.

34. David Scott, "Colonial Governmentality," 200.

GLOSSARY

Aratchy/Aratchchi: A village official or a low-level attendant or assistant.

Adigar: Minister or chief officer of the kingdom of Kandy.

Almirah: Cabinet; chest of drawers.

Ayurveda: A traditional system of medicine which arrived in Lanka from India.

Burgher: Eurasian; those descended from Europeans, and especially the Portuguese and Dutch in this period.

Chetty/Chettie: A community from South India, usually associated with banking and moneylending.

Dagaba: A Buddhist monument built on a relic, which is usually dome-shaped; also *cetiya*.

Devale: Temple.

Dhonie/Doney: Small sailing vessel with one mast.

Disavanie/Disavony: A province in the kingdom of Kandy administered by a *disave* or *disapati*.

Ganinnanse: Aristocratic monk who had not received higher ordination.

Ganga: River.

Fanams: A copper currency used in Ceylon, twelve fanams were equivalent to a rix dollar.

Kachcheri: Government agent's office.

Kavi: Verse, to be sung and recited.

Korale: A unit of administration within a disavani; administered by a *korala*.

Lascoreen/Lascarin: Local soldier or troops.

Malabar: A term used by Europeans in the island to denote the community later designated as Tamils.

Mudliyar: Administrator of a *korale* under British rule.

Maha Mudliyar: Primary "native" assistant to the British governor of the island.

Moor: Muslim, or of Arab origin.

Muhandiram/Mohandiram: A title of rank assigned to a native in the bureaucracy of the British colonial state, and sometimes the assistant of a Mudliyar.

Maha Vidane: See *vidane*.

Nayak/Nayakkar: The line of kings descended from South India, who ruled the kingdom of Kandy, and their relatives.

Nikaya: A Buddhist fraternity.

Parsi/Parsee: South Asian community, said to have descended from Zoroastrians.

Pata rata: Literally, low country, or lowlands.

Pattini: A deity to Sinhala Buddhists, also worshipped in India, linked to smallpox.

Perahera: A Buddhist procession.

Pettah: An area outside the fort (of Colombo).

Raja/Rani: A South Asian ruler, *Raja* is the male form and *Rani* the female.

Rajakariya: Service to the king or the temple; taken on board by the British as compulsory service.

Rattemahatmaya: The chief of a district or *rata*.

Sandesa/Sandesya: A poem, usually written in the genre of a messenger bird in flight, including the instructions given to it and what is seen by it.

Sannasa: A grant or official deed relating to land.

Sangha: Buddhist clergy.

Sasana: The community of Buddhist monks, and the teaching of the Buddha and practice of the religion.

Sepoy: A soldier from the Indian mainland.

Star Pagodas: A gold currency in use in India, which increased in value in relation to Ceylonese copper currency in the early period of British rule.

Tri Sinhala: Term denoting the three principalities of Sinhala – *Ruhunu*, *Pihiti* and *Maya*, that have been seen to make up a united polity.

Uda rata: Literally, up country, or highlands.

Vidane: A headman or local village official, who could be given a title of honor and be called *vidane aratchchi*; those with higher status were *Maha vidane*.

Vihara: Buddhist temple.

BIBLIOGRAPHY

MANUSCRIPT SOURCES

British Library, London

Jorgi Astaka, Jonston Astaka, John Armour Astaka, Deanwood Astaka, and *Sir Robert Brownrigg Astaka.* Or 6601 (11)
Dalada Astaka. Or 6602 (8).
Dathagotrapradipaya. Or 6606 (27).
Mayura Sandesa. Or 6611 (137).
Pol Wismaya. Or 6611 (200).

Colombo Museum, Colombo

Sinhala Rajavamsa. AR 14.
Sinhala Rajavamsa. AN 14.
Rajavamsa. AN 15.
Ahalepola Daruwan Marima. SN 69N5.
Ahalepola Hatana. SN 69.
Ingrisi Hatana. K 11.
Narendra Charitavalokana Dipikava. 1926 V.7.
Ruvanvali Dagab Varnava. AR 9.
Ruvanvali Vistaraya. 1899/7/7.

India Office Records, Asia, Pacific and Africa
Collections, British Library, London (IOC)

F/4/245; F/4/339; F/4/354; F/4/515; F/4/527; F/4/880; F/4/1013; F/4/1250; F/4/1377; F/4/1461; F/4/1523; F/4/1566; F/5/1594; F/4/1599; F/4/1698; F/4/1719; F/4/1773.

The National Archives, Kew, London (TNA)

Files and papers in CO/54.

New York Public Library, New York

Alexander Johnston papers.

Rhodes House Library, Oxford

Letters from W. Huskisson to R. Dundas, wishing to know of his letters on the subject of trade between the Cape and Ceylon are being considered. Mss Afr. b.4.155; Mss Afr. b.4.177, Rhodes House Library, University of Oxford.

Sri Lanka National Archives, Colombo and Kandy, Sri Lanka (SLNA/ SLNAK)

Files and papers in Lots 1, 5, 6, 9, 10, 19, 21, 25 and 47.

Wellcome Library, London

Udarata janasanganayaka gana hasya kavi. WS 258.
Kolamba sita mahanuvarata mahaparaak tanima. WS 1654 (v).

NEWSPAPERS

Asiatic Annual Register.
The Asiatic Journal.
The Ceylon Almanac.
The Ceylon Examiner.
The Ceylon Government Gazette.
The Ceylon Times.
Colombo Journal.
The Colombo Observer.
Mirror of Parliament.
Proceedings of the Ceylon Agricultural Society.
Saturday Magazine.

WORKS PUBLISHED BEFORE 1900

Anon. "Address by the Hon Justice Stark, delivered at the opening of the general meeting of the Asiatic Society of Ceylon, Thursday, May 1, 1845." *Journal of the Ceylon branch of the Royal Asiatic Society* 1 (1845): 1–5.
Anon. *Analyses of New Works of Voyages and Travels Lately Published in London.* London: B. McMillan, 1809.

Anon. "An Historical Account of the Island of Ceylon." *Asiatic Annual Register*, 1799 95–121.

Anon. "On the Character and Cerebral Development of the Inhabitants of Ceylon." *Phrenological Journal* 7 (1831–32): 634–52.

Anon. "Papers Regarding the Practicability of Forming a Navigable Passage between Ceylon and the Mainland of India." Communicated by Lieut.-Colonel W. Monteith. *Journal of the Royal Geographical Society* 3 (1833): 1–25.

Anon. "Remarks upon the Comparative Healthfulness and other Local Advantages of Nuwera Eliya in the Island of Ceylon and the Neilgherry Hills in Hindoostan." *Colombo Journal* (1832): 472–23.

Baker, Samuel. *Eight Years in Ceylon*. London: Longmans, Green, and Co., 1855.

Bell, H.C.P. *Archaeological Survey of Ceylon, Kegalle District*. Colombo: Ceylon Government Press, 1892.

Bennett, J. W. *Ceylon and its Capabilities: An Account of its Natural Resources, Indigenous Productions and Commercial Facilities*. London: W. H. Allen & Co., 1843.

———. "The Cocoa-Nut Tree." *Ceylon Miscellany* 3 (1842): 250–67.

Bertolacci, Anthony. *A View of the Agricultural, Commercial and Financial Interests of Ceylon* London: Black, Parbury and Allen, 1817.

Brodie, A. Oswald. "Topographical and Statistical Account of the District of Noowerakalawiya." *Journal of the Ceylon Branch of the Royal Asiatic Society* 9 (1856–58): 136–61.

Buultjens, A. E., trans. "Governor Van Eck's Expedition against the King of Kandy, 1765." *Journal of the Royal Asiatic Society, Ceylon Branch* 16 (1899): 36–78.

Callaway, John, *A Vocabulary in Cingalese and English*. Colombo: Wesleyan Mission Press, 1820.

Casie Chetty, Mervyn, "Introduction." In Simon Casie Chetty, *Ceylon Gazetteer*, i–vi.

Casie Chetty, Simon, ed., *Ceylon Gazetteer*. First published 1834; republished, Delhi: Navrang, 1989.

———. "The Pursuit of Natural History in Ceylon." *Young Ceylon* 1 (1850): 105ff.

Champion, W. "Remarks on the State of Botany in Ceylon." *Ceylon Miscellany* 2 (1843): 18–31.

Chapman, I. J. "Some Additional Remarks upon the Ancient City of Anurajapura or Anuradhapura." *Journal of the Royal Asiatic Society of Great Britain and Ireland* 12 (1850): 164–78.

———. "Some Remarks upon the Ancient city of Anarajapura." *Transactions of the Royal Asiatic Society* 3 (1832): 463–95.

Christie, Thomas. *An Account of the Ravages Committed in Ceylon by Small-pox, Previously to the Introduction of Vaccination*. Cheltenham: J. and S. Griffith, 1811.

Clarke, J.B.B., ed. *An Account of the Religious and Literary Life of Adam Clarke by a Member of his Family*. London: T. S. Clarke, 1833.

Clough, Benjamin. *Compendious Pali Grammar with a Copious Vocabulary in the Same Language*. Colombo: Wesleyan Mission Press, 1824.

———. *A Dictionary of the English and Singhalese and Singhalese and English*

Languages, under the Patronage of the Government of Ceylon. 2 vols. Colombo: Wesleyan Mission Press, 1821.

Cordiner, James. *A Description of Ceylon: Containing an Account of the Country, Inhabitants and Natural Productions.* London: Longman, Hurst, Rees and Orme, 1807.

D'Oyly, John. "A Sketch of the Constitution of the Kandyan Kingdom, By late Sir John D'Oyly, Communicated by Sir A. Johnston." *Transactions of the Royal Asiatic Society* 3 (1835): 191–252.

Davy, John. *An Account of the Interior of Ceylon and of its Inhabitants.* London: Longman, Hurst, Rees, Orme & Brown, 1821.

———. "A Description of Adam's Peak." *The Quarterly Journal of Science and the Arts* 5 (1818): 25–30.

De Bussche, L. *Letters on Ceylon, Particularly Relative to the Kingdom of Kandy.* London: Printed for S. Passey, High-Street, Newington Butts, 1826.

De Butts, Augustus. *Rambles in Ceylon.* London: W. H. Allen, 1841.

De Vos, F., ed. and trans. "A Short History of the Principal Events that Occurred in the Island of Ceilon, Since the Arrival of the First Netherlanders in the Year 1602, And, Afterwards from the Establishment of the 'Honourable Company' in the Same Island, till the Year 1757." *Journal of the Royal Asiatic Society, Ceylon Branch* 11 (1889): 1–150.

Ferguson, John. *Ceylon in 1883: Describing the Progress of the Island since 1803.* London: John Haddon, 1893.

Fernando, M. E., ed. *Vadiga Hatana* (also titled *Ahalepola Varnava*). Colombo: Luxman Press, n.d.

Forbes, Jonathan. *Eleven Years in Ceylon: Comprising Sketches of the Field Sports and Natural History of that Colony and an Account of its History and Antiquities.* London: R. Bentley, 1840.

Gordon, Peter. *India; or Notes on the Administration of the Establishments in India.* Calcutta: n.p., 1828.

Graham, Maria. *Journal of a Residence in India.* Edinburgh: Archibald Constable and Company, 1812.

Hamilton, V. M., and S. M. Fasson. *Scenes in Ceylon: Twenty Cartoons with Descriptions in Verse.* London: Chapman & Hall, 1881.

Hamilton, Walter. *A Geographical, Statistical and Historical Description of Hindostan.* London: John Murray, 1820.

Hardy, Robert Spence. *The British Government and the Idolatry of Ceylon.* Colombo: Wesleyan Mission Press, 1839.

———. *Eastern Monachism.* London: Partridge and Oakey, 1850.

Heber, Reginald, and Amelia Heber. *Narrative of a Journey through the Upper Provinces of India from Calcutta to Bombay, 1824–1825, with Notes upon Ceylon Written by Mrs Heber.* London: John Murray, 1828.

Hodgson, Brian. "Sketch of Buddhism, Derived from the Bauddha Scriptures of Nipal." *Transactions of the Royal Asiatic Society of Great Britain and Ireland* 2 (1829): 222–57.

Holman, James. *A Voyage Round the World from 1827 to 1832.* London: Smith, Elder, 1834.

Ievers, R. W. *Manual of the North-central Province of Ceylon.* Colombo: George J. A. Skeen, 1899.

Jerdan, William. "Alexander Johnston." In *National Portrait Gallery of Illustrious and Eminent Personages of the nineteenth century,* 3:1–12. London: Fisher, Son, & Jackson, 1832.

Johnston, Arthur. *Narrative of the Operations of a Detachment in an Expedition to Candy in the Island of Ceylon in the Year 1804.* London: C. and R. Baldwin, 1810.

Kinnis, J. *A Letter to the Inhabitants of Ceylon on the Advantages of Vaccination.* Ceylon: Cotta Church Mission Press, 1837.

———. *A Report on the Small-pox as it Appeared in Ceylon in 1833–34.* Colombo: Ceylon Government Press, 1835.

Knighton, William. *Forest Life in Ceylon.* London: Hurst and Blackett, 1854.

Knox, Robert. *An Historical Relation of the Island of Ceylon* (1681). Edited by J.H.O. Paulusz. Dehiwela, Sri Lanka: Tisara Prakasakayo, 1989.

Lawrie, A. C. *A Gazetteer of the Central Provinces of Ceylon.* Colombo: G.J.A. Skeen, 1896.

Le Mesurier, C. J. R. *Manual of the Nuwara Eliya District of the Central Province of Ceylon.* Colombo: G.J.A. Skeen, Government Printer, 1893.

Lewis, R. E. "The Rural Economy of the Singhalese, More Particularly with Reference to the District of Saffragam, with Some Account of their Superstitions." *Journal of the Ceylon Branch of the Royal Asiatic Society* 4 (1848–49): 31–52.

Lyttleton, William. *A Set of Views in the Island of Ceylon.* London: E. Orme, 1819.

Marshall, Henry. *Ceylon: A General Description of the Island and its Inhabitants.* London: W. H. Allen, 1846.

———. *Contribution to a Natural and Economical History of the Coco-Nut Tree.* Edinburgh: Printed by John Stark, 1836.

McKenzie, Colin. "Remarks on Some Antiquities on the West and South Coasts of Ceylon, Written in the Year 1796." *Asiatic Researches* 6 (1799): 433–34.

Moon, Alexander. *A Catalogue of the Indigenous and Exotic Plants Growing in Ceylon, Distinguishing the Several Esculent Vegetables, Fruits, Roots and Grains Together with a Sketch of the Divisions of Genera and Species in Use amongst the Sinhalese, also an Outline of the Linnean System of Botany in the English and Singhalese Languages for the Use of the Singhalese.* Colombo: Wesleyan Mission Press, 1824.

Percival, Robert. *An Account of the Island of Ceylon containing its History, Geography, Natural History, with the Manners and Customs of its Various Inhabitants.* London: C. and R. Baldwin, 1803.

Pridham, Charles. *An Historical and Statistical Account of Ceylon and its Dependencies.* London: T. and W. Boone, 1849.

Selby, H. C. *Report of the Trial of John Mackenzie Ross for Libel.* Colombo: Government Press, 1839.

Sirr, Henry Charles. *Ceylon and the Cingalese.* London: W. Shoberl, 1850.

Skeen, William. *Adam's Peak: Legendary, Traditional and Historic Notices of the Samanala and Sri Pada with a Descriptive Route from Colombo to the Sacred Footprint.* Colombo: W.L.H. Skeen, 1870.

Skinner, Thomas. *Fifty Years in Ceylon: An Autobiography.* London: W. H. Allen and
    Co., 1891.
Tennent, James Emerson. *Ceylon: An Account of the Island, Physical, Historical, and
    Topographical, with Notices of its Natural History, Antiquities, and Productions.*
    London: Longman, Green, Longman and Roberts, 1860.
Tillekeratne, D. A. "Gajaman Nona." *The Orientalist* 3 (1888–89): 62–64.
Tillekeratne, John F. "The Life of Karatota Kirti Sri Dhammarama, High Priest of
    Matara in the Southern Province of the Island of Ceylon." *The Orientalist* 3
    (1888–89): 204–8.
Trimen, Henry. Obituary of Seneviratne, *Journal of Botany* 32 (1894): 255–56.
Tolfrey W., and A. Armour, eds. *The Singhalese Translation of the New Testament of
    Our Lord and Saviour Jesus Christ.* Colombo: Wesleyan Mission Press, 1817.
Turnour, George. "Account of the Tooth Relic of Ceylon." *Journal of the Asiatic Society
    of Bengal* 6 (1837): 856–68.
———. *Epitome of the History of Ceylon.* Colombo: Cotta Mission Press, 1836.
———. "Examination of the Pali Buddhistical Annals." *Journal of the Asiatic Society of
    Bengal* 6 (1837): 501–28.
———. "Examination of the Pali Buddhistical Annals, no. 2." *Journal of the Asiatic
    Society of Bengal* 7 (1838): 686–701.
———. "Examination of the Pali Buddhistical Annals, no. 3." *Journal of the Asiatic
    Society of Bengal* 7 (1838): 789–817.
———. "Examination of the Pali Buddhistical Annals, no. 4." *Journal of the Asiatic
    Society of Bengal* 7 (1838): 919–33.
———. "Examination of Some Points of Buddhist Chronology." *Journal of the Asiatic
    Society of Bengal* 5 (1836): 521–36.
Upham, Edward. *The History and Doctrine of Buddhism, Popularly Illustrated with
    Notices of the Kappooism, or Demon Worship, and of the Bali, or Planetary Incan-
    tations of Ceylon.* London: R. Ackerman, 1829.
———. *Proposals for Publishing by Subscription the Sacred and Historical Books of
    Ceylon.* Pamphlet in Cambridge University Library, n.d.
Upham, Edward, ed. and trans. *The Mahavansi, the Raja-ratnacari and the Raja-vali,
    forming the Sacred and Historical Books of Ceylon.* London: Parbury, Allen, and
    Co., 1833.
Valentia, George. *Voyages and Travels to India, Ceylon, the Red Sea, Abyssinia and
    Egypt.* London: W. Miller, 1809.
Mrs. Walker. "Journal of Mrs. Walker and Col. Walker to Adam's Peak in 1833." In
    Companion to the Botanical Magazine, ed. William Jackson Hooker, 1 (1835): 3–14.

                    WORKS PUBLISHED AFTER 1900

Anon. "George Turnour (1799–1843)." *Ceylon Literary Register* 3 (1933–34): 477–79.
Anon. "Sir Alexander Johnston's Memorandum." *Ceylon Literary Register,* 3d series,
    1 (1931): 39–43.
Anon. "Upham's Sacred and Historical Books of Ceylon." *Ceylon Literary Register* 1
    (1931): 381–82.

Anon. "Visitors' Guide to the Royal Botanical Gardens, Peradeniya." From the author's visits in 2005 and 2008, in author's collection.

Abeyawardana, H.A.P. *Boundary Divisions of Mediaeval Sri Lanka.* Polgasovita: Academy of Sri Lankan Culture, 1999.

———. *Heritage of Sabaragamuwa.* Ratnapura: The Sabragamuwa Bank in association with the Central Bank of Sri Lanka, 2002.

Alavi, Seema. *Sepoys and the Company: Tradition and Transition in Northern India, 1770–1830.* Delhi and Oxford: Oxford University Press, 1998.

———, ed. *The Eighteenth Century in India.* New Delhi: Oxford University Press, 2002.

Almond, Philip. C. *The British Discovery of Buddhism.* Cambridge: Cambridge University Press, 1988.

Ameresekere, H. E. "The Statue of Kusta Raja at Weligama." *Ceylon Literary Register* 2 (1932): 486–88.

Appadurai, Arjun. *Fear of Small Numbers: An Essay on the Geography of Anger.* Durham, N.C.: Duke University Press, 2006.

Arasaratnam, Sinnappah. *Ceylon and the Dutch, 1600–1800: External Influences and Internal Change in Early Modern Sri Lanka.* Aldershot: Variorum, 1996.

———. "Ceylon in the Indian Ocean Trade: 1500–1800." In *India and the Indian Ocean, 1500–1800,* ed. A. Das Gupta and M. N. Pearson, 224–39. Calcutta; Oxford: Oxford University Press, 1987.

———. "The Kingdom of Kandy: Aspects of its External Relations and Commerce, 1658–1710." *Ceylon Journal of Historical and Social Studies* 3 (1960): 90–127.

———. "Sri Lanka's Trade, Internal and External in the Seventeenth and Eighteenth Centuries." In *History of Sri Lanka,* ed. K. M. De Silva, vol. 2:399–416. Peradeniya, Sri Lanka: University of Peradeniya, 1995.

Armitage, David, and Sanjay Subrahmanyam, eds. *The Age of Revolutions in a Global Context, c. 1760–1840.* Basingstoke: Palgrave Macmillan, 2010.

Arnold, David. *Colonizing the Body: State Medicine and Epidemic Disease in Nineteenth-century India.* Berkeley and Los Angeles: University of California Press, 1993.

———. *The Tropics and the Travelling Gaze: India, Landscape and Science, 1800–1856.* Seattle: University of Washington Press, 2006.

Axelby, Richard. "Calcutta Botanic Garden and the Re-ordering of the Indian Environment," *Archives of Natural History* 3 (2008): 150–63

Ballantyne, Tony. *Orientalism and Race: Aryanism in the British Empire.* Basingstoke: Palgrave Macmillan, 2006.

Bandaranayake, Senake. "The External Factor in Sri Lanka's Historical Formation." *The Ceylon Historical Journal* 25 (1978): 74–94.

Bandaranayake, Senake, and Gamini Jayasinghe. *The Rock and Wall Paintings of Sri Lanka.* Colombo: Lake House Bookshop, 1986.

Barnett, L. D. "Alphabetical Guide to Sinhalese Folklore from Ballad Sources." *Indian Antiquary* 5 (1916): 1–116.

Barrow, Ian. *Surveying and Mapping in Colonial Sri Lanka.* Delhi; Oxford: Oxford University Press, 2008.

Bayly, C. A. "Afterword." *Modern Intellectual History* 4 (2007): 163–69.

———. *Empire and Information: Intelligence Gathering and Social Communication in India, 1780–1870*. Cambridge: Cambridge University Press, 1999.

———. *Recovering Liberties: Indian Thought in the Age of Liberalism and Empire*. Cambridge: Cambridge University Press, 2012.

———. *The Origins of Nationality in India: Patriotism and Ethical Government in the Making of Modern India*. Delhi: Oxford University Press, 1998.

Bayly, Susan. *Caste, Society and Politics from the Eighteenth Century to the Modern Age*. Cambridge: Cambridge University Press, 1999.

Beven, F. "The Press." In *Twentieth Century Impressions of Ceylon: Its History, People, Commerce, Industries and Resources*, ed. A. Wright, 301–19. London: Lloyd's Greater Britain Pub. Co., 1907.

Bhattacharya, Sanjoy, Mark Harrison, and Michael Worboys. *Fractured States: Smallpox, Public Health and Vaccination Policy in British India, 1800–1947*. New Delhi: Orient Longman, 2005.

Bingham, P. M. *History of the Public Works Department*. 2 vols. Colombo: Govt. Printer, 1921–23.

Birla, Ritu. *Stages of Capital: Law, Culture and Market Governance in Late Colonial India*. Durham, N.C.: Duke University Press, 2009.

Blackburn, Anne M. *Buddhist Learning and Textual Practice in Eighteenth-century Lankan Monastic Culture*. Princeton: Princeton University Press, 2001.

———. "Localizing Lineage: Importing Higher Education in Theravadin South and Southeast Asia." In *Constituting Communities: Theravada Buddhism and the Religious Cultures of South and Southeast Asia*, ed. John Clifford Holt, Jacob N. Kinnard, and Jonathan Walters, 131–49. Albany: State University of New York Press, 2003.

———. *Locations of Buddhism: Colonialism and Modernity in Sri Lanka*. Chicago: University of Chicago Press, 2010.

Blackton, Charles S. "The Action Phase of 1915," *The Journal of Asian Studies* 29 (1970): 235–54.

Blussé, Leonard. *Visible Cities: Canton, Nagasaki and Batavia and the Coming of the Americans*. Cambridge, Mass.: Harvard University Press, 2008.

Bose, Sugata. *A Hundred Horizons: The Indian Ocean in an Age of Global Empire*. Cambridge, Mass.: Harvard University Press, 2006.

Bose, Sugata, and Ayesha Jalal. *Modern South Asia: History, Culture and Political Economy*. New York: Routledge, 1998.

Brohier, R. L. *Lands, Maps, Surveys: A Review of the Evidence of the Land Surveys as Practised in Ceylon*. 2 vols. Colombo: Ceylon Government Press, 1950.

Chakrabarty, Dipesh. *Provincializing Europe: Postcolonial Thought and Historical Difference*. Princeton: Princeton University Press, 2000.

Charney, Michael W. *Powerful Learning: Buddhist Literati and the Throne in Burma's Last Dynasty, 1752–1885*. Ann Arbor: University of Michigan Press, 2006.

Chatterjee, Kumkum. *The Cultures of History in Early Modern India: Persianization and Mughal Culture in Bengal*. Delhi; Oxford: Oxford University Press, 2009.

Chatterjee, Partha. *The Nation and its Fragments: Colonial and Postcolonial Histories.* Princeton: Princeton University Press, 1993.

Chatterji, Joya. *Bengal Divided: Hindu Communalism and Partition, 1932–1947.* Cambridge: Cambridge University Press, 1994.

Chaudhuri, K. N. *Asia before Europe: Economy and Civilisation of the Indian Ocean from the Rise of Islam to 1750.* Cambridge: Cambridge University Press, 1990.

———. *Trade and Civilisation in the Indian Ocean: An Economic History from the Rise of Islam to 1750.* Cambridge: Cambridge University Press, 1985.

Chichester, H. M. "Maitland, Sir Thomas (1760–1824)." Revised by Roger T. Stearn, *Oxford Dictionary of National Biography.* Oxford: Oxford University Press, 2004; online ed., http://www.oxforddnb.com/view/article/17835 (accessed July 26, 2011).

Clapham, J. H. "The last years of the Navigation Acts." *English Historical Review* 25 (1910): 480–501.

Codrington, H. W., ed. *Diary of Mr. John D'Oyly, with Introduction and Notes.* Colombo: Colombo Apothecaries' Co., 1917.

Cohn, Bernard. *An Anthropologist Among the Historians and Other Essays.* Delhi; Oxford: Oxford University Press, 1987.

Coleman, Simon, and John Eade, eds. *Reframing Pilgrimage: Cultures in Motion.* London: Routledge, 2004.

Coleman, Simon, and John Elsner. *Pilgrimage, Past and Present: Sacred Travel and Sacred Space in World Religions.* Cambridge, Mass.: Harvard University Press, 1995.

Comaroff, John. "Ethnicity, Nationalism and the Politics of Difference in an Age of Revolution." In *The Politics of Difference: Ethnic Premises in a World of Power*, ed. Edwin N. Wilmsen and Patrick McAllister, 162–83. Chicago: University of Chicago Press.

Coperahewa, Sandagomi. "Purifying the Sinhala Language: The *Hela* Movement of Munidasa Cumaratunga (1930s to 1940s)." *Modern Asian Studies* 46 (2012): 857–91.

Cooper, Frederick. *Colonialism in Question: Theory, Knowledge, History.* Berkeley and Los Angeles: University of California Press, 2005.

Cooter, R., M. Harrison, and S. Sturdy, eds. *Medicine and Modern Warfare.* Amsterdam; Atlanta: Rodopi, 1999.

Crosby, A. W. *Ecological Imperialism: The Biological Expansion of Europe, 900–1900.* Cambridge: Cambridge University Press, 1986.

Daniel, E. Valentine. "Afterword: Sacred Places, Violent Spaces." In *Sri Lanka: History and the Roots of Conflict*, ed. Jonathan Spencer, 226–46. London; New York: Routledge, 1990.

De Silva, Colvin R. *Ceylon under British Occupation, 1795–1833.* 2 vols. Colombo: The Colombo Apothecaries' Co. 1953–62.

De Silva, K. M. "The Colebrooke-Cameron Reforms." *Ceylon Journal of Historical and Social Studies* 2 (1959): 245–56.

———. *A History of Sri Lanka.* Delhi: Oxford University Press, 1981.

———. *Letters on Ceylon, 1846–1850: The Administration of Viscount Torrington and the 'Rebellion' of 1848.* Kandy: K.V.G. de Silva, 1965.

———. *Social Policy and Missionary Organizations in Ceylon, 1840–1855*. London: Longmans for the Royal Commonwealth Society, 1965.

De Silva, W. A. "A Contribution to Sinhalese Plant Lore." *Journal of the Royal Asiatic Society, Ceylon Branch* 12 (1891): 113–44.

———. "A Probable Origin of the Name Kushtharajagala." *Journal of the Royal Asiatic Society, Ceylon Branch* 28 (1919): 86.

———. "Sinhalese Vittipot (Books of Incidents) and Kadaimpot (Books of Division Boundaries)." *Journal of the Royal Asiatic Society Ceylon Branch* 30 (1927): 303–25.

Dening, Greg. *Islands and Beaches: Discourses on a Silent Land, Marquesas, 1774–1880*. Carlton, Vic.: Melbourne University Press, 1980.

Desmond, R. *The European Discovery of the Indian Flora*. Oxford: Oxford University Press/Royal Botanic Gardens, 1992.

Dewaraja, Lorna. *The Kandyan Kingdom of Sri Lanka, 1707–1782*. Colombo: Lake House Investments, 1988.

———. *The Muslims of Sri Lanka: One Thousand Years of Ethnic Harmony, 900–1915*. Colombo: Lanka Islamic Foundation, 1994.

———. *A Study of the Political, Administrative, and Social Structure of the Kandyan Kingdom*. Colombo: Lake House Investments, 1972.

Dharmadasa, K.N.O. *Language, Religion and Ethnic Assertiveness: The Growth of Sinhalese Nationalism in Sri Lanka*. Ann Arbor: University of Michigan Press, 1992.

———. "'The People of the Lion': Ethnic Identity, Ideology, and Historical Revisionism in Contemporary Sri Lanka." *Ethnic Studies Report* 10 (1992): 37–59.

———. "The Sinhala-Buddhist Identity and the Nayakkar Dynasty in the Politics of the Kandyan Kingdom, 1793–1815." *Ceylon Journal of Historical and Social Studies* 6.1 (1976): 1–23.

Dharmawansa, P.R.A. "Peradeniya." *Ceylon University Magazine* (March 1957): 38–40.

Dirks, Nicholas. *Castes of Mind: Colonialism and the Making of Modern India*. Princeton: Princeton University Press, 2001.

Disanayaka, J. B. *Paintings of Kelani Vihara*. Translated by Kusum Disanayaka. Nugegoda: Godage Press, 2004.

Dodson, Michael. *Orientalism, Empire and National Culture: India, 1770–1880*. Basingstoke: Palgrave Macmillan, 2007.

Drayton, R. *Nature's Government: Science, Imperial Britain and the "Improvement" of the World*. New Haven: Yale University Press, 2000.

Duncan, James S. *The City as Text: The Politics of Landscape Interpretation in the Kandyan Kingdom*. Cambridge: Cambridge University Press, 1990.

———. "Embodying Colonialism: Domination and Resistance in Nineteenth-Century Ceylonese Coffee Plantations." *Journal of Historical Geography* 28.3 (2002): 317–38.

———. *In the Shadow of the Tropics: Climate, Race and Biopower in Nineteenth-Century Ceylon*. Aldershot: Ashgate, 2007.

———. "The Struggle to be Temperate: Climate and 'Moral Masculinity' in Mid-Nineteenth Century Ceylon." *Singapore Journal of Tropical Geography* 21.1 (2000): 34–37.

Edmond, Rod, and Vanessa Smith, eds. *Islands in History and Representation*. London: Routledge, 2003.

Edney, Matthew. *Mapping an Empire: The Geographical Construction of British India.* Chicago: University of Chicago Press, 1997.

Effendi, Mohammed Sameer Bin Haji Ismail. *Personages of the Past: Moors, Malays and other Muslims of the Past of Sri Lanka.* Colombo: Rainbow Printers, 1982.

Elsner, Jaś, and Joan-Pau Rubies, eds. *Voyages and Visions: Towards a Cultural History of Travel.* London: Reaktion Books, 1999.

Ferguson, J. "The Coconut Palm in Ceylon: Beginning, Rise and Progress of its Cultivation." *Journal of the Royal Asiatic Society, Ceylon Branch* 19 (1906): 39–70.

Fernando, B. W. *Ceylon Currency: British Period, 1796–1936.* Colombo: Ceylon Government Press, 1939.

Foucault, M. "Governmentality." In *Power: Essential Works of Foucault, 1954–1984*, ed. James D. Faubion, vol. 3, 201–22. London: Penguin, 2002.

Frost, Mark. "Cosmopolitan Fragments from a Splintered Isle: Colombo and 'Ceylonese' Nationalism." In *Ethnicities, Diasporas and 'Grounded' Cosmopolitanisms in Asia*, ed. Joel Kahn, 59–69. Singapore: Asia Research Institute Monograph Series, 2004.

———. "'Wider Opportunities': Religious Revival, Nationalist Awakening and the Global Dimension in Colombo, 1870–1920." *Modern Asian Studies* 36 (2002): 937–67.

Gamage, H. H., ed. *Andare Saha Gajaman Nona.* Colombo: Indika Publishers, 1988.

Gandhi, Gopalkrishna, ed. *Gandhi and Sri Lanka, 1905–1947.* Ratmalana, Sri Lanka: Vishva Lekha, 2002.

Geiger, W., trans. *Culavamsa, Being the Most Recent Part of the Mahavamsa*, and from the German into English by C. M. Rickmers. London: Pali Text Society, 1929–30.

Godakumbura, C. E. "History of Archaeology in Ceylon." *Journal of the Ceylon Branch of the Royal Asiatic Society* 13 (1969): 1–38.

———. *Sinhalese Literature.* Colombo: Colombo Apothecaries' Co. 1955.

Gombrich, Richard Francis, and Gananath Obeyesekere. *Buddhism Transformed: Religious Change in Sri Lanka.* Princeton: Princeton University Press, 1988.

Gooneratne, Brendon, and Yasmine Gooneratne. *This Inscrutable Englishman, Sir John D'Oyly, Baronet, 1774–1824.* London; New York: Cassell, 1999.

Gooneratne, Yasmine. *Relative Merits: A Personal Memoir of the Bandaranaike Family of Sri Lanka.* London: Hurst, 1986.

Goonesekera, S. *Mount Lavinia: The Governor's Palace.* Colombo: Paradise Isle Publication, 2006.

Goonewardene, K. W. "The Accession of Sri Vijaya Rajasimha." In *Sesquicentennial Commemorative Volume of the Royal Asiatic Society of Sri Lanka 1845–1995*, ed. G.P.S.H. de Silva and C. G. Uragoda, 441–67. Colombo: Sridevi Printers, 1995.

Granville, William. "Deportation of Sri Vikrama Rajasinha, Journal of Reminiscences Relating to the Late King of Kandy When on His Voyage from Colombo to Madras in 1816, a Prisoner-of-War on Board His Majesty's Ship 'Cornwallis.'" *Ceylon Literary Register* 3 (1933–34): 487–504; 543–50.

Grove, Richard. *Green Imperialism: Colonial Expansion, Tropical Island Edens and the Origins of Environmentalism, 1600–1860.* Cambridge: Cambridge University Press, 1995.

Guha, Ranajit. *An Indian Historiography of India: A Nineteenth Century Agenda and its Implications*. Calcutta: Published for Centre for Studies in Social Sciences, Calcutta, by K. P. Bagchi & Co. 1998.

Gunawardana, R.A.L.H. "Colonialism, Ethnicity and the Construction of the Past: The Changing 'Ethnic Identity' of the Last Four Kings of the Kandyan kingdom." In *Pivot Politics: Changing Cultural Identities in Early State Formation Processes*, ed. Martin van Bakel, Renee Hagesteijn, and Pieter van de Velde, 197–211. Amsterdam: Het Spinhuis, 1994.

———. "The People of the Lion: Sinhala Identity and Ideology in History and Historiography." *Sri Lanka Journal of the Humanities* 1 and 2 (1979): 1–36.

Gunawardana, Sirancee. *Palm-leaf Manuscripts of Sri Lanka*. Ratmalana, Sri Lanka: Vishva Lekha, 1997.

Guruge, Ananda. *From the Living Fountains of Buddhism: Sri Lankan Support to Pioneering Western Orientalists*. Colombo: Ministry of Cultural Affairs, 1984.

———. *Return to Righteousness: A Collection of Speeches, Essays and Letters of the Anagarika Dharmapala*. Colombo: Ministry of Education and Cultural Affairs: 1965.

Hallisey, Charles. "Roads Taken and Not Taken in the Study of Theravada Buddhism." In *Curators of the Buddha: The Study of Buddhism under Colonialism*, ed. Donald S. Lopez, 31–61. Chicago: University of Chicago Press, 1995.

Harris, E. *Theravada Buddhism and the British Encounter: Religious, Missionary and Colonial Experience in Nineteenth-Century Sri Lanka*. London: Routledge, 2006.

Harrison, Mark. "Disease and Medicine in the Armies of British India, 1750–1830: The Treatment of Fevers and the Emergence of Tropical Therapeutics." In ed. *British Military and Naval Medicine, 1600–1830*, ed. Geoffrey L. Hudson, 87–120. Amsterdam; New York: Rodopi, 2007.

———. *Public Health in British India: Anglo-Indian Preventive Medicine, 1859–1914*. Cambridge: Cambridge University Press, 1994.

Head, Raymond. *Catalogue of Paintings, Drawings, Engravings and Busts in the Collection of the Royal Asiatic Society*. London: Royal Asiatic Society, 1991.

Headrick, Daniel. *Tools of Empire: Technology and European Imperialism in the Nineteenth Century*. New York; Oxford: Oxford University Press, 1981.

———. *When Information Came of Age: Technologies of Information in an Age of Reason and Revolution, 1700–1850*. Oxford; New York: Oxford University Press, 2000.

Hevavasam, P.B.J. *Matara Yugaye Sahityadharayan ha Sahitya Nibandhana*. [The Literary Works and Writers of the Matara Era]. Sinhala, Colombo: Lamkandhuve Mudranalaye, 1996.

Hewavitarana, P. A., ed. *Narendra Caritavalokana Pradipikava* by Yatanvala Mahathera of Asgiri Vihara. Sinhala. Colombo: Mahabodhi Press, 1926.

Ho, Engseng. *The Graves of Tarim: Genealogy and Mobility Across the Indian Ocean*. Berkeley and Los Angeles: University of California Press, 2006.

Hofmeyr, Isabel. "The Black Atlantic Meets the Indian Ocean: Forging New Paradigms of Transnationalism for the Global South." *Social Dynamics* 33 (2007): 3–32.

Holt, John Clifford. "Pilgrimage and the Structure of Sinhalese Buddhism." *Journal of the International Association of Buddhist Studies* 5 (1982): 23–40.

———. *The Religious World of Kirti Sri: Buddhism, Art, and Politics in Late Medieval Sri Lanka.* New York; Oxford: Oxford University Press, 1996.

Holt, John Clifford, J. N. Kinnard, and J. S. Walters, eds. *Constituting Communities: Theravada Buddhism and the Religious Cultures of South and South-east Asia.* Albany: State University of New York Press, 2003.

Jalal, Ayesha. *The Sole Spokesman: Jinnah, the Muslim League and the Demand for Pakistan.* Cambridge: Cambridge University Press, 1985.

Jayawardena, Kumari. "Economic and Political Factors in the 1915 Riots." *The Journal of Asian Studies* 29 (1970): 223–33.

———. *Perpetual Ferment: Popular Revolts in Sri Lanka in the Eighteenth and Nineteenth Centuries.* Colombo: Social Scientists' Association, 2010.

Jeganathan, Pradeep. "Authorizing History, Ordering Land: The Conquest of Anuradhapura." In *Unmaking the Nation: The Politics of Identity and History in Modern Sri Lanka,* ed. Pradeep Jeganathan and Qadri Ismail, 106–36.

Jeganathan, Pradeep, and Qadri Ismail, eds. *Unmaking the Nation: The Politics of Identity and History in Modern Sri Lanka.* Colombo: Social Scientists' Association, 1995.

Jennings, W. Ivor. *The Kandy Road,* ed. H. A. I. Goonetileke. Pamphlet. Peradeniya: University of Peradeniya, 1993.

Jones, Margaret. *Health Policy in Britain's Model Colony: Ceylon, 1900–1948.* Hyderabad: Orient Longman, 2004.

———. *The Hospital System and Health Care: Sri Lanka, 1815–1960.* New Delhi: Orient Blackswan, 2009.

Kannangara, A. P. "The Riots of 1915 in Sri Lanka: A Study in the Roots of Communal Violence." *Past and Present* 102 (1984): 130–64.

Kannangara, P. D. *The History of the Ceylon Civil Service, 1802–1833: A Study of Administrative Change in Ceylon.* Dehiwela, Sri Lanka: Tisara Prakasakayo, 1966.

Kapila, Shruti. "Race Matters: Orientalism and Religion, India and Beyond, c. 1770–1880." *Modern Asian Studies* 41 (2007): 471–513.

———, ed. "An Intellectual History for India." Special section, *Modern Intellectual History* 4 (2007): 3–169.

Kaviraj, S. "On the Construction of Colonial Power: Structure, Discourse, Hegemony." In *Contesting Colonial Hegemony: State and Society in Africa and India,* ed. D. Engels and S. Marks, 19–54. London: British Academic Press, 1994.

Kennedy, Dane. *Magic Mountains: Hill Stations and the British Raj.* Berkeley and Los Angeles: University of California Press, 1996.

Kenny, Judith K. "Climate, Race and Imperial Authority: The Symbolic Landscape of the British Hill Stations in India." *Annals of the Association of American Geographers* 85 (1995): 694–714.

Larson, Pier M. *Ocean of Letters: Language and Creolization in an Indian Ocean Diaspora.* Cambridge: Cambridge University Press, 2009.

Lieberman, Victor. *Strange Parallels: South-east Asia in Global Context, c. 800–1830.* Cambridge: Cambridge University Press, 2003.

Liyanaratne, Jinadasa. *Buddhism and Traditional Medicine in Sri Lanka.* Kelaniya: Kelaniya University Press, 1999.

———. "Indian Medicine in Sri Lanka." *Bulletin de l'École française d'Extrême-Orient* 76 (1987): 201–15.

———. "Studies in Medical Herbaria." *Transactions of the Ceylon College of Physicians* 97 (1973): 97–108.

———, ed. *Bhessajjamanjusa.* Oxford: Pali Text Society, 1996.

Lopez, Donald S., ed. *Curators of the Buddha: The Study of Buddhism under Colonialism.* Chicago: University of Chicago Press, 1995.

Majeed, Javed. *Ungoverned Imaginings: James Mill's "History of British India" and Orientalism.* Oxford: Clarendon Press, 1992.

Malalgoda, K. *Buddhism in Sinhalese Society, 1750–1900: A Study of Religious Revival and Change.* Berkeley and Los Angeles: University of California Press, 1976.

Marikar, A.I.L. *Glimpses from the Past of the Moors of Sri Lanka.* Colombo: Moors Islamic Culture Home, 1976.

Marshall, Peter, ed. *The Eighteenth Century in Indian History: Evolution or Revolution?* New Delhi: Oxford University Press, 2003.

Meegama, S. A. *Famine, Fevers and Fear: The State and Disease in British Colonial Sri Lanka.* Colombo: Sarasavi Printers, 2012.

Mendis, G. C. *Ceylon, Today and Yesterday: Main Currents of Ceylon history.* Colombo: Lake House Investments, first published 1957, 3d ed., 1995.

———, ed. *The Colebrooke-Cameron Papers: Documents on British Colonial Policy in Ceylon, 1796–1833.* London: Oxford University Press, 1956.

Mendis, V.L.B. *Foreign Relations of Sri Lanka, from Earliest Times to 1965.* Dehiwela, Sri Lanka: Tisara Prakasakayo, 1983.

Metcalf, Thomas R. *Imperial Connections: India in the Indian Ocean Arena, 1860–1920.* Berkeley and Los Angeles: University of California Press, 2007.

Mignolo, Walter D. "The Many Faces of Cosmo-polis: Border Thinking and Critical Cosmopolitanism." *Public Culture* 12 (2000): 721–48.

Mishra, R. L. *The Mortuary Monuments in Ancient and Medieval India.* Delhi: B. R. Publishing Corporation, 1991.

Mrazek, Rudolf. *Engineers of Happy Land: Technology and Nationalism in a Colony.* Princeton: Princeton University Press, 2002.

Mudiyanse, Nandasena. "Correspondence between Siam and Sri Lanka in the 18th century." *The Buddhist* 44 (1973): 15–22.

Munasinghe, Indira. *The Colonial Economy on Track: Roads and Railways in Sri Lanka, 1800–1905.* Colombo: Social Scientists' Association, 2002.

Nell, Andreas. "Some Trees and Plants Mentioned in the Mahavamsa." *The Ceylon Historical Journal* 2.3 and 4 (1953): 258–64.

Nevill, Hugh. *Sinhala Verse (Kavi).* Collected by Hugh Nevill, 1869–86; edited by P.E.P. Deraniyagala. Ceylon National Manuscripts Series, vols. 4–6. 3 vols. Colombo: Dept. of National Museums, 1954–55.

Nissan, Elizabeth. "History in the Making: Anuradhapura and the Sinhala Buddhist Nation." *Social Analysis* 25 (1989): 64–77.

———. "The Sacred City of Anuradhapura: Aspects of Sinhalese Buddhism and Nationalism." Ph.D. diss., University of London, 1985.

Obeyesekere, Gananath. "Buddhism, Nationhood and Cultural Identity: A Question of

Fundamentals." In *Fundamentalisms Comprehended*, ed. Martin E. Marty and R. Scott Appleby, 231–53. Chicago: University of Chicago Press, 1995.

———. "The Great Tradition and the Little in the Perspective of Sinhalese Buddhism." *Journal of Asian Studies* 22 (1963): 139–53.

———, ed. *Bandaravaliya saha Kadaimpot*. Sinhala, Colombo: Godage Publishers, 2005.

———, ed. "Buddhism Nationhood and Cultural Identity: The Premodern and Precolonial Formations." *I.C.E.S. Ethnicity Course Lecture Series*. Colombo: International Centre for Ethnic Studies, 2005.

———, ed. *Rare Historical Manuscript Series 1*. Colombo, International Centre for Ethnic Studies and Godage Publishers, n.d.

———, ed. *Ravana Rajavaliya saha Upat Katha*. Sinhala, Colombo: Godage Publishers, 2005.

———, ed. *Vanni Rajavaliya*. Sinhala, Colombo: Godage Publishers, 2005.

———, ed. *Vanni Upata, Vanni Vittiya saha Vanni Kadaimpot*. Sinhala, Colombo: Godage Publishers, 2005.

Paranavitana, K. D., and C. G. Uragoda. "Medicinalia Ceylonica: Specifications of Indigenous Medicines of Ceylon Sent by the Dutch to Batavia in 1746." *Journal of the Royal Asiatic Society, Ceylon Branch*, n.s. 53 (2007): 1–55.

Paranavitana, Senarat. *The God of Adam's Peak*. Ascona, Switzerland: Artibus Asiae Publishers, 1958.

Parthasarathi, Prasannan. *The Transition to a Colonial Economy: Weavers, Merchants and Kings in South India, 1720–1800*. Cambridge: Cambridge University Press, 2001.

Peebles, Patrick. *The Plantation Tamils of Ceylon*. London: Leicester University Press, 2001.

Perera, J.A.W. "Sri Wickrama Rajasinha's Exile in India, 1816–1832." *Young Ceylon* (April 1936): 424–27.

Petch, T. *Bibliography of Books and Papers Relating to Agriculture and Botany to the End of the Year 1915*. Colombo: H. R. Cottle, 1925.

———. "The Early History of Botanic Gardens in Ceylon with Notes on the Topography of Ceylon." *Ceylon Antiquary and Literary Register* 5.3 (1920): 119–24.

———. "The Early History of Botanic Gardens in Ceylon." *Ceylon Antiquary and Literary Register* 7.2 (1921): 63–73.

———. "Sinhalese Plant Names." *Ceylon Antiquary and Literary Register* 7 (1922): 169–79.

Pethiyagoda, Rohan. "The Family de Alwis Seneviratne of Sri Lanka: Pioneers in Biological Illustration." *Journal of South Asian Natural History* 4.2 (1998): 99–110.

———. *Pearls, Spices and Green Gold: An Illustrated History of Biodiversity Exploration in Sri Lanka*. Colombo: Wildlife Heritage Trust, 2007.

Pieris, Paul E. "Two Bhikkhus in England, 1818–1820." *Journal of the Royal Asiatic Society, Ceylon Branch* 38 (1948): 91–92.

———. *Sinhale and the Patriots, 1815–1818*. Colombo: Colombo Apothecaries' Co. 1950.

———. *Tri Sinhala: The Last Phase, 1796–1815*. Cambridge: W. Heffer & Sons, 1939.

——— ed. *Letters to Ceylon, 1814–1824, Being Correspondence Addressed to Sir John D'Oyly*. Cambridge: W. Heffer & Sons, 1938.

————, trans. "An Account of Kirti Sri's Embassy to Siam in 1672 Saka (1750 A.D.)."
    Journal of the Royal Asiatic Society, Ceylon branch 18 (1903): 17–45.

Pieris, Ralph. *Sinhalese Social Organization: The Kandyan Period.* Colombo: Ceylon
    University Press Board, 1956.

Pollock, Sheldon. "Cosmopolitan and Vernacular in History." *Public Culture* 12 (2000):
    591–625.

Prakash, Gyan. "Writing Post-Orientalist Histories of the Third World: Perspectives
    from Indian Historiography." *Comparative Studies in Society and History* 31 (1990):
    383–408.

Raheem, Ismeth. *A Catalogue of an Exhibition of Paintings, Engravings and Drawings
    of Ceylon by Nineteenth-Century Artists.* Colombo: British Council, 1986.

Raj, Kapil. *Relocating Modern Science: Circulation and the Construction of Knowledge
    in South Asia and Europe, 1650–1900.* Basingstoke: Palgrave Macmillan, 2007.

Ramaswamy, S. "Visualizing India's Geo-body: Globes, Maps, Bodyscapes." *Contribu-
    tions to Indian Sociology* 36.1 and 2 (2002): 157–95.

Rangarajan, M. *Fencing the Forest: Conservation and Ecological Change in India's
    Central Provinces, 1860–1914.* New Delhi: Oxford University Press, 1996.

Rasanayagam, S., trans. "Tamil Documents in the Government Archives." *Historical
    Manuscripts Commission, Ceylon* 3 (1937): 1–13.

Raven-Hart, R. "The Great Road." In *Sir Paul Pieris: Felicitation Volume Presented by
    Friends and Admirers,* ed. S. Paranavitana and J. de Lanerolle, 135–39. Colombo: Sir
    Paul Pieris Felicitation Committee, 1956.

Reid, Anthony, ed. *The Last Stand of Asian Autonomies: Responses to Modernity in the
    Diverse States of South-east Asia and Korea, 1750–1900.* Basingstoke: Macmillan,
    1997.

Roberts, Michael. "For Humanity. For the Sinhalese. Dharmapala as Crusading Bosat,"
    The Journal of Asian Studies 56 (1997): 1006–32.

————. "From Southern India to Lanka: The Traffic in Commodities, Bodies and Myths
    from the Thirteenth Century Onwards." *South Asia: Journal of South Asian Stud-
    ies* 3.1 (1980): 37–47.

————. *Sinhala Consciousness in the Kandyan Period, 1590s to 1815.* Colombo: Vijitha
    Yapa Publications, 2004.

Roberts, Michael, Ismeth Raheem, and Percy Colin-Thome. *People in Between: The
    Burghers and the Middle Class in the Transformations within Sri Lanka, 1790s–
    1960s.* Ratmalana, Sri Lanka: Sarvodaya Book Pub. Services, 1989.

Rogers, John D. "Caste as a Social Category and Identity in Colonial Lanka." *Indian
    Economic and Social History Review* 41.1 (2004): 51–77.

————. "Colonial Perceptions of Ethnicity and Culture in Early Nineteenth-Century
    Sri Lanka." In *Society and Ideology: Essays in South Asian History,* ed. Peter Robb,
    97–109. Delhi: Oxford University Press, 1993.

————. "Early British Rule and Social Classification in Lanka." *Modern Asian Studies*
    38 (2004) 625–47.

————. "Historical Images in the British period." In *Sri Lanka: History and the Roots of
    Conflict,* ed. J. Spencer, 87–106. London; New York: Routledge, 1990.

———. "Post-Orientalism and the Interpretation of Pre-modern and Modern Political Identities: The Case of Sri Lanka." *Journal of Asian Studies* 53 (1994): 10–23.

Rutnam, James T. "Ancient Nuwara Eliya." *Ceylon Fortnightly Review*, Part 1: "The Ramayana Legend," 10.4 (June 21, 1957), 25, 27 and 36; Part 2: "A Link with King Dutugemunu," 10.6 (July 19, 1957), 19, 21, and 33; Part 3: "A Lithic Record," 10.8 (August 23, 1957), 17, 19.

———. *The Early Life of Sir Alexander Johnston, Third Chief Justice of Ceylon*. Colombo: Law and Society Trust of Sri Lanka, 1988.

Said, Edward. *Orientalism*. New York: Pantheon Books, 1978.

Samaranayake, D.P.R. *Sulurajavaliya*. Maradana: Free Press, 1959.

Samaraweera, Vijaya. "Ceylon's Trade Relations with Coromandel during Early British Times, 1796–1837." *Modern Ceylon Studies* 3 (1972): 1–17.

———. "The Cinnamon Trade of Ceylon." *Indian Economic and Social History Review* 8 (1971): 415–42.

———. "Economic and Social Developments under the British, 1796–1832." In *University of Ceylon, History of Ceylon*, ed. K. M. De Silva, 3:48–65. Colombo: University of Ceylon, 1973.

———. "Governor Sir Robert Wilmot Horton and the Reforms of 1833 in Ceylon." *The Historical Journal* (1972): 209–28.

Sartori, Andrew. *Bengal in Global Concept History: Culturalism in the Age of Capital*. Chicago: University of Chicago Press, 2008.

Sawers, Simon. "Memoranda and Notes on the Kandyan Law of Inheritance, Marriage, Slavery Etc. by Simon Sawers." In Fredric Austin Hayley, *The Laws and Customs of the Sinhalese or Kandyan Law*, appendix 1, 1–33. Colombo: H. W. Cave & Co. 1923.

Schrikker, Alicia. *Dutch and British Colonial Intervention in Sri Lanka, 1780–1815: Expansion and Reform*. Leiden; Boston: Brill, 2007.

Scott, David. "Colonial Governmentality." *Social Text* 43 (1995): 191–220.

Senadhira, Adrian. *History of Scientific Literature of Sri Lanka*. Colombo: Foremost Productions Ltd., 1995.

Senaveratna, John M. *Sri-Pada (Adam's Peak): Historic Pilgrimages to the Sacred Footprint*. Colombo: U.N.P. Journal, 1950.

Seneviratna, Anuradha. *Sunset in a Valley, Kotmale: Record of a Lost Culture and Civilisation*. Nugegoda, Sri Lanka: Sarasavi Publishers, 2001.

Seneviratne, H. L. "'The Alien King.'" *Ceylon Journal of Historical and Social Studies* 6.1 (1976): 55–61.

Siddhartha, Rambukwelle, ed. and trans. "Letters of J. D'Oyly." *Historical Manuscript Commission of Ceylon* 2 (1937): 1–18.

Singh, Upinder. *The Discovery of Ancient India: Early Archeologists and the Beginnings of Archaeology*. Delhi: Permanent Black; Bangalore: Distributed by Orient Longman, 2004.

Sivasundaram, Sujit. "Appropriation to Supremacy: Ideas of the 'Native' in the Rise of British Imperial Heritage." In *From Plunder to Preservation: Britain and the Heritage of Empire c. 1800–1940*, ed. Peter Mandler and Astrid Swenson. Oxford: Oxford University Press/ British Academy, forthcoming.

———. "Buddhist Kingship, British Archaeology and Historical Narratives in Sri Lanka, c. 1750–1850." *Past and Present* 197 (2007): 111–42.

———. "Ethnicity, Indigeneity and Migration in the Advent of British Rule to Sri Lanka." *American Historical Review* 115 (2010): 428–52.

———, ed. "Global Histories of Science." Special section, *Isis* 101 (2010): 95–158.

———. "Sciences and the Global: On Methods, Questions and Theory." *Isis* 101 (2010): 146–58.

———. "Tales of the Land: British Geography and Kandyan Resistance in Sri Lanka, c. 1803–1850." *Modern Asian Studies* 41 (2005): 925–61.

———. "'Trading Knowledge': The East India Company's Elephants in India and Britain." *The Historical Journal* 48 (2005): 27–63.

Skaria, A. *Hybrid Histories: Forests, Frontiers and Wildness in Western India.* Delhi; Oxford: Oxford University Press, 1999.

Somasunderam, Ramesh. "British Infiltration of Ceylon (Sri Lanka) in the Nineteenth Century: A Study of the D'Oyly Papers between 1805 and 1818." Ph.D. diss., The University of Western Australia, 2008.

Spencer, Jonathan, ed. *Sri Lanka: History and the Roots of Conflict.* London; New York: Routledge, 1990.

Strathern, Alan. *Kingship and Conversion in Sixteenth-century Sri Lanka: Portuguese Imperialism in a Buddhist Land.* Cambridge: Cambridge University Press, 2007.

Subrahmanyam, Sanjay. "Noble Harvest: Managing the Pearl Fishery of Mannar, 1500–1925." In *Institutions and Economic Change in South Asia,* ed. Burton Stein and Sanjay Subrahmanyam, 134–72. Delhi: Oxford University Press, 1996.

———. *Penumbral Visions: Making Polities in Early Modern South India.* Ann Arbor: University of Michigan Press, 2001.

Tagliacozzo, Eric. *Secret Trades, Porous Borders: Smuggling and States Along a Southeastern Frontier, 1865–1914.* New Haven: Yale University Press, 2005.

Tambiah, H. W. "The Alexander Johnston Papers." *Ceylon Historical Journal* 3.1 (1953–54): 18–30.

Tammita-Delgoda, SinhaRaja. *Ridi Vihare: The Flowering of Kandyan Art.* Pannipitiya, Sri Lanka: Stamford Lake, 2006.

Thant Myint-U. *Where China Meets India: Burma and the New Crossroads of Asia.* London: Faber and Faber, 2011; New York: Farrar, Straus and Giroux, 2011.

Thomas, Nicholas. *Colonialism's Culture: Anthropology, Travel and Government.* Oxford: Polity Press, 1994.

Trautmann, Thomas. *Aryans and British India.* Berkeley and Los Angeles: University of California Press, 1997.

Travers, Robert. *Ideology and Empire in Eighteenth-Century India: The British in Bengal.* Cambridge: Cambridge University Press, 2007.

Turner, L.J.B. "The Town of Kandy about the Year 1815 A.D." *The Ceylon Antiquary* (1918): 76–83.

———, ed. "A Constitution of the Kandyan Kingdom by John D'Oyly." *Ceylon Historical Journal* 24 (1929): 1–241.

Uragoda, C. G. *A History of Medicine in Sri Lanka From the Earliest Times to 1948.* Colombo: Sri Lanka Medical Association, 1987.

Van Lohuizen-de Leeuw, J. E. "The Kustaraja Image: An Identification." In *Paranavitana Felicitation Volume on Art & Architecture and Oriental Studies,* ed. N. A. Jayawickrama, 253–61. Colombo: M. D. Gunasena & Co. 1965.

Vanden Driesen, Ian H. *The Long Walk: Indian Plantation Labour in Sri Lanka in the Nineteenth Century.* New Delhi: Prestige Books, 1997.

Vanderstraaten, J. L. "A Brief Sketch of the Medical History of Ceylon." *Journal of the Royal Asiatic Society, Ceylon Branch* 9 (1886): 306–35.

———. *History of Medical Service in Ceylon.* Colombo: Colombo Apothecaries' Co., 1901.

Vaughan, Megan. *Creating the Creole Island: Slavery in Eighteenth-Century Mauritius.* Durham, N. C.: Duke University Press, 2005.

Vertovec, Steven, and Robin Cohen, eds. *Conceiving Cosmopolitanism: Theory, Context, and Practice.* Oxford: Oxford University Press, 2002.

Vink, Markus P. M. "Indian Ocean Studies and the New Thalassology." *Journal of Global History* 2 (2007): 41–62.

Walters, Jonathan S., and Michael B. Colley. "Making History: George Turnour, Edward Upham and the 'Discovery' of the *Mahavamsa.*" *The Sri Lanka Journal of the Humanities* 22 (2006): 135–67.

Ward, Kerry. *Networks of Empire: Forced Migration in the Dutch East India Company.* Cambridge: Cambridge University Press, 2009.

Washbrook, David. "Progress and Problems: South Asian Economic and Social History, c. 1720–1860." *Modern Asian Studies* 22 (1988): 57–96.

———. "South India, 1770–1840: The Colonial Transition." *Modern Asian Studies* 38 (2004): 479–515.

Webb, James L. A. *Tropical Pioneers: Human Agency and Ecological Change in the Highlands of Sri Lanka, 1800–1900.* Athens, Ohio: Ohio University Press, 2002.

Wenzlhuemer, Roland. "Indian Labour Immigration and British Labour Policy in Nineteenth-Century Ceylon." *Modern Asian Studies* 41 (2007): 575–602.

Wickramasinghe, Nira. "Many Little Revolts or One Rebellion? The Maritime Provinces of Ceylon/Sri Lanka between 1796 and 1800." *South Asia: Journal of South Asian Studies* 32 (2009): 170–188.

———. *Sri Lanka in the Modern Age: A History of Contested Identities.* London: C. Hurst & Co.; New Delhi: Distributed in South Asia by Foundation Books, 2006.

Wickremaratne, L. A. "Education and Social Change, 1832 to c. 1900." In *History of Ceylon,* vol. 3, ed. K. M. De Silva, 165–86. Peradeniya: University of Ceylon, 1973, 165–86.

Wickremasinghe, Sirima. "Ceylon's Relations with South-east Asia, with Special Reference to Burma." *Ceylon Journal of Historical and Social Studies* 3 (1960): 38–58.

Wickremeratne, U. C. *The Conservative Nature of the British Rule of Sri Lanka: With Particular Emphasis on the Period, 1796–1802.* New Delhi: Navrang, 1996.

———. "The English East India Company and Society in the Maritime Provinces of

Ceylon, 1796–1802." *Journal of the Royal Asiatic Society of Great Britain and Ireland* 103 (1971): 139–55.

———. "Lord North and the Kandyan Kingdom, 1798–1805." *Journal of the Royal Asiatic Society of Great Britain and Ireland* 105 (1973): 31–42.

Wickremesekera, Channa. *Kandy at War: Indigenous Military Resistance to European Expansion in Sri Lanka, 1594–1818.* New Delhi: Manohar, 2004.

Willis, J. C. "The Royal Botanic Gardens of Ceylon and their History." *Annals of the Royal Botanical Gardens, Peradeniya* 1.1 (1901): 1–15.

Wilson, Jon E. *The Domination of Strangers: Modern Governance in Eastern India, 1780–1835.* Basingstoke: Palgrave Macmillan, 2008.

———. "Early Colonial India beyond Empire." *The Historical Journal* 50 (2007): 951–70.

Young, R. F., and G.P.V. Somaratna. *Vain Debates: The Buddhist-Christian Controversies of Nineteenth-Century Ceylon.* Vienna: De Nobili Research Library, 1996.

INDEX